Blockchain Applications - Transforming Industries, Enhancing Security, and Addressing Ethical Considerations

Edited by Vsevolod Chernyshenko and Vardan Mkrttchian

Published in London, United Kingdom

Blockchain Applications - Transforming Industries, Enhancing Security, and Addressing Ethical Considerations
http://dx.doi.org/10.5772/intechopen.100780
Edited by Vsevolod Chernyshenko and Vardan Mkrttchian

Contributors
Aryan Chaudhary, Keshav Kaushik, Sunil Kumar, Ruben Gevorgyan, Yenok Hakobyan, Serge Chernyshenko, Vardan Mkrttchian, Valentin Afanasyev, Vsevolod Chernyshenko, Asadi Srinivasulu, Anand Kumar Gupta, Kamal Kant Hiran, Tarkeswar Barua, Goddindla Sreenivasulu, Sivaram Rajeyyagari, Madhusudhana Subramanyam, Antonio Merchán Murillo, Yibin Dong, Seong K. Mun, Yue Wang, Alan Ma, Leyla Gamidullaeva, Ivan Karelin, Svetlana Zinchenko, Dominique Bernard Kanga, Azouazi Mohamed, Yassine El Ghoumrari Mohammed, Daifa Abderrahmane, Luiz Cruz Villares, Joana R. Pereira, Giselle Garcia, Armando de Jesús Plasencia Salgueiro, Arlety García García, Jerome Verny, Wei Guan, Javier Ibáñez Jiménez, Eva María Ibáñez Jiménez, Thillaiarasu Nadesan, Doddi Srilatha, Hanlie Smuts, André Schreuder, Reyan Abdalaziz Ahmed M. Zein, Hossana Twinomurinzi

First published in London, United Kingdom, 2023 by IntechOpen
IntechOpen is the global imprint of INTECHOPEN LIMITED, registered in England and Wales, registration number: 11086078, 5 Princes Gate Court, London, SW7 2QJ, United Kingdom

British Library Cataloguing-in-Publication Data
A catalogue record for this book is available from the British Library

Additional hard and PDF copies can be obtained from orders@intechopen.com

Blockchain Applications - Transforming Industries, Enhancing Security, and Addressing Ethical Considerations
Edited by Vsevolod Chernyshenko and Vardan Mkrttchian
p. cm.
Print ISBN 978-1-80356-053-3
Online ISBN 978-1-80356-054-0
eBook (PDF) ISBN 978-1-80356-055-7

For EU product safety concerns:
IN TECH d.o.o., Prolaz Marije Krucifikse Kozulić 3, 51000 Rijeka, Croatia,
info@intechopen.com or visit our website at intechopen.com.

Meet the editor

Dr. Vsevolod Chernyshenko is a distinguished expert in blockchain technologies and system modeling. He has a Ph.D. in Computer Sciences and has successfully initiated and participated in numerous projects, displaying exceptional skills as a concept designer, data architect, and agile manager. His expertise extends to blockchain, data processing, nonlinear systems modeling, and fintech. Currently holding positions at the Financial University under the Government of the Russian Federation, Dr. Chernyshenko continues to contribute to the evolution of blockchain technologies by supervising young researchers and collaborating with academia in quantitative finance, business analysis, and data modeling. He leads research projects, develops distributed database solutions and predictive models for oil cargo transportation analytics, and supervises a startup focused on AI-based forestry assessment.

Vardan Mkrttchian graduated from the Faculty of Cybernetics, State Engineering University of Armenia as an electronics engineer and received his D.Sc (Engineering) in Control Systems from Lomonosov Moscow State University (former USSR). Dr. Mkrttchian taught undergraduate and graduate courses on control systems and information sciences and technology at Astrakhan State University, Russian Federation for 6 years. Currently, he is a full professor in the CAD and Economy departments of Penza State University, Russia. He is currently the chief executive of HHH University, Australia, and a team leader of international academics. He also serves as executive director of the HHH Technology Incorporation. Dr. Mkrttchian has authored more than 400 refereed publications and 20 books. He is also editor-in-chief of the *International Journal of Applied Research in Bioinformatics* (IJARB).

Contents

Preface

Blockchain Applications – Transforming Industries, Enhancing Security, and Addressing Ethical Considerations explores the exciting and rapidly evolving field of blockchain technology and its diverse applications across industries. This volume is a collection of thought-provoking chapters that delve into the potential of blockchain to revolutionize sectors such as health care, energy, finance, supply chain, and more.

In this book, you will embark on a journey through the multifaceted world of blockchain. From exploring the role of blockchain in cybersecurity and its impact on the audit profession to investigating its potential in tourism, nature conservation, and energy markets, the book provides insights into the innovative ways this technology is reshaping our society.

The volume also examines the integration of artificial intelligence with blockchain, the challenges and legal implications associated with non-fungible tokens (NFTs), and the ethical considerations surrounding the use of blockchain in various contexts. We explore the concept of decentralized autonomous organizations (DAOs), the use of deep learning techniques for early prediction and detection of diseases, and the significance of trust and privacy in longitudinal personal health records.

Each chapter offers a unique perspective, with authors presenting their research findings, insights, and practical applications of blockchain technology. We believe this collection will not only contribute to the scholarly understanding of blockchain but also provide valuable insights to professionals, policymakers, and technology enthusiasts.

As an editor, I am grateful to the contributors who have shared their expertise and passion for blockchain. Our team hopes this volume sparks further exploration, discussion, and innovation in the field. Our aim is to offer a comprehensive overview of blockchain applications, shedding light on their transformative potential while addressing the challenges and ethical considerations that arise along the way.

We invite you to delve into the pages of this book and discover the diverse applications of blockchain technology that are shaping the future of industries worldwide.

Enjoy the journey!

Vsevolod Chernyshenko
Financial University under the Government of the Russian Federation,
Moscow, Russian Federation

Vardan Mkrttchian
HHH University,
Sidney, Australia

Section 1

Introduction to Blockchain Technology

Chapter 1

Blockchain for Cyber-Physical Systems

Doddi Srilatha and Thillaiarasu Nadesan

Abstract

Cyber-physical systems (CPSs) are the intelligent systems that offer an interaction among computational, software, and networking resources in a continuous and dynamic fashion. Future systems are likely to be created and developed using CPSs, which have been recognized as a significant area of research. The electric power grid, energy systems, body area networks, modern vehicles, smart homes, cooperative robotics, and smart transportation are the examples for CPS. The security aspects of CPSs can be enhanced with blockchain (BC) technology. For instance, with the combination of CPSs and blockchain, a peer-to-peer energy market is made possible where machines may automatically buy and sell energy based on parameters specified by the user. In this chapter, we summarize recent developments in the creation and applications of CPS, the state-of-the-art and pertinent concepts, numerous CPS applications that have employed blockchain, relevant solutions, and open challenging issues.

Keywords: cyber-physical systems, blockchain, power grid, internet of things, transportation systems, vehicular systems

1. Introduction

Cyber-physical systems (CPSs) is a sort of architecture that combines sensing and communication technology to give the society a number of advantages. To put it another way, a computerized process system (CYPS) is an engineering structure whose physical system or procedure is made up of cybernetic elements such as computing hardware and a communication network [1]. All the mechanisms of CPS are same tightly combined with respectively other, by which we can understand that the functionality of single module depends on the additional factor. The CPSs have a sequential development in current years in the fields of energy, smart homes, smart vehicles, health, transportation, and industrial Internet of Things (IIoT). The main expanses of investigate to be considered for scheming such schemes to be keen, effective and flexible, remain constancy, dependability, robustness, safety and confidentiality.

Rapid development of technologies also made such schemes for thoughtful and profound risks. If such dangers remain not attained, then we would misplace all the assistances that the CPS provides. Blockchain establishes trust among the nodes to generate new basics for most dispersed arrangements. The important technology to allow decentralization that plays a significant part in CPS field is blockchain.

Blockchain (BC) is a protected numerical ledger of dealings that not only maintains records of the economic world but similarly in additional areas that maintain historical sign of the dealings that has charge. This is the core skill of Bitcoin. Financial establishments, for a long time, consume negotiated around the need aimed at the circulated decision-making procedure, but never took it on to the next step, till the beginning of crypto-currencies fueled through the BC skill [2].

Firstly, the blockchain skill was mainly used for defending the economic transactions, smart contracts, storage space systems, and notary. However, the assistances were not sticked to particular needs but there were also other requests such as supply chain, health care, and transportation. The manufacturing comprehended that blockchain can expand the effectiveness. This made an active part of effort, where researchers and professors are now observing at other submissions where blockchain can be exploited.

2. Blockchain model

The hack of done a billion of Yahoo files [3], the Equifax, the increase in information breaches [4], and then ransomware occurrences are some of the cyber-attacks conveyed in recent years. Every day, more than a million cyber-threats are published, and through 2020, approximately 200 million IoT strategy [5] determination requires protection. Nearly business analysts predicted that in a few years, this figure will reach 29 billion.

BC is a distributed system that organizes payments and is to be taken into consideration as a possible option to defend cyber-attacks since it employs network participant consensus to generate trust. When opposed to centralized systems, which become inefficient as the number of linked devices rises, distributed systems provide a number of advantages. Currently [6], blockchain is surrounded by a strong and quickly expanding ecosystem, and there are more applications for safeguarding transactions than ever before. Blockchain was initially designed for purely digital applications, but as time has gone on, it has also proven useful for applications that integrate digital and physical elements.

2.1 Blockchain clarification

BC is essentially, unchallengeable file to which novel transactions with time stamps are continuously added and organized into hash chains called blocks [7]. This protocol's most important feature determines how a system of users, recognized as miners, may come to an understanding about the blockchain's present status. The BC designs come in a variety of forms, including private (i.e., permissioned) and public (i.e., permission-less).

- A community blockchain is single that anybody may connect; typically, they are permission-less, giving all users an identical level of privileges.

- A closed blockchain where confidentiality is valued and each joining node is pre-nominated is known as a private blockchain. All network users do not have equal access to them since they are permitted.

Example of popular permission-less blockchain procedure is Bitcoin [8]. It randomly chooses a new miner who has the capability to obligate or insert a fresh block to the blockchain on average once per minute. Who adds the next transaction and how it is done are the two main issues to be resolved. For the same, there are two techniques: proof-of-work (PoW) and proof-of-stake (PoS).

Deliberate the circumstance where person 1 wishes to pay person 2 in basic words. Person 1 declares its purpose first and then authenticates the operation by signing it with a cryptographic key. The legitimacy of the digital signatures and the accessibility of possessions remain subsequently verified by network administrators or miners. The additional transactions are included in the blockchain after the process is finished.

Individually block comprises a single code-named hash, which comprises the hash of earlier block and is used to attach the blocks composed in an exact order. To establish the credibility as a leader any miner should perform a set of computations. These calculations solve the problem of mapping data of any size to an immovable size. In any system, a spearhead can remain selected in one of the following traditions:

Once everybody has confirmed this, they will choose that specific mine job as their leader. The PoW method is computationally intensive since numerous miners try to answer the puzzles at the same time until one of them prospers.

The distinctive level assessment of BC is exposed in the **Figure 1**. Here, once a deal is demanded, an information construction is allocated to support a set of transactions for all nodes on the network. Earlier calculation everything to the BC construction all the nodes perform the block confirmation process. Once nodes perform block verification, they will receive the PoW rewards.

Similarly, each new node linking the disseminated arrangement of BC becomes a fully copy of the BC. It is directed to each point confidentially the blockchain outline when another block is made. At this time, each node will confirm the block also check whether the statistics expressed there is correct. If the whole thing is normal, the block will be additional to each node's local blockchain.

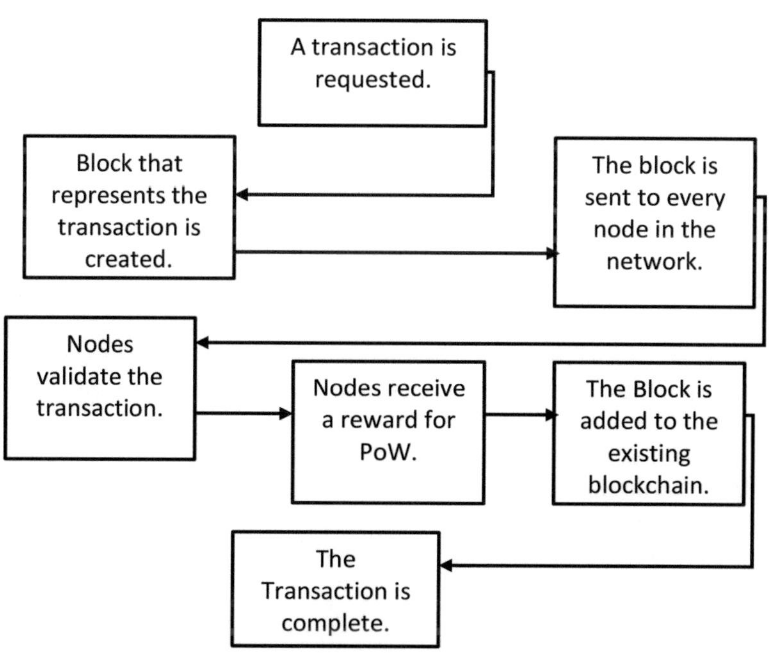

Figure 1.
The blockchain process.

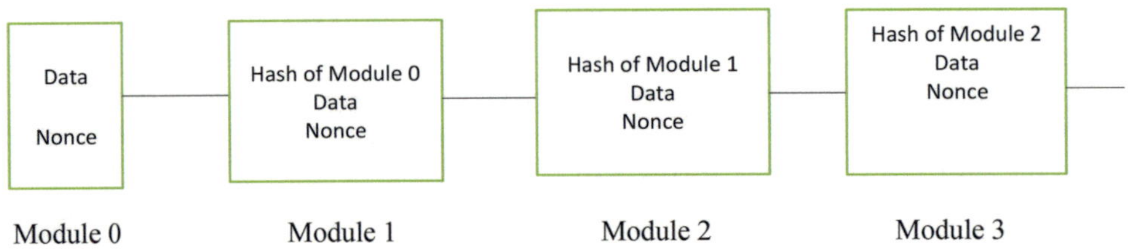

Figure 2.
POW of blockchain mechanism process.

In PoS, a leader with the maximum quantity of stake in the system is designated. The number of coins that the miner owns determines the amount of the stack in the network. Honestly, this is based on the theory that it is most likely a miner who is very interested in the network, and the break of the system implicitly receives the leader through appending his slab to the leader block. **Figure 2** demonstrates the illustration of deal records of BC.

Using a cryptographic hash algorithm, the block's principal job is to keep track of a sequence of confirmed transactions. The essential characteristics make the hash function effective:

- It produces a fixed-length output regardless of the distance of the contribution.
- It is available, which implies that given an input, it always produces the similar result.
- It is irreversible; therefore, it is unbearable to get the identical input from the output.
- Any little changes to the initial input result in new output.
- The hash calculations are quick and incur little overhead.

Once a miner finds a random number that causes the hash value of its block to fall beneath the trouble threshold, the block will eventually be measured effective and then ready for transmission to the network, and the miner will be rewarded for his efforts.

An attacker can change the database's contents and produce additional sets of transactional data to establish alternative chain of records, which is a possible attack scenario. However, the achievement of varying anything in the chain will lead to domino consequence; thereby, it invalidates all the modules that trail. After the whole block has been invalidated, the network miners must once again search for a time being that produces a hash key underneath the desired struggle.

This demonstrates that of all current technologies, blockchain is the most innovative technology that is also the most effective and secure.

The open-source, permission-less public designs, like those used by Bitcoin and Ethereum, let anybody download the code, provide evidence of work, as well as earn the ability to approve network transactions. Open and transparent architecture describes this style. On the other side, private blockchain organizations, like R3 [9] and EWF [10], function as a team. It has a permissioned read and write architecture and a semi-distributed design. This kind of architecture is quicker and uses people with confirmed credentials who have been pre-approved.

3. CPS related to blockchain

Through the increasing popularity of processers in the earlier few eras, accounts consume changed after primarily paper credentials to digital versions shaped and then maintained on computers.

This is single of the numerous cyber requests, the ones permitted through processors. Even though these documents are made and kept electronically, information must still be entered manually. It can be claimed that people are still the major source of information gathering in these applications since money transfers, health records, and coverage records are only a few of the numerous instances in this category.

The sensors increasingly taking the place of people as the main data collecting source in many organizations, thanks to the rise of IoT and the growth of sensing technologies over the last several years. These systems, also known as CPS, bring together physical processes, software, and information to create a cohesive system with design, analysis, and abstraction capabilities. A number of disciplines are involved in the study, which has as its fundamental elements used for dynamic analysis.

BC financial transactions have been extensively studied and recorded. All of these technological advancements have made it possible to transfer money directly to approve individuals without using centralized authority. The use of smart contracts on the BC reduces the likelihood of delays, suppression, or other outside influences. This is unbreakable, ensures total financial stability, and maintains track of the conditions of the contract. It is also easier to track and manage online identity when utilizing blockchain. Blockchain is employed as a low-cost notary system, as detailed in Ref. [11], preventing various frauds by producing distinctive certificates that would be simple to validate. **Table 1** represents BC technology applications.

3.1 Health and medical level BC applications

BC is currently existence cast-off in community strength and health investigates grounded on patient data and application management. Assessment criteria depending on viability, planned capabilities, and acquiescence may be rummage-sale to evaluate BC-based decentralized systems in the healthcare sector [12]. This is the fundamental advantage of blockchain, which is crucial for health data since it makes it difficult to modify or erase a record while leaving a data trace of the effort. Blockchain is being used by several nations to safeguard clinical trial and health archives by associating access to the information with authorization locations. BC

Schemes	Submissions	Shared inspiration
Medical level	Health equipment, networks for medical analysis	Medical and healthcare facilities of the highest calibers.
Transportation	Transportation networks, railroad technology, aviation, and airspace management	Zero vehicular deaths, less cramming, and less delays.
Automotive	Organization nursing & management.	Extreme performance & produce.
Power grids	Building construction, supply of electricity, & environmental conservation	Advantages to the environment include delivery of power without blackouts.

Table 1.
Applications of blockchain technology.

also provides safety overslide, allowing the use of barcodes to scan medicines and helping them enter secure digital blocks when ownership changes, thereby reducing the risk of counterfeiting. There is supply chain access to real-time records. Blockchain can be used for a variety of applications, specifically information sharing, admittance regulator, medical analysis, management, and supply restraint management [13].

3.1.1 Healthcare information organization

Research on the subject of interoperability between blockchain applications in healthcare can be originated here [14]. The MedRec model [15] provides a proof-of-work that allows blockchain designs and other decentralization standards to anchor and communicate with other medical record technologies. The architecture is organized using Ethereum smart contracts, which also provides a log that keeps track of medical data and allows users to see entire data, review care records, and exchange information. An innovative approach for structuring open APIs and system structure transparency, as well as conditioning with the supplier's existing framework, has also been described in this study [16]. An adaptive method for processing large-scale personal health data using a tree-like method is proposed [17]. A blockchain-based program has been adopted to secure documents stored in the cloud. A further platform built on the blockchain for healthcare data is BlockHIE [18], which includes off-chain archiving with on-chain validation to guarantee privacy as well as authenticity for records. With the use of blockchain technology, healthcare data may be transferred in a secure and auditable way [19]. Additionally, it employs blockchain to manage access-based health data [20, 21]. **Figure 3** represents the scope of CSP. Lastly, a blockchain-based technology for maintaining patient and physician databases in the field of healthcare analytics is shown. The benefits of a parallel healthcare system are brought in by a methodology that is given in Ref. [22] and is built on artificial systems, computational trials, and then a simultaneous implementation utilizing blockchain technology. To handle health data at the individual and institutional levels, several experts have suggested using private blockchain solutions [23].

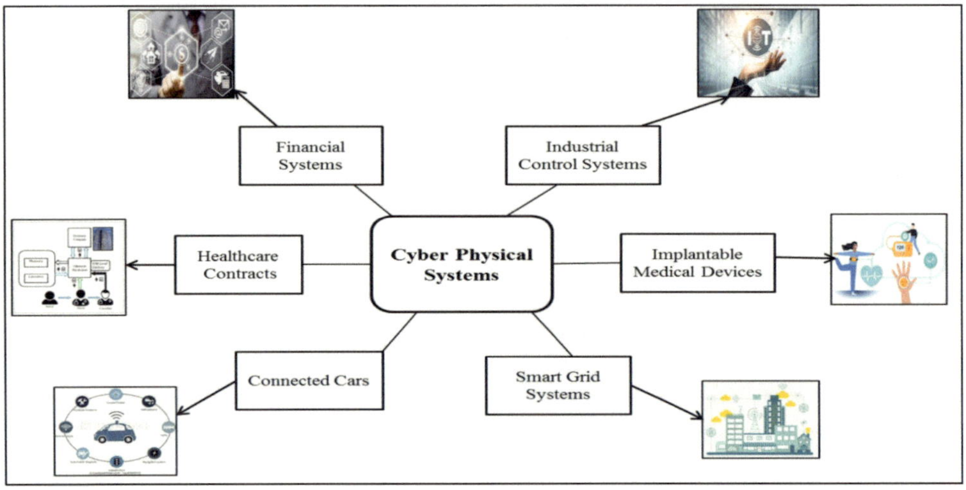

Figure 3.
The scope of CPS.

3.1.2 Implantable health care device safety

Advanced health care systems are essential for enduring nursing, giving them a major health care knowledge [24–26] and transforming health care data into an important source of health care knowledge [26, 27]. We spoke about the study on managing health information in the preceding segment. In this unit, we will be learning the latest developments in health information sharing. According to statistics, the US's GDP and per capita health care expenditures are higher (a record 2.5 trillion U.S. dollars (17.3%), per capita expenditure in 2009 was approximately 8050 U.S. dollars) [28], with a growth rate of 1 to the normal growth rate in 2019. 6.3% per year. By 2019, it will reach 19.6% of GDP [28].

Medical equipment's are anticipated to have a market share of $186 billion by the end of 2019 [28], manufacture medical equipment is one of the largest marketplaces in this industry. The U.S. Department of Commerce (DOC) recognizes that medical device exports in 2015 exceeded $44 billion [29], which was mainly driven by key innovations from more than 6500 medicinal expedient businesses in the US country. **Table 2** presents BC forth coming challenges in medical applications.

3.2 Blockchain claims in manufacturing control schemes

Manufacturing Regulate Schemes are used to monitor &switch physical objects that may be found in a variety of different sectors, since organization-critical nuclear facilities to everyday irrigation schemes (ICS).

Performance measure	Objective	Forthcoming guidelines
Medical-level applications with patient interaction [14, 30]	Information sharing, interpreting, and application	Sophisticated information analytics and strong care organization.
Medical health care records management [15]	Controls patient admission to medical records though enabling thorough document review, care traceability, and data exchange.	Assemble the needs for customized integration to create an available functioning.
BC modeling [18]	Medical data sharing for automated medicinal annals and individual clinical records.	For privacy and authenticity, on-chain validation and off-chain storing are both used.
Healthcare analytics [31]	Health data collection, storage, and exchange	Analytics for healthcare using BC and AI.
BC and (IoT) level applications [32]	Incorporating cloud-level Bigdata.	Need an agreement model, lower block processing, and transaction validation computing costs.
MedShare [33]	Model for data exchange across cloud service providers.	Reduced latency facilitates data giving out and advancement
Information authentication with privacy [16]	A batching approach for dispensation personal health information using Hyper ledger fabric and trees.	Information about one's own health as well as medical information.
Confidentiality desecration [34]	Statement, backup and retrieval, and anonymization of data.	Advancement of raw data, protect it.

Table 2.
Offerings the numerous BC use cases, plan contests, and upcoming instructions in medical.

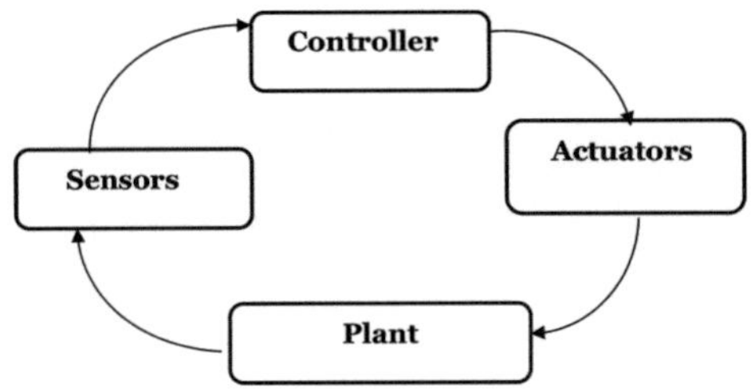

Figure 4.
Industrial control systems.

As demonstrated in **Figure 4**, ICS uses sensors to perceive and gather data, and then transmit that data to the controller, which then uses the actuator to deliver feedback.

The following are the essential elements of an ICS environment:

- A plant with operational technologies for data collecting, communication, and local processing. This is referred to as a sensor. It is a tool for measuring actual physical amount. Cameras, accelerators, gyroscopes, radar, lidar, and other devices are examples of sensors.
- A computer device that can be programmed to carry out activities using programming logic analysis to be done.
- A control loop gives the controller the ability to carry out different operations by deciphering signals from the sensors. The actuator, which is a component of this system, modifies the physical quantity that the system is observing. Actuators often include devices such as motor drivers, LEDs, lasers, speakers, switches, valves.

The future transformation of industrial systems will heavily rely on the IIoT. Similar to ICS, the intelligence and connectivity are delivered through sensors and actuators with networking and computation capabilities. It is predicted that billions of data-generating gadgets will already be online and linked to the Internet. Numerous applications, including infrastructure, transportation, and agriculture, will profit from this. In this type of system, transactions involve reading data from various sensors, with time and space tags that indicate where and when the measurement occurred; then, these data are distributed to various network participants. It is important to keep a historical record of these transactions because they are used to influence key decisions. This assumes that the record has not been manipulated and will be tracked when such attempts occur.

A powerful and effective new group of mechanisms for dealings created by corporeal resources is introduced by blockchain. The grouping of BC and the IoT provides us with a multifunctional, truly distributed point-to-point system, and it is possible to interact clearly and reliably with distributed sensors [35]. To implement access control regulations, a BC-based system for safe shared verification is required. To provide both privacy and security, this scheme employs a triangulation with combined quality sign, multi-receiver encoding, and communication verification code [36]. In order to offer a reliable mechanism for the identification and verification of devices, blockchain is utilized to construct virtual zones. The Trust Bubble, a decentralized system made up of several virtual spaces, ensures robust device identification and authentication while safeguarding the accessibility and integrity of data [37–39]. **Table 3** presents ICS design challenges with analysis.

Parameter field	Objectives	Forth coming instructions
IoT maintenance [35]	Load information from your iPhone or mobile and meter.	Necessitates a lot of memory and is not quick.
Smart Home [40]	Smart phones use symmetric cryptography and lightweight protection.	Investigate implications in further IoT fields.
Industrial IIoT [36]	Exploit the financial gains from credit banks.	Plans created for risky circumstances with good or bad recognition ratings.
EVs with fog computing [37]	To produce the evidence of work, data submission periodicity, and energy contribution are utilized.	For center-less trust, cooperative intelligence, as well as spatiotemporal sensibility, hybrid cloud, and edge calculating.
Distributive switch scheme for advantage calculation [38]	Higher-level executing strategic decision-making	Accountable for process control at the executive level.
BSeIn [39]	Access control and safe authentication process for Industry 4.0	Incorporating value systems inside organizations
Foams of trust [41]	Safe virtual spaces where objects may recognize and trust one another.	Collaboration among virtual zones.
BCencounters IoT [42]	Scalable IoT permissions	IoT scenarios need for technology that is flexible.
Device organization arrangement on blockchain [43], confidentiality conserving [44].	Distribution of device data while compromising privacy	Data anonymization is a possibility, and there are other difficulties and possibilities as well, such as regulation, fault tolerance, non-reputation, and trust [45]

Table 3.
ICS design challenges.

One may automate labor-intensive job processes when all the gadgets in an IoT network are interconnected and equipped with decision-making capacities. Nevertheless, this necessitates that a historical log of these activities and the information that inspired them be kept. According to ongoing scientific study, like the one cited above, the use of BCs in the IoT space will satisfy the demand for cryptographic verifiability, leading to significant improvements across many sectors.

3.3 Applications in transportation

Prospective transportation would depend heavily on autonomous cars, which will also be important for societal advancement. These automobiles have a significant impact on traffic management, and comfort while driving and messaging and road safety. An autonomous, reliable, and decentralized intelligent transportation system, which makes better use of the structure and advantages of traditional intelligent transportation systems (ITS), is especially suitable for crowd sourcing innovation. **Figure 5** presents the ITS national architecture proposed by department of transportation [46]. Physical, data, network, consensus, incentive, and application layers are the seven layers of the conceptual paradigm for ITS. In a heterogeneous intelligent transport system, dispersed key management is also used [47]. In order to cut down on key transfer time, it incorporates dynamic key management and key transfer across heterogeneous networks.

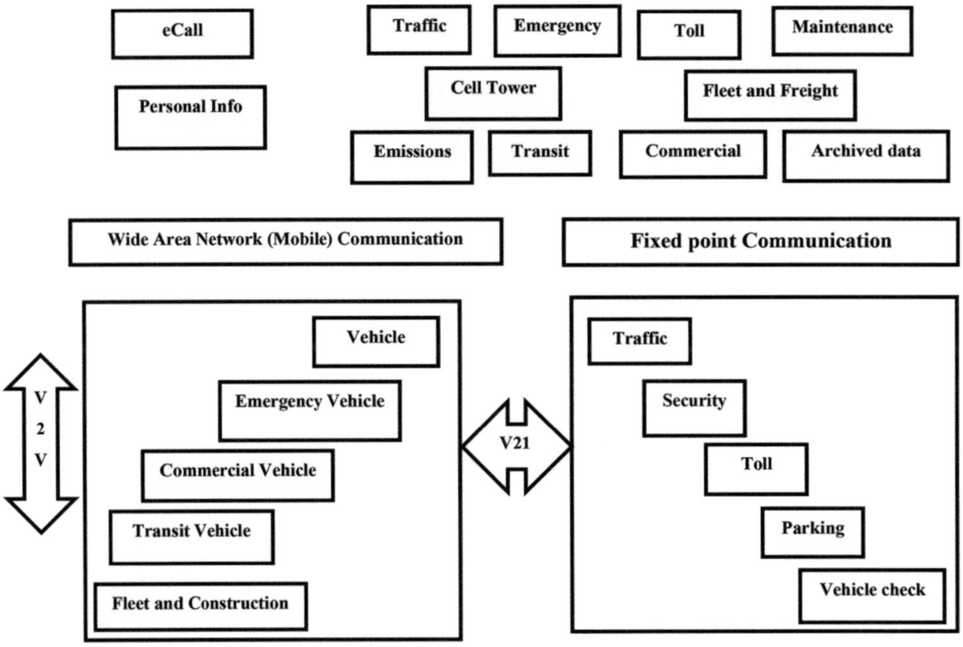

Figure 5.
ITS-based US DOT architecture.

The connected automobiles have in-built sensor capabilities that allow them to monitor their surroundings and provide a thorough 360-degree perspective of what is around. They include things such as cameras, proximity sensors, light and radio-frequency detection sensors, navigation systems, and webcams, to mention a few. In the event of a collision, they have the capacity to synchronize data from several sensors—a process known as sensor fusion—with real-time data to keep the cars and infrastructure components informed. Advanced driver aid technologies including adaptive cruise control, lane departure warning, and collision avoidance systems have increased as result of these features. In order to achieve all these functions, these vehicles are also equipped with communication equipment and protocols for exchanging information between all objects in the vehicle network. Dedicated Short-Range Communication (DSRC) is currently approved ITS 5.9 GHz band protocol for vehicle safety (V2V) applications [48]. Similarly, based on software, the key question to be answered is software updates when new features are added. **Table 4** presents the BC design challenges and future directions in transportation sector.

3.4 Applications in smart grid systems

The smart grid networks that deliver power among supply and consumption in an optimal manner. This is achieved by integrating data technology, telecommunications, and energy into existing electrical systems. Smart grid systems integrate sensors and software into existing networks and provide information to public utilities and private users, who can use this information to react quickly to changes. The smart grids not only improve the effectiveness and dependability of the electricity supply, but they also act as a catalyst for the integration of renewable energy sources into current networks, lowering carbon emissions.

Application domain	Objectives/use cases	Future directions
Intelligent Transport Systems [46]	Conceptual framework for intelligent transportation networks with seven layers	Examine the justification, innovative business methods, and real-world application situations.
Distributed key management [47]	Reduces the time it takes to transmit keys during the handover of automobiles by making use of the dynamic transactional collecting period.	Blockchain-based maintenance of pseudonyms.
Charge it up [49]	For postponement, latency, safety, and cost in smart mobility networks, use the state channel.	State channels can be used by intelligent mobility devices for connection & control records.
Reward-based systems [50]	Reliability of vehicle behavior, as well as legal and unlawful conduct of vehicles.	Activity with many vehicles in a dubious situation.
TangleCV [51]	Security using a decentralized trust system	Vehicles entering and exiting the network.
Trustbit [52]	Intelligent vehicle communication using a reward-based scheme.	More use cases on communication level.
Intelligent vehicle trust point [53]	Crypto ID to assure vehicle reliability.	Using Bitcoin to make payments at petrol stations.
Identification of vehicles [54]	Blockchain-based connectivity that is secure.	For the blockchain validations, do somewhat expensive hash operations.
Software update system [55]	System for secure wireless (SW) updates.	Verify the findings with a larger dataset.
CUBE [56]	Framework for information security.	Utilize artificial intelligence (AI) to thwart nefarious assaults.

Table 4.
BC use cases, design challenges, and future directions in transportation sector.

There are multiple options for network modernization through BC. The existing power grid cannot withstand cyber-attacks on distributed energy sources and network peripherals. The business model reduces costs through cutting out third parties [57, 58]. Many methods to increasing arbitrage chances to harvest and sell energy at separate level are studied and used. Information about energy use is distributed gathered through smart metering equipment. Utilizing self-enforcing smart agreements, the predicted energy adaptability at the customer level may be managed programmatically. We can learn how energy demand and manufacture can remain matched at a smart grid level and how to integrate the reward and penalty mechanism to balance energy demand by monitoring the flexibility among energy consumption and then the demand answer signal. Smart contracts built on blockchain technology give security and flexibility, an unchangeable record of transactions, and the ability to automate processes and conduct micro-transactions quickly and cheaply. The design and modeling of the local energy market is implemented in a private blockchain, providing real-time pricing data with manual agents. It simulates the best decision based on the predicted production capacity, thereby automatically executing well-founded cost decisions. In similar work, the manufacture and consumption load curves are converted to distance keeping inserts in order to find the right speed. It embeds protection while using blockchain to make computations for bid negotiations publicly transparent. On electric cars, this research was expanded upon and tested [59]. Data exchange is made possible, while sensitive user data is protected in a related work that proposes a blockchain-based privacy-preserving payments system for car to grid networks [60].

A strategy uses an instrument to integrate Bitcoin features into the market for renewable energy [61]. A robot that provides consumers with sales advice is also part of this project. An effective energy management system is created using modern technologies in the replicable smart area concept. Join a platform built with blockchain technology and the IoT [62]. A blockchain-based approach controls efficient aggregation and protects privacy for power grid interactions in smart communities [63]. By examining the user's energy consumption profile, a blockchain-based solution is here offered to prevent application usage patterns. Similar to this, the security and privacy issues with smart grid are reduced by using a sovereign blockchain that offers transparency and provenance [64]. Coworkers may negotiate energy costs discreetly

Application domain	Objectives/use cases	Future directions
Modernize Grid [57]	Blockchain—based smart, investment management, access control, and business flow.	Less centralization of the system is desirable.
Smart energy grid [58]	Energy exchanges among energy suppliers and individual consumers.	Consideration should be given to rural residents.
Smart grid resilience [66]	Keep track of real-time loads, while agreements carry out dispersed sales and purchases from clients.	Applications should be simulated in a realistic setting.
Decentralized management of demand response [67]	Energy demand and supply matching using consensus-based validation.	Multi-stakeholder market places are implemented.
Blockchain based smart contracts [68]	Shortened payment times and fewer middlemen required.	Micro grids will improve the energy systems' resilience.
Privacy Preserving smart grid tariff decisions [69]	Provides dependability, verifiability, and transparency.	Deployment with rigor.
Electric vehicle charging [59]	Identify the least expensive charging station in a given area.	Large-scale electric car scalability issues and managing the payment phase.
Payment mechanism for vehicle to grid network [60]	Data sharing and privacy protection in grid networks connecting vehicles.	Various pricing policies and privacy requirements.
Crypto-trading energy market [61]	Robotic adviser to improve trading of energy.	Energy users will connect to intelligent grid systems digitally.
Smart city through IoT [62]	Using distributed storage to keep track of every transaction.	Replication throughout several cities
Efficient aggregation for power grid communications [63]	Improved computing speed to protect user privacy	Lessen the computation complexity that authentication causes, specifically at system startup.
Grid-monitoring [64]	A invention that enables users to keep an eye on the electricity without outside interference.	Application of the suggested model.

Table 5.
BC use cases, design challenges, and future directions in smart grid.

and safely using a proof of concept for a decentralized energy trading system that makes use of blockchain technology, multiple signatures, and anonymous encrypted message streams [65]. To enhance the system's overall performance, certain network members in this initiative restricted energy output and sales. The blockchain is used to manage transactions. **Table 5** presents the use cases, design challenges, and future directions in smart grid domain.

4. Blockchain limitations and future directions

As a large number of scientific studies discussed in this chapter have shown, blockchain technology has become very popular in recent years, which may change the way people work and connect, and lay the basis for new requests using network devices. However, it has certain boundaries, such as:

- It does not scale through the amount of associated devices because it is restricted through the block size and the time required to calculating the hash.

- In some cases, transaction fees need to be paid or around additional reward device for miners.

- Although not as centralized as the single bank perception, it still relies on several large organizations such as miners.

- The computing also storage necessities of blockchain members are very general because they must keep the whole ledger and contribute in the deal review procedure as an endorser or miner. These boundaries have not made blockchain an ideal skill for large systems connecting IoT devices. In this case, Tangle was launched in 2017 as a skill for verifying transactions then protecting applications related to the Internet of Things (IoT) [70].

Examples include low resource consumption, extensive interoperability, billions of nano-dealings, and data integrity because it is quicker, more energy and resource competent, and more resistant to quantization.

Tangle is an advanced transaction system and statistics argument layer intended to protect requests related to the Internet of Things. It is founded on a focused acyclic graph named Tangle, which is a typical information construction method. It aims to overcome approximately of the shortcomings of blockchain. In the tangled network, apiece transaction must be run PoW to confirm the previous two transactions. The basic theory is that the more transactions verified in parallel, the faster the network will expand. Tangle has the characteristics of scalability, resource optimization, data transmission safety, and quantum training. Tangle emerged as a third-generation crypto-currency, which does not require any additional costs to verify transactions, but is still safe. **Table 6** compares blockchain and entanglement.

Blockchain is less de-centralized than Tangle. Blockchain will likely link several IoT technologies to a single gateway, which will then take part in the blockchain network. This is known as a grouped or semi-decentralized method. It supports the idea that a small IoT device may participate directly in the tangle network.

Blockchain	Tangle
Blockchain is made up of a number of nodes, or frames of transactions, each of which is attached to the one before it in a lengthy chain that is always growing. It has the ability to circle back.	A collection of data components that only flow in one way make up a tangle. It cannot ever go backwards.
Ownership is semi-distributed and decentralized.	Really dispersed ownership and decentralized.
Due to its frame process, which entails the solution of a mathematical problem and verification by group consensus, blockchain claims a substantial level of safety.	To complete its own operation and subsequently establish a data node, a Tangle device simply has to validate and approve two prior ones. The tangle is less safe than blockchain because of this less reliable process.
With more operations competing for less available block spaces, transaction performance decreases as the size of the network grows. Blockchain requires a lot of processing power as a result.	As the number of users rises, tangle adaptability grows, making it lightweight and necessitating less-processing resources.
High power demands result in high-energy needs.	Low power usage reduces the need for energy.
Blockchain is not sustainable since it requires about 10 minutes to complete a transaction.	Compared to blockchain, it is speedier and more scalable because of its low overhead PoW.
Transaction fees are charged by miners.	Transaction fees are absent since there is no idea of miners.
Since it employs an elliptic curve signature technique, it is not quantum resistant.	Security from quantum mechanics due to the usage of hash-based signatures.

Table 6.
Blockchain and tangle comparison.

5. Conclusions

This bankruptcy affords a holistic review of diverse CYPS in which dispersed database methods, together with blockchain or tangle, remain used. CYPSs and structures that manipulate also screen the bodily global round us. Blockchain and their inherent aggregate of consent algorithms, disbursed information loading, and then stable protocols may be applied to construct robustness and dependability in those structures. This broad side defines how packages, together with a clever grid, independent automobiles, and IoT devices, have advanced with the aid of using dispensing the function of facts validation throughout the community peers, thereby disposing of the dangers related to a centralized construction. This bankruptcy surveys improvements, use belongings, layout tasks, and destiny instructions in blockchain studies throughout the health care, clever grid, independent vehicle, and business manufacturing technique packages and demonstrates how those packages consume advanced from this skill. This bankruptcy defines blockchain skill, that is, a communal database that raises best with the aid of using joining new information, verifies customers with sturdy cryptography, and leverages financial incentives to inspire mistrustful strangers to control and stable updates. This investigation bankruptcy gives the blessings and drawbacks of this progressive skill. This bankruptcy additionally defines a scientific version that may remain used as a useful resource to decide if a selected utility container gets advantage from this skill. The version became examined on packages, specifically the linked automobile file and college database. For future work, we propose to furnish a complete list of applications and challenges of BC technology in smart grids, healthcare, and IoT.

DOI: http://dx.doi.org/10.5772/intechopen.110394

Author details

Doddi Srilatha[1,2*] and Thillaiarasu Nadesan[1]

1 School of Computing and Information Technology, REVA University, Bengaluru, India

2 Department of CSE-AIML, Sreenidhi Institute of Science and Technology, Hyderabad, India

*Address all correspondence to: doddisrilatha@gmail.com

References

[1] National Institute of Standards and Technology. Available online: https://www.nist.gov/el/cyber-physicalsystems [Accessed: February 21, 2019]

[2] Markham JW. A Financial History of the United States: From Enron-Era Scandals to the Subprime Crisis (2004-2006); from the Subprime Crisis to the Great Recession (2006-2009). Abingdon, UK: Routledge; 2015

[3] Smith KT, Smith M, Smith JL. Case studies of cybercrime and its impact on marketing activity and shareholder value. Academic of Marketing Studies Journal. 2011;**15**:67

[4] Gressin S. The Equifax Data Breach: What to Do. Washington, DC, USA: Federal Trade Commission; 2017

[5] Evans D. The Internet of Things How the Next Evolution of the Internet Is Changing Everything. Cisco White Paper 2011. Available online: http://www.cisco.com/web/about/ac79/docs/innov/IoT_IBSG_0411FINAL.pdf [Accessed: January 1, 2020]

[6] Haber S, Stornetta WS. How to time-stamp a digital document. In: Proceedings of the Conference on the Theory and Application of Cryptography, Santa Barbara, CA, USA, 11-15 August 1990. Berlin/Heidelberg, Germany: Springer; 1990. pp. 437-455

[7] Coron JS, Dodis Y, Malinaud C, Puniya P. Merkle-Damgård revisited: How to construct a hash function. In: Proceedings of the Annual International Cryptology Conference, Santa Barbara, CA, USA, 14-18 August 2015. Berlin/Heidelberg, Germany: Springer; 2005. pp. 430-448

[8] Nakamoto S. Bitcoin: A Peer-to-Peer Electronic Cash System. Berlin, Germany: ResearchGate; 2008

[9] Ghafarian A, Seno SAH. Exploring digital forensics tools in backtrack 5.0 r3. In: Proceedings of the International Conference on Security and Management (SAM). Las Vegas, NV, USA: University of Mashhad; 2014

[10] Goranovic A, Meisel M, Fotiadis L, Wilker S, Treytl A, Sauter T. Blockchain applications in microgrids an overview of current projects and concepts. In: Proceedings of the IECON 2017-43rd Annual Conference of the IEEE Industrial Electronics Society. Beijing, China: IEEE; 2017. pp. 6153-6158

[11] Arredondo A. Blockchain and Certificate Authority Cryptography for an Asynchronous on-Line Public Notary System. Ph.D. Thesis. Austin, TX, USA: The University of Texas; 2018

[12] Zhang P, Walker MA, White J, Schmidt DC, Lenz G. Metrics for assessing blockchain-based healthcare decentralized apps. In: Proceedings of the 2017 IEEE 19th International Conference on e-Health Networking, Applications and Services (Healthcom). Dalian, China: IEEE; 2017. pp. 1-4

[13] Hölbl M, Kompara M, Kamišalic A, NemecZlatolas L. A systematic review of the use of blockchain in healthcare. Symmetry. 2018;**10**:470

[14] HIMSS. What is Interoperability? Available online: https://www.himss.org/what-interoperability [Accessed: November 14, 2019]

[15] Ekblaw A, Azaria A, Halamka JD, Lippman A. A case study for Blockchain

in healthcare: "MedRec" prototype for electronic health records and medical research data. In: Proceedings of the IEEE Open and Big Data Conference. Vol. 13. Washington, DC, USA: IEEE; 2016. p. 13

[16] Liang X, Zhao J, Shetty S, Liu J, Li D. Integrating blockchain for data sharing and collaboration in mobile healthcare applications. In: Proceedings of the 2017 IEEE 28th Annual International Symposium on Personal, Indoor, and Mobile Radio Communications (PIMRC). Montreal, QC, Canada: IEEE; 2017. pp. 1-5

[17] Esposito C, De Santis A, Tortora G, Chang H, Choo KK. Blockchain: A panacea for healthcare cloud-based data security and privacy? IEEE Cloud Computers. IEEE. 2018;**5**:31-37

[18] Jiang S, Cao J, Wu H, Yang Y, Ma M, He J. Blochie: A blockchain-based platform for healthcare information exchange. In: Proceedings of the 2018 IEEE International Conference on Smart Computing (SMARTCOMP). Sicily, Italy: IEEE; 2018. pp. 49-56

[19] Theodouli A, Arakliotis S, Moschou K, Votis K, Tzovaras D. On the design of a Blockchain-based system to facilitate healthcare data sharing. In: Proceedings of the 2018 17th IEEE International Conference on Trust, Security and Privacy in Computing and Communications/12th IEEE International Conference on Big Data Science and Engineering (TrustCom/BigDataSE). New York, NY, USA: IEEE; 2018. pp. 1374-1379

[20] Peterson K, Deeduvanu R, Kanjamala P, Boles K. A blockchain-based approach to health information exchange networks. In: Proceedings of the NIST Workshop Blockchain Healthcare. Vol. 1. Los Angeles, CA, USA; 2019. pp. 1-10. Available from: https://www.healthit.gov/sites/default/files/12-55-blockchain-based-approach-final.pdf

[21] Al Omar A, Rahman MS, Basu A, Kiyomoto S. Medibchain: A blockchain based privacy preserving platform for healthcare data. In: Proceedings of the International Conference on Security, Privacy and Anonymity in Computation, Communication and Storage. Guangzhou, China, Berlin/Heidelberg, Germany: Springer; 2017. pp. 534-543

[22] Wang S, Wang J, Wang X, Qiu T, Yuan Y, Ouyang L, et al. Blockchain-powered parallel healthcare systems based on the ACP approach. IEEE Transactions on Computational Social Systems. 2018;**28**:942-950

[23] Bhuiyan MZ, Zaman A, Wang T, Wang G, Tao H, Hassan MM. Blockchain and big data to transform the healthcare. In: Proceedings of the ACM International Conference on Data Processing and Applications. Guangzhou, China: Association for Computing Machinery; 2018. pp. 62-68

[24] Rathore H, Al-Ali AK, Mohamed A, Du X, Guizani M. A novel deep learning strategy for classifying different attack patterns for deep brain implants. IEEE Access. 2019;**7**:24154-24164

[25] Rathore H, Fu C, Mohamed A, Al-Ali A, Du X, Guizani M, et al. Multi-layer security scheme for implantable medical devices. In: Neural Computing and Applications. Berlin/Heidelberg, Germany: Springer; 2018. pp. 1-14

[26] Rathore H, Mohamed A, Al-Ali A, Du X, Guizani M. A review of security challenges, attacks and resolutions for wireless medical devices. In: Proceedings of the 13th IEEE International Conference on Wireless Communications and Mobile Computing (IWCMC).

Valencia, Spain: IEEE; 2017. pp. 1495-1501

[27] Rathore H, Wenzel L, Al-Ali AK, Mohamed A, Du X, Guizani M. Multi-layer perceptron model on chip for secure diabetic treatment. IEEE Access. 2018;**6**:44718-44730

[28] Medtech. The U.S. Market for Medical Devices: Opportunities and Challenges for Swiss Companies. Bern, Switzerland: Medtech; 2017

[29] SelectUSA. Medical Technology Spotlight. Available online: https://www.selectusa.gov/medicaltechnology-industry-united-states [Accessed: January 11, 2017]

[30] Gordon, W.J.; Catalini, C. Blockchain technology for healthcare: Facilitating the transition to patient-driven interoperability. Computational and Structural Biotechnology Journal 2018, 16, 224-230

[31] Le Nguyen T. Blockchain in healthcare: A new technology benefit for both patients and doctors. In: Proceedings of the 2018 Portland International Conference on Management of Engineering and Technology (PICMET). Honolulu, HI, USA: IEEE; 2018. pp. 1-6

[32] Karafiloski E, Mishev A. Blockchain solutions for big data challenges: A literature review. In: Proceedings of the IEEE EUROCON 2017-17th International Conference on Smart Technologies. Ohrid, Macedonia: IEEE; 2017. pp. 763-768

[33] Xia QI, Sifah EB, Asamoah KO, Gao J, Du X, Guizani M. MeDShare: Trust-less medical data sharing among cloud service providers via blockchain. IEEE Access. 2017;**5**:14757-14767

[34] Yue X, Wang H, Jin D, Li M, Jiang W. Healthcare data gateways: Found healthcare intelligence on blockchain with novel privacy risk control. Journal of Medical Systems. 2016;**40**:218

[35] Huh S, Cho S, Kim S. Managing IoT devices using blockchain platform. In: Proceedings of the 19th International Conference on IEEE Advanced Communication Technology (ICACT). PyeongChang, Korea: IEEE; 2017. pp. 464-467

[36] Li Z, Kang J, Yu R, Ye D, Deng Q, Zhang Y. Consortium blockchain for secure energy trading in industrial internet of things. IEEE Transactions on Industrial Informatics. 2018;**14**:3690-3700

[37] Liu H, Zhang Y, Yang T. Blockchain-enabled security in electric vehicles cloud and edge computing. IEEE Network. 2018;**32**:78-83

[38] Stanciu A. Blockchain based distributed control system for edge computing. In: Proceedings of the IEEE 21st International Conference on Control Systems and Computer Science (CSCS). Bucharest, Romania: IEEE; 2017. pp. 667-671

[39] Lin C, He D, Huang X, Choo KK, Vasilakos AV. BSeIn: A blockchain-based secure mutual authentication with fine-grained access control system for industry 4.0. Journal of Network and Computer Applications. 2018;**116**:42-52

[40] Dorri A, Kanhere SS, Jurdak R, Gauravaram P. Blockchain for IoT security and privacy: The case study of a smart home. In: Proceedings of the 2017 IEEE International Conference on Pervasive Computing and Communications Workshops (PerCom Workshops). Kona, HI, USA: IEEE; 2017. pp. 618-623

[41] Hammi MT, Hammi B, Bellot P, Serhrouchni A. Bubbles of trust: A decentralized blockchain-based authentication system for IoT. Computers & Security. 2018;**78**:126-142

[42] Novo O. Blockchain meets IoT: An architecture for scalable access management in IoT. IEEE IOT Journal. 2018;**5**:1184-1195

[43] He Q, Xu Y, Liu Z, He J, Sun Y, Zhang R. A privacy-preserving IoT device management scheme based on blockchain. International Journal of Distributed Sensor Networks. 2018;**14**:11

[44] Ouaddah A, Elkalam AA, Ouahman AA. Towards a novel privacy-preserving access control model based on Blockchain technology in IoT. In: Europe and MENA Cooperation Advances in Information and Communication Technologies. Cham, Switzerland: Springer; 2017

[45] Dorri, A.; Kanhere, S.S.; Jurdak, R. Blockchain in IoT: Challenges and solutions. arXiv 2016, abs/1608.05187.2016:1-13

[46] Yuan Y, Wang FY. Towards blockchain-based intelligent transportation systems. In: Proceedings of the 2016 IEEE 19th International Conference on Intelligent Transportation Systems (ITSC). Rio de Janeiro, Brazil: IEEE; 2016. pp. 2663-2668

[47] Lei A, Cruickshank H, Cao Y, Asuquo P, Ogah CP, Sun Z. Blockchain-based dynamic key management for heterogeneous intelligent transportation systems. IEEE Internet of Things Journal. 2017;**4**:1832-1843

[48] Morgan YL. Notes on DSRC and WAVE standards suite: Its architecture, design, and characteristics. IEEE Communications Surveys and Tutorials. 2010;**12**:504-518

[49] Pedrosa AR, Pau G. ChargeltUp: On blockchain-based technologies for autonomous vehicles. In: Proceedings of the ACM 1st Workshop on Cryptocurrencies and Blockchains for Distributed Systems. Munich, Germany: Association for Computing Machinery; 2018. pp. 87-92

[50] Singh M, Kim S. Blockchain based intelligent vehicle data sharing framework. arXiv. abs/1708.09721.2017:1-4

[51] Rathore H, Samant A, Jadliwala M, Mohamed A. TangleCV: Decentralized technique for secure message sharing in connected vehicles. In: Proceedings of the ACM Workshop on Automotive Cybersecurity. Richardson, TX, USA: Association for Computing Machinery; 2019. pp. 45-48

[52] Singh M, Kim S. Trust bit: Reward-based intelligent vehicle commination using blockchain. In: Proceedings of the 2018 IEEE 4th World Forum on Internet of Things (WF-IoT). Singapore: IEEE; 2018. pp. 62-67

[53] Singh M, Kim S. Branch based blockchain technology in intelligent vehicle. Computer Networks. 2018;**145**:219-231

[54] Galvo JAL, Mathar R. Secure Blockchain-based communication scheme for connected vehicles. In: Proceedings of the 2018 IEEE European Conference on Networks and Communications (EuCNC). Ljubljana, Slovenia: IEEE; 2018. pp. 347-351

[55] Steger M, Dorri A, Kanhere SS, Römer K, Jurdak R, Karner M. Secure wireless automotive software updates using blockchains: A proof of concept. In: Springer Advanced Microsystems for Automotive Applications 2017. Berlin/Heidelberg, Germany: Springer; 2018. pp. 137-149

[56] CUBE. Autonomous Car Network Security Platform Based on Blockchain. White Paper, Cube. 2017. Available online: https://cubeint.io/wp-content/uploads/2019/10/Cube-Whitepaper-Centered-v2-3.pdf [Accessed: January 3, 2020]

[57] Basden J, Cottrell M. How Utilities Are Using Blockchain to Modernize the Grid. Boston, MA, USA: Harvard Business; 2017

[58] Alessandra P, Scarpato N, Di Nunzio L, Francesca F, Mario R. Smarter city: Smart energy grid based on blockchain technology. International Journal on Advanced Science, Engineering and Information Technology. 2018;**8**:298-306

[59] Knirsch F, Unterweger A, Engel D. Privacy-preserving blockchain-based electric vehicle charging with dynamic tariff decisions. Computer Science – Research and Development. 2018;**33**:71-79

[60] Gao F, Zhu L, Shen M, Sharif K, Wan Z, Ren K. A blockchain-based privacy-preserving payment mechanism for vehicle-to-grid networks. IEEE Network. 2018;**32**:184-192

[61] Mannaro K, Pinna A, Marchesi M. Crypto-trading: Blockchain-oriented energy market. In: Proceedings of the IEEE AEIT International Annual Conference. Cagliari, Italy: IEEE; 2017. pp. 1-5

[62] Lazaroiu C, Roscia M. Smart district through IoT and blockchain. In: Proceedings of the 2017 IEEE 6th International Conference on Renewable Energy Research and Applications (ICRERA). San Diego, CA, USA: IEEE; 2017. pp. 454-461

[63] Guan Z, Si G, Zhang X, Wu L, Guizani N, Du X, et al. Privacy-preserving and efficient aggregation based on blockchain for power grid communications in smart communities. IEEE Communications Magazine. 2018;**56**:82-88

[64] Gao J, Asamoah KO, Sifah EB, Smahi A, Xia Q, Xia H, et al. Gridmonitoring: Secured sovereign blockchain based monitoring on smart grid. IEEE Access. 2018;**6**:9917-9925 [CrossRef]

[65] Aitzhan NZ, Svetinovic D. Security and privacy in decentralized energy trading through multi-signatures, blockchain and anonymous messaging streams. IEEE Transactions on Dependable and Secure Computing. 2016;**15**:840-852

[66] Mylrea M, Gourisetti SNG. Blockchain for smart grid resilience: Exchanging distributed energy at speed, scale and security. In: Proceedings of the IEEE Resilience Week (RWS). Wilmington, DE, USA: IEEE; 2017. pp. 18-23

[67] Pop C, Cioara T, Antal M, Anghel I, Salomie I, Bertoncini M. Blockchain based decentralized management of demand response programs in smart energy grids. Sensors. 2018;**18**:162

[68] Cohn A, West T, Parker C. Smart after all: Blockchain, smart contracts, parametric insurance, and smart energy grids. The Georgetown Law Technology Review. 2017;**1**:273-304

[69] Knirsch F, Unterweger A, Eibl G, Engel D. Privacy-preserving smart grid tariff decisions with blockchain-based smart contracts. In: Springer Sustainable Cloud and Energy Services. Berlin/Heidelberg, Germany: Springer; 2018. pp. 85-116

[70] Popov S. The Tangle. White Paper. 2018. Available online: https://assets.ctfassets.net/r1dr6vzfxhev/2t4uxvsIqk0EUau6g2sw0g/45eae33637ca92f85dd9f4a3a218e1ec/iota1_4_3.pdf [Accessed: December 16, 2019]

Chapter 2

Blockchain-Enabled Smart Legal Contracts

Alan Ma

Abstract

A smart legal contract is a binding contract in which some or all the contractual terms are defined in and/or performed automatically by a computer program. It runs on a blockchain platform and carries the features of the blockchain of being automatically self-executed, and immutable, providing permanent records with real-time information, and reducing cumbersome documentation using high processing power. In major jurisdictions around the world, it is generally recognised that the smart legal contract is capable of having contractual force just like a traditional natural language contract. It has the potential to have entire complex commercial contracts written in and executed by computer codes. This chapter explains the concept and operation of the smart legal contract. Its advancement as an integral part of legal practices and a mainstream area of law is described in chronological order. The judgement of the first significant case relating to the use of smart contracts is unpacked. Whilst the acceptance of smart contracts by legal practices has gained pace, novel legal issues have been emerging in this area of law. This chapter identifies and proposes solutions to key legal issues arising from the operation of computer code and the resolution of disputes of smart legal contracts.

Keywords: blockchain, smart legal contract, digital platform, digital dispute resolution, arbitration

1. Introduction

Blockchain is a distributed ledger comprised of immutable blocks chained together to create an encrypted history of transactions. It carries the functionalities of automatically self-executed, avoiding human errors, immutable, providing permanent records with real-time information, and reducing cumbersome documentation using high processing power. Its embedded feature of transparency and integrity shows the reliable nature of the technology. These features are attractive for commercial use that involves a large amount of repetitive and similar transactions for example in the logistics sector for management of supply chains [1, 2], the financial services industry [3], the insurance industry to deal with payment of claims [4, 5], the handling of health care data [6, 7], the UK court justice system [8, 9], and the list is growing fast.

A smart legal contract is a computer program that defines obligations between parties and is performed, solely or in part, automatically by software algorithms.

It runs on a blockchain platform and carries blockchain features. It has the potential to replace the traditional natural language contract system so that the benefits offered by blockchain technology are fully utilised.

Commercial contracts have always been using natural language, and their development represents human civilisation in doing business [10]. It remains a scientific fantasy that offers no practical use to society if the smart contract is not recognised and enforceable as a way of entering contracts between parties. The smart contract is now developed in such a way that it is no longer the law of the future. However, as its applications become widespread, novel legal issues emerge. Each of these issues creates barriers to adoption in industries and practices. Identification and resolutions of these issues are essential for maximising the benefits of the technologies.

The objectives of this chapter are threefold. Firstly, there have been diverse discussions on the validity and enforceability of contracts written in computer code. This chapter seeks to clarify and unify the descriptions of blockchain-enabled smart contract that gives legal effects. Secondly, it provides an understanding of the development history of smart contracts to give further development of this practice area a sense of direction. The third objective is to identify the legal issues that arise from this novel practice area where the technology behind it is in a nascent state with vast scope for development and reaching maturity.

Section 2 of this chapter gives an account of the advance of the smart contract from its origin as a concept to today's status of being recognised by the legal profession as enforceable and does not require any law reforms for its application in England and Wales. The first reported court case that has significant impacts on the use of smart contracts is unpacked considering the current status of its recognition and applications. Section 3 identifies and proposes solutions to key legal issues arising from the operation of computer code. The current digital dispute resolution rules that apply to smart legal contracts are examined in Section 4.

At the outset, it is noteworthy that there are various types of emerging technologies, including artificial intelligence (AI) and blockchain. The current recognised practice of smart legal contracts is enabled by blockchain without any deployment of AI. This book chapter does not cover any discussions of the concept of combining AI and blockchain technology that potentially offers other versions of a smart contract.

It should also be noted that this book chapter builds on, but far from exclusively, the materials discussed and reported in the author's published work [11, 12]. Readers will realise that this chapter represents substantial work, in breadth and depth, beyond those reported in the previous publications.

2. The status of smart legal contracts

2.1 The origin

The concept of smart contracts was created by a computer scientist, Nick Szabo. In his publication in 1997 [13], he defined a smart contract as "A set of promises specified in digital form, including protocols which the parties perform on those promises." Szabo illustrated his definition by reference to a vending machine. A consumer inserts coins into the machine (satisfying the condition of the contract), and the vending machine automatically dispenses the chocolate bar (meeting the terms of the "contract"). The transaction is facilitated by software, which enables the transfer of output (the chocolate bar) on the occurrence of input (the correct payment). A smart contract is, therefore, a computer program that contains certain inputs and executes a set of instructions to come to one of many pre-determined outcomes upon the occurrence of a triggering event. It is a computer program that defines obligations between parties, and it runs on a blockchain. Since then, whilst the definition of a smart contract has been refined, many debates have been on whether the smart contract is a breakthrough discovery or a mere fantasy. Amongst these discussions, the terms, smart contract and smart legal contract, are now defined by the Law Commission of England and Wales (the "Law Commission") [14]. The smart contract is defined as "Computer code that, upon the occurrence of a specified condition or conditions, is capable of running automatically according to pre-specified functions". Whereas the smart legal contract is defined as "a legally binding contract in which some or all of the contractual terms are defined in and/or performed automatically by a computer program."

2.2 Smart contract v traditional natural language contract

Contracts define the rights and obligations between the parties. The legal principles of contract law are well established, and their application is pervasive in our daily life. Contracts written in the natural language can be traced back to the Middle Ages and they form the bedrock of doing business. No alternative forms had been considered until the birth of smart contracts. With such attractiveness emerging from blockchain technology, can smart contracts replace natural language contracts entirely? The question begs a comparison of the smart contract with the contract written in natural language.

A short answer is no, not at present. The smart contract has its limitations on handling only the pre-determined outcomes programmed in the software. By contrast, the nature of natural language contracts is flexible and ambiguous. It allows room for courts to interpret contracts and exercise judgement to maintain justice and fairness. Smart contracts cannot handle the nuances and intricacy of various clauses in complex commercial contracts.

Figure 1 sets out, in unidirectional, the various stages of the development of smart contracts with the eventuality of replacing natural language with computer code. Stage 0 is the traditional natural language contract that we have been using since

0	I	II	III	IV
Natural language traditional contract	Simple contract with encoded payment mechanism	Natural language contract with encoded performance mechanism	Switching between natural language contract and coded computer software	Replacing natural language with computer codes

Figure 1.
A unidirectional illustration of the smart contract development.

ancient times. It is possible to draw up a traditional contract with certain functions encoded in digital form. The transformation starts in Stage I by writing the straightforward rules-based functions responding to specific inputs, like payment is processed when the goods are delivered and verified. The logic is if the delivery of goods is fulfilled, the payment will be released in a simple contractual arrangement for the payment of a purchase.

Stage II extends the rules-based functions to more complex cases that define parties' performance obligations, like the automatic deduction of liquidated damages for late delivery of work when the level of damages is pre-estimated and agreed upon before the work starts. Therefore, the logic is the payment function will not be triggered until the goods are delivered. However, if the delivery is late, the sum equal to the liquidated damages will be deducted automatically.

Stage III is more advanced with interactions between the natural language contract and coded computer software. Following the above example, will the delay be caused by the occurrence of an event that excuses the late delivery and avoids the application of liquidated damages? The ultimate Stage IV is a complete replacement of a natural language contract with computer code.

2.3 Making an inroad

Even with the limitations that not all of them are fully automated by computer codes, smart contracts are efficient in many ways and generate positive impacts on the contracting life cycle. Optimum benefits can only be gained if smart contracts are recognised and used by the legal professions. In England and Wales, over a period of four years, blockchain-enabled smart contract progressed from being the least accepted emerging technology to being recognised as an instrument fulfilling the same function as the traditional natural language contract. **Figure 2** shows the rise of

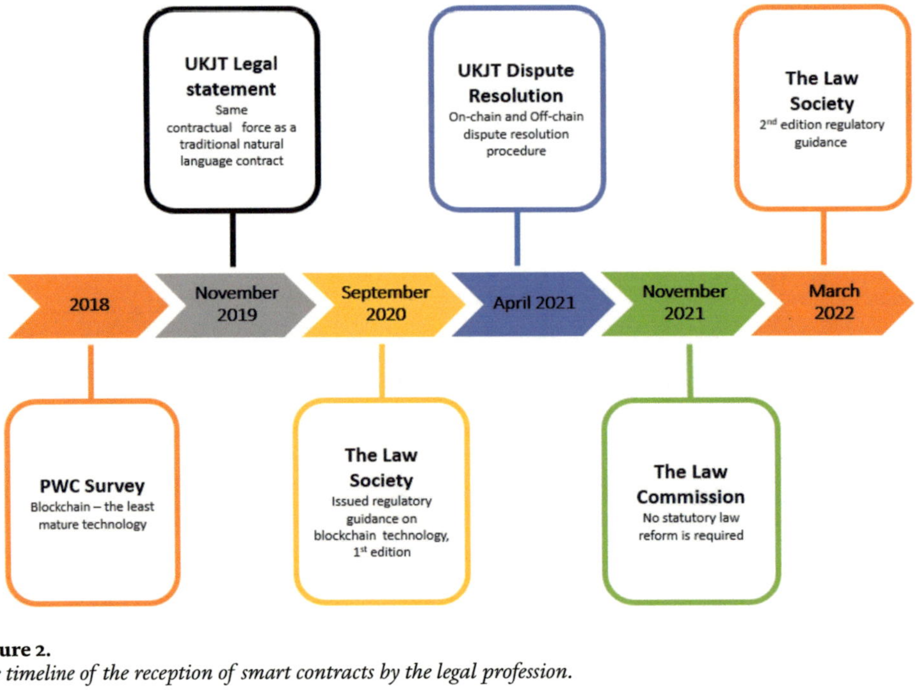

Figure 2.
The timeline of the reception of smart contracts by the legal profession.

smart contracts from an emerging technology that aids to define and perform contractual obligations to becoming an integral part of legal practices and a mainstream area of law.

In 2018, the legal profession did not embrace blockchain technology with much enthusiasm. This is reflected in the survey done by PWC on the top 100 law firms regarding the use of emerging technologies in legal practice [15]. The survey reveals that blockchain technology was not interested/aware and it was classified as being under research by law firms. The technology was the least mature in application amongst other digital and emerging technologies at the time, like Artificial Intelligence, Robotic Automation, and Predictive Analytics. Since then, the reception of smart contracts by the legal industry has grown from zero to formally embraced as an instrument that is capable of having contractual force just like a traditional natural language contract. This is demonstrated by a series of publications from the government-backed UK Jurisdiction Taskforce, the Law Society, and the Law Commission of England and Wales.

In 2019, the UK Jurisdiction Taskforce (UKJT) published a legal statement on smart contracts [16]. Based on extensive consultation from a broad range of experts: technologists, legal practitioners, judiciary, and academics, the legal statement states that a smart contract satisfies the legal requirements of creating an agreement between parties and is capable of having contractual force just like the traditional natural language contract. The rights and obligations written in computer code are binding and enforceable under the well-established English contract law.

In 2020, the Law Society of England and Wales published regulatory guidance on blockchain technology [17]. The guidance provides a rich source of information on the best practices for legal practitioners working on transactions involving smart legal contracts. This represents a formal recognition of the accepted use of smart contracts. A significant implication of this publication is that every practising lawyer in England and Wales must be conversant with the application and operation of smart contracts as per the professional code of conduct. Keeping the regulations in sync with the other advancements, the second edition of the report was published in March 2022 [18].

In 2021, on consideration of whether statutory reform is required to deal with the increasing use of smart contracts, the Law Commission carried out an extensive study on the application of existing English law to the smart contracts [14]. It should be noted that each country has its own legal system. In general, the legal systems in countries around the world can be classified as either common law or civil code. In common law countries, the system is based on case law derived from judicial decisions; whereas the civil code is built on codified statutes [19].

The Law Commission concluded that "the current legal framework is clearly able to facilitate and support the use of smart legal contracts". They confirmed that statutory law reform is not needed as common law in general can respond consistently and flexibly to new commercial mechanisms. Judges can apply and adapt existing principles to new situations as they arise. The Law Commission's study supports the earlier UKJT legal statement that a smart contract can be enforced and used as a natural language contract.

2.4 Smart contracts and civil code

The above-mentioned reports and statements represent the current status of legal opinion and practices in England and Wales. The Law Commission report [14] reveals that in the United States of America, which is common law jurisdiction, several states, including Arizona, Nevada, Ohio, and Tennessee have provided legislation to protect the legal effects of smart legal contracts. The report also shows that smart legal contracts are recognised, with or without legislation, in Australia, China, Dubai, Estonia, India, New Zealand, Sweden, and Switzerland. These jurisdictions include both common law and civil codes countries. On this basis, it appears that it does not matter whether the jurisdiction is based on common law or the civil codes insofar as the smart legal contracts are concerned. Indeed, the issues in implementing smart contracts in any legal system should be the same irrespective of the structure, practice, and overarching principles of the legal system. However, there is not enough evidence to draw any conclusion that it is no difference in the application of smart legal contracts given that there are some fundamental differences between common law and civil code jurisdictions in the context of contract law, including the requirement and consideration in the formation and conceptual interpretation.

2.5 Forms of smart legal contracts

Based on the degree of automation in a given contract, the Law Commission characterises smart legal contracts in three forms as follows:

Form 1 - Natural language contract with automated performance: a contract in natural language but includes agreement from certain aspects of the contract to be performed using a computer program designed for this purpose. In Form 1, the role of the code is limited to performing obligations, and the parties' obligations are defined in the natural language contract. This is currently the most common form of smart legal contract being used.

Form 2 - Hybrid contract: In a hybrid contract, the contractual terms are defined both in natural language and code. The code defines contractual obligations as well as performing them. This leads to the situation where some or all of the contractual obligations are performed automatically as per the code written.

Form 3 - Solely coded contract. In this form, all the contractual obligations between parties are defined and performed automatically by computer code. e. This represents a total replacement of the natural language contract. As explained above, the current technology behind the smart contract does not reach this stage. The Law Commission classifies that this form represents the extreme of the spectrum and considers the use of solely coded contract as "low".

The subsequent discussions will follow the three forms as characterised by the Law Commission.

2.6 Case law

The development of smart legal contracts in the common law jurisdictions will be influenced and steered by court decisions on any cases that come to light. Quoine Pte Ltd. v B2C2 Ltd. [20] is the first significant case that concerns cryptocurrency trading and has implications for the interpretation of smart contracts. The decisions of the Singapore High Court and Court of Appeal have been widely reported and analysed in various jurisdictions across the globe.

The case concerns contracts made by two computer programs using deterministic algorithms that are executed automatically. In a trading of cryptocurrency, B2C2 traded Ethereum in exchange for Bitcoin with Quoine on Quoine's automated trading platform. B2C2's computer program initiated the trade and made an offer to sell in

Ethereum and the automatic function of Quoine's platform accepted that offer, leading to an exchange of cryptocurrency. It credited B2C2's account with the proceeds of the trades without human intervention. However, caused by an error in the automated contracting system, the exchange rate calculated by the algorithm was 250 times higher than the true value of the currency. When the technical officer later manually reviewed the trades, he discovered the error and reversed the trade. B2C2 disagreed and claimed that Quoine was in breach of the contract between the parties. Quoine's position was that if there was a contract, such contract was vitiated because of the mistake.

In its judgement, the Singaporean court confirmed that the automation process gives rise to a binding contract for the sale of cryptocurrency. The parties held their programs as a mechanism for reaching an agreement and were bound by the agreements that were entered into by those programs. This part of the decision confirmed the contracts entered automatically by computer codes, like the smart legal contracts, create legal effects and are enforceable.

In considering the unilateral mistake made by Quoine's computer program, the courts concluded that the intention of the party should be assessed at the time when the computer code was written, and what is relevant is the state of the mind of the programmer. Applying the principles developed by the successive cases based on traditional natural language contracts, the courts found, in the instant case, that unilateral mistake did not vitiate the contract as it did not go to the terms of the contract.

In assessing the remedy of the breach of contracts, considering the automatic self-execution of the trading system and the cryptocurrency volatility of the market in which the system operated; the court decided that a remedy mechanism of reversing the trades at prevailing rates was not appropriate. This approach was the same to that of the breach of traditional natural language contracts when operating in a volatile environment. Given the common feature of computer programs, like smart contracts, for being self-execution, automotive and, high-speed that by the time when a breach occurred in a specific transaction is discovered, many other subsequent transactions would be completed, it is envisaged that the application of specific performance as a remedy will be limited.

Quoine v B2C2 is the leading case to date concerning contracts entered by computer programs without human intervention. It showed how the courts apply well-established legal principles to deal with novel issues arising from the use of emerging technologies. No doubt, as the adoption of smart legal contracts becomes more and more common, further court decisions will provide further and better guidance by the legal precedents. This is how the contract law in the natural language has been developing since ancient times and the experiences and wisdom generated by the case law will guide and steer the maturity of the law of smart contracts.

3. Specific legal issues

3.1 Additional steps and process

The usual procedure for parties entering a commercial contract is that the parties negotiate the heads of terms on the commercial arrangement of the transaction. They seek legal advice on the risks and any appropriate compromises to close the deal. The lawyers' task is to find out their clients' needs and protect their clients' interests and draft a contract for agreement with the opposing party. The terms and the clauses, which reflect the intention of the parties, should be clear without ambiguity.

Traditionally, the negotiation, instructions to lawyers, and the drafting of a contract are carried out in natural language. The formation and execution of a smart legal contract between parties involves additional steps and processes. Coupled with its self-execution and immutable nature, novel legal issues emerge. This section focuses on certain specific key legal issues that emerge from the use of smart legal contracts as discussed below. Other issues that are in common with digital technologies like data privacy and cyber security are not considered.

3.2 Interpretation of computer codes

Language is often susceptible to more than one possible meaning. When things go wrong, parties argue about the meaning of the language used in the way that suits them. It is not uncommon that the courts are asked to interpret what was agreed between the parties. The court applies an objective test to identify the intention of the contracting parties. The court refers to "what a reasonable person having all the background knowledge which would have been available to the parties would have understood them to be using the language in the contract to mean" [21].

In all the 3 forms of a smart legal contract, computer codes have a role to play. Form 1 does not pose any issues as the code is used as a tool to assist the execution of the performance; it does not define the obligations between parties. As for the other two forms, an issue arises as to how to interpret the code that defines the contractual obligations between parties which are based on the intention of the parties at the time the contract was entered. As in the interpretation of the contractual terms in natural language, the Law Commission adopts the objective test. The question to ask is what the term would mean to a "reasonable coder", a person with the knowledge and understanding of code at the time when the codes were written.

It should be noted that the "reasonable coder" approach is consistent with the current approach adopted by the courts to call expert evidence to assist on the meaning of contractual terms drafted in a foreign language. It is also observed that the reasonable coder approach is consistent with that taken by the courts in Quoine v B2C2, as discussed above.

3.3 Third-party coders

Forms 2 and 3 of the smart legal contracts require computer code input. The additional process of "translating" the intention of the parties by writing the terms and conditions of the agreement in computer code introduces a number of risk elements that require additional terms for parties to agree upon. It should be noted that the contractual relationship in a transaction extends beyond the parties themselves as they either jointly or independently instruct a third-party coder. The risks arising from the addition coder are:

- Loss in the "translation". Errors are caused by the communication between the coder and the client because either the coder fails to understand the instructions from the parties, or the parties fail to provide clearly defined instructions.

- Errors in the data input or human intervention in the operation.

- Defective coding and software, there is no bug-free software.

- Events that are out of the control of the parties.
- Writing blockchain software that fits for purposes requires specialist skill, the exercise of which require reasonable skills and care. The coder could be negligent in undertaking the tasks.
- An obligation defined by the natural language narrative as well as by the computer code may be in conflict. A clause is required to stipulate which one takes precedent.

3.4 Oracles

The current status of smart legal contracts is that the percentage of forming and performing a contract purely using computer codes is low (i.e Form 3). For the other two forms, interaction with external sources is facilitated by using oracles to retrieve off-chain data and information and push them to the chain. For example, in a situation to decide whether a threshold of adverse weather condition is reached on a specific day in order to decide whether a contractual event is triggered. Data from the metrology office can be injected into the chain via the oracle.

Risks arise from a failure of the oracle or inputting incorrect data in their smart legal contract. Terms and conditions on the reliability, accuracy, and timing of the external data sources are required to be included in any contracts between the parties as users and the external data sources provider.

3.5 Self-execution nature and immutability

The superior ability of smart legal contracts is their self-execution and immutability. Once the chain is activated, the performance of the contracts is automatic without human intervention, and the records are immutable and permanent. This means that the coded terms cannot be changed once the smart contract has been entered into the distributed ledger on the blockchain platform. The implication is that it would be difficult to amend a smart legal contract should an error is found during the process. The high-speed ability powered by the computer processing power means that an error in one block could propagate to many subsequent transactions causing widespread disruptions. Parties are required to get it right the first time to avoid undesirable possible disastrous consequences.

3.6 Blockchain platform

Smart legal contracts run on a blockchain platform. The digital platform stores information and records transactions. It drives self-execution of the blocks and the distribution of ledgers to the users. Generally, blockchain platforms fall into two types of ecosystems, (a) open (public, permission) and (b) close (private, non-permission). Most business-to-business commercial transactions operate in a closed environment, where participants are joined using a private key 28 and all users can be identified. As in any shared access, users are operating on mutual trust and cooperative natures. It is well settled that under English law, obligations imposed by good faith are not enforceable. If the parties wish to impose such a duty, they must do so expressly [22].

To set out explicit contractual terms for the shared users of a platform, the starting point is to examine the arrangement of the organisations that develop and operate blockchain platforms. These organisations are separate entities. They joined together and formed a blockchain consortium. Currently, there are in general 3 types of consortia models (a) contractual consortium model, (b) joint venture model, and (c) developer agreement and participant agreement models. A comprehensive discussion on the advantages and disadvantages of the different business models for forming and operating a consortium is given in the Law Society for England and Wales' guidance on blockchain [18]. It is concluded that there is no preferred model as the choice is subjective to the specific requirements and aspirations of the consortium members in a particular sector.

In using any of the models, contractual arrangements must be put in place to govern and regulate the obligations and commitment of all parties concerned in two generic groups: (a) amongst the consortium members regulating the structure and governing principles in the development and running of the platform - the consortium agreement, and (b) the participation agreement that regulates the rules of joining the platform in respect of the use of the platform.

The challenges and issues associated with the consortium governance and the participation agreements are of no difference to those that apply to a joint industry multi-stakeholder enterprise project. The current practice is that contractual arrangements are dealt with in the traditional way by a raft of natural language instruments, like formal contracts supplemented by protocols, policies, standards and regulations.

At the top level of the consortium governance, the key issues are decision-making authority, funding, costs, income allocation, legal entity structures, risk allocation, and identification and ownership of intellectual property. At the day-to-day operation level, issues of the duties and the identity of the information manager, the agreement with the software providers or developers regarding their obligations, the performance level, and availability of the network, and specification of the intellectual property ownership.

The participant agreement sets out the rules for joining the platform. Attention should be given to the allocation of responsibility for the operation of the platform and coordination of the data, allocation of liability, risk and responsibility for errors, access to data in the system, data privacy and cybersecurity.

3.7 Jurisdiction and governing law

The law that governs a contract determines its validity, effect and discharge, directly affecting the rights and obligations of the parties. As explained in Section 2.3 above, a country's legal system is unique. The applicable law that governs a contract formed and performed in one country is not necessarily to be the same as that in another country.

Jurisdiction refers to the legal authority of a court or tribunal to exercise justice in matters. In other words, which court or tribunal has the power to hear and resolve disputes arising between the parties. Jurisdiction and governing law are usually discussed together in the context of cross-border transactions. They are related but governing law is not the same as jurisdiction. The legal system of a country (i.e., a jurisdiction) can apply the governing law of a different country.

In the absence of explicit provisions in a contract that stipulate the governing law and jurisdiction, the issues arising from the choice of law and jurisdiction are dealt with in accordance with the principles developed under the practice area of private international law. In the context of the smart legal contract, can the same principles apply to the traditional natural language contracts apply?

The Law Commission [14] identified certain jurisdictional issues unique to smart legal contracts, like the physical location of the defendant, the contract's place of formation, the third-party coders who concluded the contracts on behalf of the parties, the facts need to be examined when determining the most significantly connected legal system, difficulty in locating the exact location where the breach concerning a digital asset rather than a physical asset in the real-world location. Further work is being commissioned by the Law Commission.

But one thing for sure is that the court will follow the explicit agreement between parties. To avoid future controversy, parties are well advised to include a governing law clause stating expressly the parties' choice of law that applies. Likewise, the contract should contain a jurisdiction clause stating that the parties have agreed to the courts/tribunals of a named country taking jurisdiction over any disputes that may arise. However, parties must be careful in specifying which law of the country will apply with due consideration of the fact that the extent of recognition of smart legal contracts is not uniformly recognised in different jurisdictions.

3.8 Insurance

Parties can manage risk by seeking appropriate insurance to cover the possible consequences of the risks identified, including service interruption, loss of income, hardware damages, and reputational damage. The novel nature of legal risks emerging from smart legal contracts calls for specialised coverage to complement the traditional insurance policies. Many unusual or specialised threats are included as optional cover and parties must ensure they understand the insurance plan. Rather than relying on the insurance policy to cover all conceivable losses, parties should have their systems in place to mitigate risks to avoid monetary burden of premium.

4. Bespoke dispute resolution rules

4.1 Introduction

In a contracting life cycle, an important piece of the jigsaw is how to resolve disputes arising from the contract. The self-execution nature of blockchain technology means that the terms of an agreement can be automatically implemented, and their fulfilment can be automatically executed. In an ideal world, disputes could be avoided altogether. However, in day-to-day commercial transactions, it is common for disputes to occur.

In April 2021, on the premises of the recognition of smart contracts by the legal industry and they are firmly established as part of the English legal system, UKJT published its Digital Dispute Resolution Rules for matters of smart contracts, distributed ledger technology, and other digital assets, ("Rules") [23]. This section examines the key features of this bespoke arbitration procedure for digital disputes like those generated from smart contracts.

4.2 The practices of arbitration

The standard process for resolving disputes between parties is through the national courts by way of litigation. The Rules specify disputes arising from smart legal contracts are to be resolved by arbitration. The use of arbitration as a dispute resolution mechanism alternative to court proceedings can be traced back to the 7th Century and international arbitration has been the preferred method of resolving cross-border disputes for years [24]. The arbitration rules and procedures are well-established.

In a nutshell, arbitration is a consensual process such that parties submit their disputes for settlement by an independent arbitral tribunal that has gone through a judicial process of hearing the evidence and arguments from both parties. The arbitral tribunal is usually made up of one or three arbitrators. Arbitrators are under a strict duty to act fairly and impartially between the parties. The award made by the arbitral tribunal is final and binding with limited grounds for appeal. In England, the basis of the appeal is either for reasons of improper administration of justice under s67 or s68 of the Arbitration Act 1996 or on a point of law under s69. The rate for a successful appeal is very low. Based on the court records between 2015 to 2018, the success rate was 0.02% during that period [25].

Through the Convention on the Recognition and Enforcement of Foreign Arbitral Awards 1958 (the New York Convention), an award made in a signatory country is enforceable in all other participating states. As of December 2021, a total of 169 countries have signed up for the New York Convention [26]. This means an arbitral award made in one contracting state is recognised and enforceable in more than 85% of countries around the world. This is a particularly attractive feature of arbitration given the global nature of blockchain networks.

There are institutions located around the world that offer the administration of arbitration, for example maintaining a panel of arbitrators for the appointment, providing fee scale, and helping with the arrangement of the arbitration process. These institutions have their sets of procedural rules, which are revised from time to time ensuring their efficiency and effectiveness. For example, during the Covid_19 pandemic, rules were revised to allow ongoing proceedings to continue [27].

The current practice of arbitration with the involvement of technology is limited to using technology to aid procedural matters, like the electronic filing of notices and documents in the arbitration proceedings, and holding virtual and/or hybrid hearings.

4.3 The new rules

A distinct advantage of arbitration over litigation is its procedural flexibility which allows tailored modification for digital disputes matter. In this way, radical reform of the national court's civil procedural rules can be avoided. The Rules aim to fill the gaps in the current practices of arbitration for resolving commercial disputes that involve novel digital technology like smart contracts. The unique features of the Rules include:

4.3.1 A rapid process

The default timeframe for the entire arbitration process is 30 days from the initiation of the arbitration proceedings to the issue of a binding decision. The procedure is designed to be short for efficiency, although the tribunal has the absolute power to modify it. The parties are not entitled to any oral hearings. The entire process can be completed on-chain, including the enforcement of the award.

4.3.2 Arbitration agreement

A cardinal principle of arbitration is that it is a consensual process - no arbitral agreement means no arbitration. The Rules provide a standard dispute resolution clause (which may be in electronic or encoded form) for parties to incorporate the Rules into a smart legal contract before a dispute has arisen. The agreed arbitration rules and procedures are stored in the block and activated if certain pre-defined events are triggered or at the request of a party.

An important consideration of using arbitration is the ability to enforce an award which is not given in the same jurisdiction via the New York Convention. However, care must be taken to comply with the requirements of the convention, one of which is that the arbitration clause must be in writing, see Article II(2) of the New York Convention 1958. To address this issue, parties should enter into a specific written agreement as in the conventional arbitration agreement.

4.3.3 The applicable law

Even if the agreement is silent on the choice of applicable law, the Rules specify that the law of England and Wales is the applicable law by default. This gives certainty and avoids controversy as to which is the governing law as discussed above.

4.3.4 Appointment of the arbitral tribunal

One feature of arbitration is that the arbitrators should be clear about the specific nature of the disputes with relevant knowledge and experience in the subject matter. Given the nascent state of smart legal contracts, an arbitrator who meets the requirements would be rare. The Rules name the Society for Computers and Law as the appointing body, but no detailed procedure for how an arbitrator is selected and appointed. So far, the appointing body does not maintain a panel of suitably qualified arbitrators. Parties can control the selection process by expressing preferences in advance as to their number, identity or qualifications but the final appointment is subject to the agreement with the appointing body. More information on the appointment of the arbitrators from the appointing body is needed to allow early adoption of the process by industry.

4.3.5 The power of the arbitral tribunal

Once the arbitral tribunal is formed, the tribunal has absolute authority over what procedure to adopt; the only restriction is that the tribunal must be sure the procedure is fair. The Rules allow for totally or partially off-chain arbitration as the tribunal sees fit.

The "off-chain" arbitration means that the dispute is to be resolved in a classic dispute resolution mechanism. As such, the arbitrator is given a private key to access the blockchain data, examine the evidence (documentary or oral hearings) and make decisions in the usual manner. It should be noted that the interplay of on-chain and off-chain arbitration requires the service of oracles, and the parties must allow for making such arrangements.

4.3.6 Implementation of a decision

The Rules allow the arbitral tribunal to implement its decision "on-chain" using a private key. This means the tribunal has the power to make changes to the smart contract as per its decision, which may reverse a transaction or allow the transaction to continue and makes changes to the digital assets. No additional steps are required to enforce the decision. Furthermore, the right to appeal under the Arbitration Act 1996 is specifically removed. It is debatable whether such a fast-track procedure with no appeal right meets the requirement of natural justices.

4.3.7 Enforcement of award

Given the applicable law is under English law, an award can be enforced through the court proceedings in England and Wales The multi-jurisdiction nature of blockchain makes the enforcement via the New York Convention attractive. However, the enforcement of a foreign award may be refused because the local legislation rules that matters relating to digital assets as illegal on the ground of public policy. Parties are advised to check the status of local law from time to time as a measure to manage such risks.

4.4 The way forward

To date, the Rules are untried with perceived uncertainties. It is only a matter of time before the first case is tested. This area of law will continue to develop, mature, and refine. The release of the Rules by UKJT, a UK government-backed body, represents the first significant step toward making progress in the resolution of digital disputes.

5. Concluding remarks

The legal professions are catching up with the advancement of technology behind smart legal contracts. Not long ago, smart contracts were considered the law of the future - it is now obsolete. The series of reports published by legislators, lawyers, and the judiciary in major jurisdictions around the world means that smart legal contracts are ready to be put into everyday use. Contracts written in computer codes are gradually becoming the norm. Businesses and individuals should be prepared for the next normal. The starting point must be to have an understanding of the development of smart legal contracts, which is explained in this chapter.

This nascent practice area generates novel legal issues. Key issues are identified in this chapter. Good practices to manage the threats from these risks are discussed. To complete the jigsaw of a contracting life cycle, a bespoke dispute resolution process for smart legal contracts is also examined.

The technology behind smart contracts inevitably continues to evolve, changing how smart contracts operate and pushing the boundaries of their applications. At the same time, novel legal issues emerge. The cycle of technological advancement, the discovery of legal issues, and overcoming them continues in the journey of revolutionising the contracting practice through smart legal contracts. The way forward must be to promote the crossflow of information and knowledge between the legal and technology professions to achieve optimum collaboration for future advancement. This book chapter is published with this spirit in mind.

Author details

Alan Ma[1,2]

1 England and Wales, and Scotland (On the Roll of the Law Society of Scotland), United Kingdom

2 Birmingham City University, England

*Address all correspondence to: siuyungalan.ma@bcu.ac.uk

References

[1] PWC. Blockchain in logistics. 2020. Available from: https://www.pwc.be/en/FY21/documents/blockchain-in-logistics.pdf. [Accessed: July 15, 2022]

[2] DHL. Blockchain in logistics: Perspectives on the upcoming impact of blockchain technology and use cases for the logistics industry. Available from: https://www.dhl.com/content/dam/dhl/global/core/documents/pdf/glo-core-blockchain-trend-report.pdf?j=112522&sfmc_sub=108088830&l=59_HTML&u=6150683&mid=7275327&jb=10. [Accessed: July 15, 2022]

[3] Statista. Blockchain applications within financial services industry worldwide as of 2021. 2022

[4] IBM. Transforming insurance management with IBM Blockchain. 2018

[5] Built-in. How using blockchain in healthcare is reviving the industry's capabilities. 2022. Available from: https://builtin.com/blockchain/blockchain-healthcare-applications-companies. [Accessed: July 15, 2022]

[6] Digital Authority Partners. Blockchain in Healthcare: An Executive's Guide for 2022. 2021. Available from: https://www.digitalauthority.me/resources/blockchain-in-healthcare/. [Accessed: July 15, 2022]

[7] HIMSS (Healthcare Information and Management Systems Society). Blockchain in Healthcare. Available from: https://www.himss.org/resources/blockchain-healthcare. [Accessed: July 15, 2022]

[8] UK Government, Ministry of Justice Digital Strategy 2025. Policy Paper, April 2022. Available from: https://www.gov.uk/government/publications/ministry-of-justice-digital-strategy-2025/ministry-of-justice-digital-strategy-2025. [Accessed: July 15, 2022]

[9] Accenture. Digitally transforming the justice system. 2021. Available from: Digitally Transforming the Justice System | Accenture. [Accessed: July 15, 2022]

[10] Mehren A. Contract. Encyclopedia Britannica. 2019. Available from: https://www.britannica.com/topic/contract-law. [Accessed: July 15, 2022]

[11] Ma A. Emerging legal issues in blockchain for construction supply chains. In: ICVISP 2020: Proceedings of the 2020 4th International Conference on Vision, Image and Signal Processing. 2020. pp. 1-7. DOI: 10.1145/3448823.3448880

[12] Ma A. The legal advancement of smart contracts, iTNOW the BCS (British Computer Society), the Chartered Institute for IT's magazine. Available from: https://www.bcs.org/articles-opinion-and-research/the-legal-advancement-of-smart-contracts/. [Accessed: July 15, 2022]

[13] Szabo N. Formalizing and securing relationships on public networks. First Monday. 1997;**2**(9). Available from: https://firstmonday.org/article/view/548/469

[14] Law Commission. Smart legal contracts, advice to government, CP563. 2021

[15] PwC. Resilience through change, 27th Annual Law Firms' Survey, London. 2018

[16] UK Jurisdiction Taskforce. Legal statement on cryptoassets and smart contracts. 2019

[17] The Law Society for England and Wales. Blockchain: Legal & Regulatory Guidance. First ed. London: The Law Society for England and Wales; 2020

[18] The Law Society for England and Wales. Blockchain: Legal & Regulatory Guidance. Second ed. London: The Law Society for England and Wales; 2022

[19] Kiralfy A, Lewis A, Glendon M. Common law. Encyclopedia Britannica. 2020. Available from: https://www.britannica.com/topic/common-law. [Accessed: July 15, 2022]

[20] B2C2 Ltd v Quoine Pte Ltd [2019] SGHC(I) 3; Quoine v B2C2 [2020] SGCA(I) 2 – SICC / Singapore Court of Appeal

[21] Lord Hoffman in Chartbrook Ltd v Persimmon Homes Ltd [2009] UKHL 38, para 14

[22] Astor Management AG & Anr v Atalaya Mining Plc & Others [2017] EWHC 425 (Comm)

[23] UK Jurisdiction Taskforce. Digital Dispute Resolution Rules. London: UK Jurisdiction Taskforce; 2021

[24] Queen Mary University of London, 2021 International Arbitration Survey: Adapting arbitration to a changing world, London 2021

[25] Sanderson B, Scott M. Appeals under the English Arbitration Act 1996, May 2018, DLA Piper publications Available from: https://www.dlapiper.com/en/uk/insights/publications/2018/05/appeals-under-the-english-arbitration-act-1996/. [Accessed: July 15, 2022]

[26] United Nations Commission on International Trade Law. Convention on the Recognition and Enforcement of Foreign Arbitral Awards (New York, 1958) (the "New York Convention"). Available from: https://uncitral.un.org/en/texts/arbitration/conventions/foreign_arbitral_awards. [Accessed: July 15, 2022]

[27] Cleary Gottlieb. International Arbitration in the Time of COVID-19: Navigating the Evolving Procedural Features and Practices of Leading Arbitral Institutions, Alert Memorandum. 2020. Available from: https://www.clearygottlieb.com/-/media/files/alert-memos-2020/international-arbitration-in-the-time-of-covid19.pdf. [Accessed: July 15, 2022]

Section 2

Legal, Regulatory and Ethical Considerations

Chapter 3

Artificial Intelligence and Blockchain: Debate around Legal Challenges

Antonio Merchán Murillo

Abstract

The rapid technological change and its development have led to an era of technology and applications that are leading us to transversal changes, based on the data that feed the Internet. This change is taking place through the Internet of Things (IoT), machine-to-machine (m2m) communications, robotics, big data, blockchain, and artificial intelligence (AI). In this context, it can be seen how artificial intelligence and blockchain are opening up, posing specific legal challenges to those imposed by technologies. Recent developments in AI are the result of increased processing power. The emergence of new traceability and authentication as blockchain can allow recording assets, participant transactions, which can provide valuable information on the origin and history. These facts should lead to study and to make regulatory proposals and to study the existing legal framework.

Keywords: artificial intelligence, ethics, legal framework, blockchain, trust

1. Introduction

Recent discussions have revolved around the need to regulate the AI sphere itself and set limits, to prevent the so-called artificial general intelligence from developing. On the other hand, blockchain is becoming popular, and its continuous attempts to implement it have curiously given confidence to the digital society of the future, although it does pose serious challenges.

Although it must be kept in mind that artificial intelligence (AI) is going to pose challenges in its application, we must not lose sight of the blockchain, which, although it focuses on validation, permanence, and achieving higher levels of certainty, control, and trust, is going to pose the challenge of acting together with AI; that is, blockchain has the mission of generating trust, transparency, and acting as a mediator. So you are going to have the challenge of making it possible for AIs to act and connect with each other.

However, we cannot forget the commercialized cloud (computing) services either. The cloud has become the perfect scenario for AI, and also for blockchain, since, for the development of these applications, not only is a significant calculation capacity necessary during learning

2. Artificial intelligence: aspects to consider

AI-based systems can simply consist of software (e.g., voice assistants, image analysis programs, search engines, voice and facial recognition systems), but AI can also be embedded in devices hardware (e.g., advanced robots, self-driving cars, drones, or Internet of Things applications).

AI is used daily, for example, to translate from one language to another, generate subtitles in videos, or block unsolicited email (spam). Far from being science fiction, AI is already part of our lives, in the use of a personal assistant to organize our workday, in the movement in a self-driving vehicle or in the songs or restaurants suggested by our phones.

AI is about developing systems capable of solving problems and performing tasks by simulating intellectual processes. The AI can be taught to solve a problem, but it can also study the problem and learn how to solve it itself without human intervention. Different systems can achieve different levels of autonomy and can act independently. In this sense, its operation and its results are unpredictable, since these systems function as "black boxes" [1].

Today, there are various definitions of artificial intelligence [2]. However, none of them has been universally accepted [3], a fact that leads us to the first challenge, making a timeless, general, and at the same time, robust definition of AI, especially when one thinks of AI and its normative regulation.

A certain issue cannot be regulated without establishing a solid definition of what is regulated. Therefore, it is essential to establish a generally accepted definition of AI that is common and flexible and does not hinder innovation, considering that AI is becoming more and more sophisticated.

The principles enunciated by UNCITRAL when establishing in their Model Laws on Electronic Commerce or on Electronic Signature the procedures and basic principles, as well as fundamental, to facilitate the use of modern techniques can serve as a starting point of communication, in order to record and communicate information in various types of circumstances, such as: nondiscrimination, neutrality with respect to technical means, and functional equivalence. These principles are widely recognized as fundamental elements of electronic commerce [4]. At the same time, they are reflected in the enunciation of the requirements that electronic communications must meet.

In this way, a common European definition must be established [5], including the definitions of its subcategories, considering the following characteristics:

a. the ability to acquire autonomy through sensors and/or through the exchange of data with its environment (interconnectivity) and the analysis of the said data;

b. the ability to learn through experience and interaction;

c. the form of the physical support of the robot;

d. the ability to adapt their behavior and actions to the environment.

2.1 Debate of regulations

In our opinion, we must consider a series of methodological and substantive questions regarding the regulation of technology in general and robotics and AI in particular. In the first place, it is worth asking whether sufficient arguments can be identified to accommodate these new technologies and therefore justify a change in the existing legal framework, that is, are existing laws sufficient to meet the regulatory challenges of the technology, and if not, should some laws be adapted to include the new technology, generally by making the language of the law more technology-neutral or rather should they be sui generis laws? With the existing administrative regulations, the lack of interoperability is guaranteed [6].

As we know, interoperability is a very important issue. When we talk about this, we mean that the processes, technologies, and protocols required to ensure the integrity of the data and the identity of the citizen, when they are transferred from one system to another, must entail, by definition, a correct interconnection of the systems and data exchange. However, this is not always carried out, let's think about the right of access.

At the technical level, although standards abound, the lack of common basic standards for some technologies must be highlighted. At the legal level, the laws that prescribe a specific predominant technology are pointed out as factors that impede progress, the difficulty of those responsible in understanding their respective frameworks of mutual trust and even in the areas of liability and compensation [7]. An example of the problem of interoperability is, and sometimes continues to be, the situation that is present in Spain at a regional level and in the EU: State authorities throughout Europe offer electronic access, focusing, above all, on the needs and national media, which has generated a complex system with different solutions, which has given rise to new obstacles to cross-border exchanges, which hamper the functioning of the single market for companies and citizens.

Second, it is necessary to clarify the direct and indirect role that ethics can play in the regulation of technology [8]. The impact of AI is cross-border. The EU is realizing this, trying to regulate its sphere and establish limits. The debates point to 11 areas: ethics, security, privacy, transparency, and accountability, work, education and capacity development, inequality and inclusion, legislation, and regulation; governance and democracy, war, and superintelligence.

These debates are justified and should be taken into account. However, they are part of a larger problem, related to the insufficient conception of artificial intelligence that society has, which makes trust difficult, and to the current laws, which have not yet recognized the specific characteristics of artificial intelligence [3]. In this way, the need arises to analyze and deepen an ethical framework so that both citizens and companies can trust the technology with which they interact, have a predictable legal environment, and have the effective guarantee that they will protect themselves and their fundamental rights and freedoms.

Let us think that technologies based on artificial intelligence influence aspects such as health, safety, productivity, or leisure, and in the medium term, they will have a great impact on energy, transport, education, and domestic activities. Regarding education, it is essential to find new models and methodologies that integrate ethical concerns in relation to the impact of artificial intelligence on humanity, especially in everything related to security, freedom, privacy, integrity, and dignity; self-determination and nondiscrimination, and the protection of personal data [9].

The complexity of AI entails the need to create an ethical and efficient framework, for which the principle of transparency must be based on, which consists of the fact that it must always be possible to justify any decision that has been adopted with the help of artificial intelligence and that can have a significant impact on the life of one or more people. On the other hand, it should always be possible to reduce the calculations of the AI system to a form understandable to humans.

3. Blockchain

Blockchain is a distributed information processing technique on which different treatments and business models can be implemented. The blockchain is a large database that is distributed among various nodes that participate in the chain. This functions as an immutable logbook that contains the complete history of all transactions that have been executed on the network5. These nodes are connected in a decentralized network, without a main computer, and they are networks called P2P (network of peers, network between equals) that communicate with each other using the same language that they transmit a message, called a token. The P2P network is a computer network in which all or some aspects work without fixed clients or servers, but rather a series of nodes that behave as equals to each other. They act simultaneously as clients and servers with respect to the other nodes of the network, allowing the direct exchange of information in any format between the interconnected computers. Usually, this type of networks is implemented as overlay networks built in the application layer of public networks as in the case of the Internet. A token (symbol, signal, or token) is a representation of the information that the network contains [10]. The information travels encrypted, due to this, it is distributed without revealing its content, as the number of transactions grows, the chain of blocks grows, and each block has its own digital footprint. Its scope is immense, and Ethereum could replace basically any intermediary, substituting products and services that depend on third parties to be totally decentralized.

3.1 Types

Depending on the permissions required to be part of a blockchain, three categories can be distinguished:

A. Public: Where anyone can download the necessary programs on their computer and set up a node and participate in the consensus process, anyone who is a party can send transactions through the Internet, which will be included in the blockchain.

B. Federated or consortium: In this class, they do not allow anyone to configure a node on their PC and participate in the transaction validation process since access permission is needed, which is usually granted to members of a certain group, for example: the group of financial entities.

C. Private: In these blockchains, the authorizations to carry out transactions are conceived by private organizations that will determine the conditions under which they will allow the reading of the transactions carried out.

3.2 Features

A. Immutable. Nobody can alter or delete the data in the registry or add new content without any validation. When a transaction occurs, all nodes on the network will have to say that it is valid, or it will not be added to the record.

B. Decentralization. There is no single person or governing authority that reviews the framework.

C. Origin: All nodes can verify the moment in which a certain asset has been registered in the block chain, who was its first owner, and all the subsequent changes of ownership that have occurred up to that moment.

D. Security. All registry data are strongly encrypted, using cryptography, which is one of the most complex mathematical algorithms in existence.

E. Each block has a unique hash ID (fingerprint) and changing it is impossible.

F. To carry out a transaction in the blockchain, the use of public and private keys is necessary.

G. Register distributed. Due to the nature of the technology, all nodes maintain the registry and therefore, the computational power is distributed among them, and the nodes act as verifiers.

H. Consensus. It is a determining factor. Without consensus, the system does not work. For the information contained in a block to be considered valid, all participants must agree. The network developer needs to implement some kind of consensus algorithm.

I. Speed. It offers a faster result. For example, a transaction might take a few minutes to complete.

4. Artificial intelligence and blockchain: the need for a joint study

Recent developments in AI are the result of increased processing power, improvements in algorithms, and exponential growth in the volume and variety of digital data. Many AI applications have begun to enter our daily lives, from machine translations to image recognition to music generation, and are increasingly being implemented in industry, government, and commerce. Connected and autonomous vehicles and AI-supported medical diagnostics are soon-to-be-common application areas.

Now, for the other to happen, there must be a communication between the machines that must be validation and with a high level of certainty and control. For this reason, we consider that joint action, if it is not obvious that it must take place, will be important. In fact, it is an issue that is being debated more and more globally.

In this context, a question of trust, transparency, reliability, speed, and effectiveness in automatic electronic transactions arises.

The emergence of new traceability and authentication systems, such as blockchain, can make it possible to record assets, transactions, and participants, which can provide valuable information about origin and history.

In this way, solutions based on blockchain can enable the rapid detection of possible illicit or defective actions within the system itself, products to illicit markets. In this way, it is evident that the deployment of digital technologies, such as blockchain, is key to the development of AI.

In this context, Regulation (EU) 2018/1807 of the European Parliament and of the Council, of November 14, 2018, regarding a framework for the free circulation of non-personal data in the European Union, arises from the need to establish administrative cooperation, based on the review of the European Interoperability Framework; however, with this standard we only intend to make you see that development is a reality.

Now, it should be borne in mind that the current regulations have not yet recognized the specific characteristics of the contracts that may arise, and neither blockchain technology nor artificial intelligence, it is true that in the cloud environment (computing), things have been developed in UNCITRAL.

For example, in the absence of direct legal regulation of AI, article 12 of the United Nations Convention on the Use of Electronic Communications in International Contracts, which establishes that a person (either a natural or legal person) on whose behalf a programmed computer was used should be responsible for any messages generated by the machine. However, in the explanatory note, UNCITRAL makes it clear that this article is an enabling provision and should not be misunderstood as transforming an automated message system or a computer into a subject of rights and duties. Electronic communications that are generated automatically by a computer or messaging system without human intervention should be interpreted as "coming from" the legal entity on behalf of which the computer or messaging system operates. The issues relating to the subject of the action that could arise in this context must be settled in accordance with rules outside the Convention, which returns us to the previous point. In any case, in our opinion, it is of vital importance to establish a legal framework, especially in an international context, in which its real applicability is included [11].

To give security to the transaction to be carried out, blockchain arises, as a decentralized technology, which entails some legal uncertainties, such as the legal nature of blockchains and shared digital records, which includes problems of judicial jurisdiction and applicable law; thus, each network node may be located in a different place as there is no "central party" responsible for the digital registry, whose nationality could serve for regulation.

As we say, as the Internet becomes part of everyday life, the need arises to study the adaptation of private international law systems to the new demands [12]. For this reason, our intention is to analyze, deepen the debate, and respond to the problems, succinctly raised, in order to contribute, as far as possible, to give certainty to the responsibility, due diligence, contracts on intelligence systems artificial, as well as the condition of artificial intelligence and the attribution of its acts of legal significance and the use of blockchain technology in the formation of smart contracts, to mention some pertinent issues, as we say, from a private international law perspective.

4.1 The existing international law

The agreements attributing jurisdiction included in the underlying contract are relevant to the extent that they are effective in accordance with the rules of our private international law system (in principle, in accordance with the provisions of Regulation 1215/2012 – Brussels Reg. I bis), which significantly restricts its operability in transactions with consumers.

Aside from these transactions, it will be necessary to observe the eventual incidence of special rules such as that of article 25.2 Brussels I bis Regulation with respect to the written form in electronic contracting, presupposition of the effectiveness of the agreements attributing jurisdiction. Likewise, the situations in which the special jurisdiction of article 7.1 Regulation 1215/2012 becomes applicable must be observed, and it may be controversial to what extent the automation of certain benefits conditions the determination of the place of fulfillment of the obligation for the purposes of that rule.

In relation to the applicable Law, we are going to bear in mind issues that have to do with the meaning of the automated fulfillment of certain commitments and its interaction with the underlying relationship between the parties. The basic criterion is that the provisions of the Rome I Regulation (including its specific consumer protection regime and the application of rules transposing the Directives on consumers) must be followed in principle in relation to the underlying transaction between the parties, without ignoring that the application of certain provisions may be a source of controversy—for example, its art. 14.2 and the specification in this framework of the place of compliance—as well as that other regulatory instruments may also be relevant.

For example, the origin criterion of the Directive on electronic commerce in relation to the contractual aspects included in the coordinated scope of the Directive when it is applicable. It will also be necessary to take into account the existence of issues subject to autonomous connection, among others, the ability to contract and, very especially, everything related to the applicable regime in terms of personal data protection, in which it will be necessary to comply with the provisions in article 3 RGPD regarding the need to comply with its provisions in the situations included within its scope of territorial application.

However, notwithstanding the foregoing, the evidentiary effectiveness to accredit transactions or other circumstances within the framework of judicial proceedings will in principle be determined, as a procedural matter, by the lex fori.

In the context of the EU, Regulation 910/2014, regarding electronic identification and trust services for electronic transactions in the internal market and which repeals Directive 1999/93/EC, is of particular relevance for these purposes.

4.2 Problems associated with AI, blockchain

All this in accordance with the principle of preexistence of existing law. Now, in the underlying contractual relationship we are going to find cases in which the technology can be given a non-nationality, as we have seen before, and we may encounter problems in the connection criteria, cross-border insolvencies that blockchain detects, etc., are problems that, if not now, they will be over time, when artificial intelligence advances and also companies evolve to the cloud.

Why?

The blockchain poses different risks because of the technology and the way of operations: One of the main problems that will affect the blockchain is the inability to control and stop its operation. In addition, the lack of control over the operation can lead to the lack of responsibility of the company that manages the platform. Let us think that, in its simplest form, blockchain is a decentralized technology or a distributed ledger in which transactions are recorded anonymously. This means that the transaction ledger is simultaneously maintained on a network of unrelated computers or servers.

Therefore, the allocation and attribution of risk and responsibility in relation to a blockchain service that is not working properly will have to be carefully analyzed, not only at the provider-customer level, but also around all the participants in the system.

It should be noted, regarding the process, that blockchain has the ability to cross-jurisdictional boundaries since the nodes in a blockchain can be located anywhere in the world.

This can pose a series of complex problems that require careful consideration in relation to citizen-State, company-State, company-company, citizen-company, company-administration, citizen-Administration relations of the same State, and of different ones. In this regard, it should be noted that, in a decentralized environment, it may be difficult to identify the appropriate set of rules to apply. Estonia does.

And it does so by proposing electronic identity as a connection criterion, since it is related to residence, in this case electronic [13], insofar as the need to link information and its management, solely, with the person who issues it, becomes essential for numerous different interactions: an organizational infrastructure (identity management) and a technical infrastructure (identity management systems), to develop, define, designate, manage, and specify authorization levels, assigning roles and identity attributes related to specific groups of people, such as company directors, employees, or customers.

The evolution of technology is creating large electronic files, with it, large commercial and state databases. A national identifier, contained in an identity card, allows capturing information about a person, which is found in different databases, so that they can be easily linked and analyzed through certain data analysis techniques. At the same time, ID cards are also getting smarter. The generation of data also has the potential to be offered in a medium where they can be directly processed. In this way, files that can be crossed and structured, as well as transferred, are created. For this reason, you have to pay special attention to any identity management system and see who are where people are.

At its simplest level, each transaction could fall within the jurisdiction of the location of each node in the network. With this, it should be noted that in an online environment, authenticating the identity of the remote party is more important than ever. It plays a key role in the fight against identity fraud and is also essential to establish the necessary trust that facilitates any type of electronic transaction.

At this point, it should be taken into account that the relationship between Law and IT goes beyond what has been seen so far [14]. That is why one of the main issues that arise, with respect to cross-border services, is the security and confidentiality of information transmitted over the Internet, which should lead us to guarantee the protection of personal data that lead to the identification of its owner.

By this, we mean that the electronic identity is an identity that is made up of information stored and transmitted to the different users of it. Let us think that the identity is a fundamental element, which links the information to its owner, located in some State, giving rise to its location, and therefore, to the effective and safe handling of the specific data that enter the cloud.

We must keep in mind that all electronic identity schemes depend on two processes: first, identity authentication and, later, identity verification. When authenticated, the identity is registered in the system and can then be used for transactions. Identity is verified at the time of each transaction, from within the cloud itself. From the information registered at that moment, the identification information arises or that will identify the person, as if it were the signature, which will be used, later, to link an individual in an inseparable way.

We observe that within the cloud, there will be two elements that will come together to facilitate the identity of the person who intends to access the cloud. These two fundamental elements are the identity fixed to the individual and another fixed to the transaction that is carried out [15]. The first will be the one that identifies the parties and, therefore, will have a direct effect on the formation and enforceability of the contract, thus determining its capacity to be contractually bound, by including elements such as the name of the legal person, its form legal, its registration number in the registry (if applicable), its registered office or address of the business center, together with the mention of its founding documents. The second would be the largest body of transaction information, and it is continually updated, based on the transactions you make in the cloud.

If the above is not enough, it will be necessary to assess the specific contract and the specific links, of the said contract, with the different countries attending, as indicated by the CJEU, in its Judgment of October 23, 2014, case C-305/13 , to the "overall assessment of all the objective elements that characterize the contractual relationship and assess the element or elements that, in his opinion, are more significant" and "in the event that it is alleged that a contract has closer ties with a country" other than the country whose law is designated by virtue of the presumption established in said section, the national court must compare the ties between the contract and the country whose law is designated by virtue of the presumption, on the one hand, and between the contract and the other country in question, on the other. The national judge must consider all the circumstances that concur, including the existence of other contracts related to the contract in question [16].

In this way, issues related to identity management must be regulated by the different legal systems that discipline the multiple activities of the specialized operators that carry out the identification tasks and the functional operators. In this context, it must be considered that the eIDAS Regulation does not impose the creation of national electronic identification schemes as such, but rather aims to guarantee their interoperability by applying the principle of mutual recognition.

Let us think that the identity is a fundamental element, which links the information to its owner, located in some State, giving rise to its location and, therefore, to the effective and safe handling of the specific data that enter the cloud. We must keep in mind that all electronic identity schemes depend on two processes: first, identity authentication and, later, identity verification. When authenticated, the identity is registered in the system and can then be used for transactions. Identity is verified at the time of each transaction, from within the cloud itself. From the information registered at that moment, the identification information arises or that will identify the person, as if it were the signature, which will be used, later, to link an individual in an inseparable way.

On the other hand, we could find ourselves in situations in which the parties intervene in unequal conditions, the applicable laws on contracts usually establish that the contract in question is a contract of adhesion. Service providers are often familiar with a limited number of local laws, and especially local laws governing contracts and the right to privacy. For that reason, they will proceed to choose an applicable law that establishes the requirements related to the protection of information that the service provider in question can or is willing to comply with, which offers rules for the elaboration of contracts that are predictable and acceptable to their purposes.

These issues may force the client company to assume certain responsibilities toward its end users in relation to the determination of the applicable type Law, which may be abusive.

Given this, as indicated by the Judgment of the CJEU, dated July 28, 2018, case C-191/15, when appreciating the abusive nature of a certain contractual clause in the framework of an action for injunction, of article 6, section 2 , of the Rome I Regulation, it turns out that the choice of the applicable law is made without prejudice to the application of the mandatory provisions provided for by the law of the country in which the consumers whose interests are defended by means of that action reside.

Such provisions may include those transposing Directive 93/13, provided that they guarantee a higher level of protection for the consumer. At this point, the main legal risk that arises for the client company is not being able to fully assess the risks linked to the contract, for example, ignorance of the weak points inherent in the technology that is being used; missing or inadequate security features; economic risks linked to the loss of data or breaches of the agreement, etc.

5. Conclusion

The need arises to analyze and deepen an adequate legal framework, so that both citizens and companies can trust the technology with which they interact, have a predictable legal environment, and have the effective guarantee that their rights will be protected and freedom.

For this reason, we propose the need to review certain criteria to prevent liability from being diluted with respect to the operations carried out by service providers, if the criteria for the applicability of the legislation are not sufficiently clear.

Additional information/notes/thanks/other declarations

The information in this chapter is based on the work published, by the author, in 2019 in the Administrative Law Review, 2019 (50). Being, therefore, an update, evolution, and revision of the conclusions reached.

Author details

Antonio Merchán Murillo
Pablo Olavide University, Spain

*Address all correspondence to: amermur@gmail.com

References

[1] UNCITRAL. Legal issues related to identity management and trust services. New York Times. 2017

[2] Čerka P, Grigienė J, Sirbikytė G. Is it possible to grant legal personality to artificial intelligence software systems? Computer Law & Security Review. 2017; **33**:685

[3] UNCITRAL. A/CN.9/960 – Work Programme of the Commission – Legal Aspects of Smart Contracts and Artificial Intelligence: Submission by the Czechia. New York: United Nations; 2018

[4] Illescas R. Law of Electronic Contracting. 2nd ed. Madrid: Civitas; 2019. p. 508

[5] European Commission. Artificial Intelligence for Europe {SWD(2018) 137 final}, Brussels, 25.4.2018 COM(2018) 237 final. 2018

[6] López de Mantarás R. Ethics in artificial intelligence. Research and Science. 2017;**48**:49

[7] Gurkaynak G, Yilmaza I, Hakseve G. Stifling artificial intelligence: Human perils. Computer Law & Security Review. 2016;**32**:758

[8] Čerka P, Grigienė J, Sirbikytė G. Liability for damages caused by artificial intelligence. Computer Law & Security Review. 2015;**31**:376-389

[9] European Commission. Proposal for a Regulation of the European Parliament and of the council laying down harmonised rules on artificial intelligence (Artificial Intelligence Act) And Amending Certain Union Legislative Acts {SEC(2021) 167 final} - {SWD(2021) 84 final} - {SWD(2021) 85 final} Brussels, 21.4.2021. COM(2021) 206 final. 2021/0106(COD). 2021

[10] De Filippi P, Wright A. Decentralized blockchain technology and the rise of lex cryptographia. 2015. Available from: https://www.intgovforum.org/cms/wks2015/uploads/proposal_background_paper/SSRNid2580664.pdf

[11] UNICTRAL. Text—Explanatory Note United Nations Commission on International Trade Law. New York: United Nations; 2007

[12] BURDEN K. EU update. Computer Law & Security Review. 2018;**34**:418

[13] Diago, MªP. La residencia digital como nuevo factor de vinculación en el Derecho Internacional Privado del Ciberespacio ¿posible conexión de futuro? Diario LA LEY. 2014;**8432**(2):2-8

[14] Viguera Revuelta R. The computer contracts. Revista de la Contratación Electrónica. 2008;**97**:61

[15] Cuthbertson A. Estonia First Country to Offer E-Residency Digital. International Business Times. 2014. Available from: https://www.ibtimes.co.uk/estonia-first-country-offer-e-residency-digital-citizenship-1468766

[16] UNCITRAL. A/CN.9/WG.IV/WP.162 - Draft Provisions on the Use and Cross border Recognition of Identity Management and Trust Services. New York: United Nations; 2020

Chapter 4

Perspective Chapter: Cryptocurrencies Effectiveness for Nature

Luiz Cruz Villares

Abstract

The rise of cryptocurrencies based on Blockchain platforms have provided multiple solutions for social and nature projects supported by concerned investors with sustainable development initiatives. Speculation and unclear uses of a cryptocurrency plays a negative role for the projects they claim to support. A positive relationship between coin investors and supported projects must position the coin value on the scale of the community involvement among the coin and project issues, thus placing the project results above speculative moves. Coin nature and social based projects may include a decentralized autonomous organization (DAO), combined with a digital currency to contribute to social and nature improvements. This organization provides a framework for the engagement of investors, beneficiaries, and implementation partners, with results measured by reliable third parties. The potential funding from non fiduciary sources for sustainable development targets may be framed under the fundraising and financial solutions models, addressing the cryptocurrency volatility risks with responsible tokenomics in attention to transaction and regulatory issues. Overall, the more clear are the object and transaction issues of a nature conservation project supported by a currency, the more successful it will be in terms of nature and social improvements and the currency valuation for all parties involved.

Keywords: blockchain, cryptocurrencies, tokens, tokenomics, nature conservation

1. Introduction

This is a study about cryptocurrencies and Blockchain (blockchain) platforms related to nature, having direct or indirect purposes to help the conservation or recovery of forest areas. The major liasons between cryptos and nature assets come in the forms of carbon credits, tree planting and nature restoration projects. Blockchain has created a new and innovative framework that provides a safe, efficient and transparent chain of transactions. Citizens wordwide daily trade digital currencies utilizing a chain of transactions registered on the web, with no central server, or proprietors, only provided by the Blockchain. Along with cryptocurrencies, blockchain has enabled the creation of a vast array of solutions for businesses and society because it provides effective uses for (1) funding sources and exchange of currency, (2) smart contracts, and (3) reporting key information for all parties involved directly and indirectly in any relationship.

Since the creation of Bitcoin [1] cryptocurrencies have been designed to be free from government and central monetary regulations[1]. Most of cryptocurrencies following Bitcoin, have been launched with unclear or weak proposals, other than speculation, rapidly becoming unvalued assets, popularly referred as "shitcoins", showing a repetive short valuaton life presenting a common "launch-peak-devaluation" pattern. Such massive devaluations have confirmed the absence of most cryptocurrencies' roles as tradeable assets and exchange of currency.

A cryptocurrency is commonly referred as a digital currency or token. Generally, a digital currency is launched through an Initial Coin Offering (ICO) in the forms of tokens, attracting initial investor interests, mainly based on speculative targets, followed by exponential increase in quotations, just to slump to very low values thereafter causing investor to dump their coins to capitalize on short-term gains [2]. This typical "boom-collapse" standard is observable in some coins launched with nature conservation objectives and related purposes, thus raising questions about the effectiveness of their roles tied to their promised objectives of nature conservation. This study presents four parts, being:

- a summarized review of cryptocurrencies volatility and the energy consumption for their mining;
- an analysis of blockchains solutions in front of nature conservation, raising questions and proposals about the effectiveness of coins linked to nature conservation purposes, especially for forest conservation initiatives, with carbon credits;
- a brief presentation of models for cryptocurrencies and nature conservation effectiveness; and
- Conclusion.

The analysis of a sample of more than 30 coins/tokens and blockchains indicates that most of them have weak assurances of the real benefits they claim to provide to nature conservation. In some cases, the issue may fall to weak information about their indicators about their related nature objects. However, the general perception of this study is that information on the majority of tokens do not specify clear and measured results of their purposes. A specific consideration is made to tokens related to carbon credits, with a brief exposition of the size and complexities associated to climate change issues. Finally, the study addresses key points and ideas associated to coins and nature, aiming at providing a mutual gain for coin valuation and social benefits and nature conservation.

2. General cryptocurrencies overview

2.1 The rise of cryptocurrencies, volatility and setbacks

Since the creation of Bitcoin in 2008, the world has seen the booming rise of decentralized, trustable platforms supporting coins, tokens and all sorts of more than

[1] As cryptocurrencies have grown more popular, initial government regulations have been imposed in some countries, for example, in China.

20.000 digital currencies (altcoins) with a total market value of nearly U$ 3 trillion at the end of 2021. However, in the first months of 2022, post Covid 19 pandemic inflationary effects and the Russian-Ukrainian war induced a massive devaluation of cryptoassets, with their value falling to just $1.3 trillion in May, 2022. Such volatility supported long time opinions about the predominant wrong uses of most cryptocurrencies, rather for speculative purposes and illicit uses, such as money-laundering and "off taxes" transactions [3]. The 2022 devaluation suggested an overall crypto market value correction for most tradeable coins and the demise of value for the majority of coins (**Figure 1**) [4].

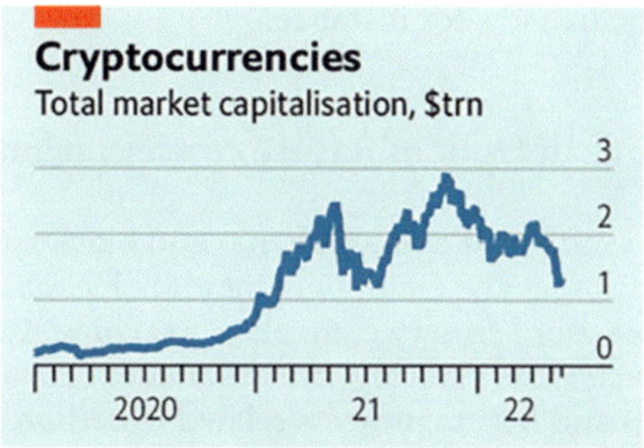

Figure 1.
Market capitalization of cryptos. Source: The economist.

Many years before the 2022 devaluation, numerous analyzes have covered this issue. Despite their astonishing price appreciation in recent years, cryptocurrencies have been subjected to accusations of pricing bubbles central to the trilemma that exists between regulatory oversight, the potential for illicit use through its anonymity within a young under-developed exchange system, and infrastructural breaches influenced by the growth of cybercriminality. Each influences the perception of the role of cryptocurrencies as a credible investment asset class and legitimate of value [5].

2.2 Energy consumption

Given the core subject about cryptocurrencies and nature, this study addresses basic environmental impacts, specifically about the energy consumption used in the mining of blocks associated to Bitcoin and altcoins transactions. Some cryptocurrencies require solving difficult cryptographic puzzles to adding transactions to a digital ledger—a blockchain—demanding verifications by algorithms. All those calculations consume energy. By May, 2022, Bitcoin currently consumed around 148 Terawatt Hours per year—0.55% of global electricity production, or roughly equivalent to the annual energy draw of countries like Malaysia or Sweden [6]. Bitcoin mining uses a proof-of-work (PoW) system. In brief, If a miner is successful, he can propose a new block of transactions to the blockchain, and receive a reward. Bitcoin mining uses a reasonable share of renewable (non fossil) energies, estimated to be between 40% and 64% at the end of 2022 [7].

Generally, all altcoins, mainly all listed in next session, are based on an alternative consensus process—proof of stake (PoS)—that presents much lower energy consumption for mining a block, comparing to PoW systems. Whereas PoW consumes 707 Kwh for a Bitcoin block mining, Ethereum, the second most popular coin (and Blockchain) requires about 64 Kwh; and secondary alternate coins consume less than

4 Kwh for mining a block. Ethereum started to move its consensus process to PoS in 2022, further reducing its block consumption to low units per Kwh as well.

Despite a positive drive to low mining energy needs, other issues such as regulations and taxations may impose a risk disincentivizing PoS coins, exactly because their ilicit uses and extreme volatilies. Given such uncertainties, these issues of energy and regulations present overall concerns for the energy balance of a cryptocurrencies for general and specific purposes, moreover, for nature conservation. Yet that "nature coins" transactions are mined through PoS systems, the underlying basis of transactions for these tokens are implicitly linked to Bitcoin, as the master cryptocurrency for exchange means, indirectly related to all coins, thus, offering questions about the "greeness" of any criptocurrency, for instance.

3. Blockchain solutions in front of nature conservation

This study assessed a sample of coins (tokens) and blockchains launched for nature conservation purposes. The choosen coins and blockchais include the ones with roles for fauna protection; fishery control; forest conservation initiatives, with and without social projects; tree planting; miscellaneous sustainability projects; and, a relevant share of forest and nature projects related to carbon sinks with carbon credits accounting. They are listed by their major roles and objectives in **Table 1**.

The sample of coins and blockchains analyzed offered relevant findings about their tokenomics[2] and roles in providing solutions related to sustainability issues. Their objectives (or promises) are widely diversed, such as purposes for fundraising, investment businesses, digital market platforms, carbon offsetting services, forest conservation, and animals monitoring. The general impression of the group of coins analyzed is that they present clear tokenomics, thus providing basic transparency to investors, however, they show a massive general devaluation pattern from its ICO to present time. In this pattern, the "boom-collapase" pattern is clearly observable with astonishing appreciations in the first months after its launch, sometimes with itermediary peaks and lows, then, followed by a massive depreciation, in many cases to values well under their initial price offering. In some cases, they show an eventual boost of appreciation in the course of their depreciation, due to massive sell-offs and other causes [8]. Such volatity supports the rather inappropriate role of these currencies to serve as store of value, means of exchange and registry, thus narrowing their use mainly for speculative moves. Rest aside their roles to provide funds for nature related projects. In this case, the analysis of information provided in the coins' websites (**Table 1**) provide a range of information suchs as the number of planted trees, metric tons of carbon storage, hectares of protected forests, animals monitored, and the like. In some cases, no information whatsoever is provided, or only vague information about "projects to be supported", under partnerships to come, and else.

The general perception of a weak object achievement by most coins related to nature is also seen from the distance between their positive market attractiveness to their effectiveness for the conservation of a piece of nature, moreover, when local populations are included, thus inducing the analysis of social and lifequality improvements. These information is not observed in any project related to forests, being, either with carbon credit metrics or not. Nevertheless, nature conservation

[2] Tokenomics is the set of elements that make a particular cryptocurrency valuable and interesting to investors. It includes issues related to the token's supply and how it is issued to subjects like its utility or objectives.

Name	Coin/Token	Blockchain	Major Purpose/objective	Website
Cecil Alliance	X		Fauna	https://www.cecilalliance.com/
Faunachain	X		Fauna	https://faunachain.io/
Reforestum	X		Fauna	https://www.cecilalliance.com/
Fishcoin	X		Fishery	https://fishcoin.co/
Ecofolio	X		Forest	https://www.ekofolio.com/
MCO2—Moss Earth	X		Forest Carbon Credits	https://nft.moss.earth/
ZCO2	X		Forest Carbon Credits	https://savethegreen.world/brasil/zco2-token
Amazonas Coin	X		Forest Carbon, Social	https://amazonascoin.com.br/
Gainforest	X		Forest Carbon, Social	https://www.gainforest.app/
Invert	X		Forest Carbon, Social	https://letsinvert.io/
KlimaDAO	X		Forest Carbon, Social	https://www.klimadao.finance/
Treedefi	X		Forest Carbon, Social	https://treedefi.com/
SOS Amazonia	X		Forest, Social	https://tokensosamazonia.com
Tupan	X		Forest, Social	https://www.tupan.io/
Wildchain	X		Nature, Fauna	https://wildchain.io
Coin Merit		X	Carbon	https://www.single.earth/
Regen Tokens		X	Carbon	https://www.regen.network/
Adaptation Ledger		X	Climate projects	https://www.adaptationledger.com
Carbonfuture		X	Climate projects	https://www.carbonfuture.earth/
DAO IPCI		X	Climate projects	https://ipci.io/
Nori		X	Climate projects	https://nori.com/
Earthledger		X	Sustainability Projects	https://earthledger.one
Opensc		X	Sustainability Projects	https://opensc.org/

Name	Coin/Token	Blockchain	Major Purpose/objective	Website
Plastic Bank		X	Sustainability Projects	https://plasticbank.com/
Poseidon		X	Sustainability Projects	https://poseidon.eco/
Cardano 1 Million Trees		X	Tree Planting	https://cardanofoundation.org/en
Forestcoin		X	Tree Planting	https://forestcoin.earth/
Global Mangrove Trust		X	Tree Planting	https://globalmangrove.org/
Landlife		X	Tree Planting	https://landlifecompany.com/
Seeds		X	Tree Planting	https://joinseeds.earth/
Teratree		X	Tree Planting	https://www.teratree.com/
Treecoin		X	Tree Planting	https://treecycle.ch/
Plantgrowsave		X	Tree Planting	https://plantgrowsave.org/

Table 1.
Blockchains and tokens assessed.

projects, especially in tropical forests, are generally the result of indigenous and/or local populations efforts to protect the forest, thus, being the human presence, part of the solution, and included in the conservation efforts with a combination of social, environmental, and economic results for a responsible project for nature. Blockchains and coins related to nature tend to overview all social roles and indicators related to nature conservation. Their essential offer tends to be resumed to viable tokenomics.

The related distant relationship between cryptos and their nature objects is also well observed in games and NFTs[3] for animals, images and other nature related objects. The gaming market has grown exponentially in the second decade of the twenty-first century. Estimates are that already 3 billion people are addicted to games, moving an industry of above U$170 billion in 2020 to reach an expected value of U$ 315 billion by 2026, doubling in size. The industry is no longer restricted to a niche, since it has become a global sector with economic dynamics that go way beyond the game's limits [9]. Some digital currencies and tokens include games as a mean to attract and retain investors to buy and stake the coin, with additional rewards coming from games, many of them related to nature objects. In this case, it is clearly observable the profile of gamers playing to win rewards with nature and fauna icons, however, with a distant relationship and effective support to effective nature objects. The "play-to-earn" mode uses nature and animals as a market attractiveness to engage players in the game, but the flows of funds to effective nature conservation projects resulting from games are to be further observed.

3.1 Blockchain and the commodification of nature

As observed, the coins and blockchain solutions for nature tends to be over appreciated and the reality of their nature related purpose, only measured by figures—stored in a blockchain—not really indicative of the complexity of a forest or natural resource situation, combining social and environmental issues. This issue is supported by a study entitled "Smart, Commodified and Encoded: Blockchain Technology for Environmental Sustainability and Nature Conservation" [10], providing complementary assessments to this issue.

This study framed a group of 27 blockchain technologies for conservation and environmental policy, part of them, here also analyzed (**Table 1**). The study reveals that a an outstanding characteristic of blockchain initiatives is "that their technical savvy, financial wizardry, and ingenious entrepreneurship are not always matched by a sophisticated understanding of the issues they support. This lack of understanding produces framings which then require blockchain-based interventions". They argue that the studied coins and blockchains represent more of "blockchain solution" than an effective solution for their related objects. Another perspective to frame this situation would be placing an intermediary activity, such as administration and finance above the effective purpose of a given project. For instance, the forest conservation activities and all issues related to nature and people involved. Blockchains for nature are likely to provide finance solutions to the forest without effective forest management and people attention. Such situatuon is also related to the issue of the "commodification of nature", which introduce a simplistic vision of nature related projects placing important technical interdependencies within the environment and the relational character of environmental goods, and it "twists the perception of the environment from systems preservation to items use or transformation" [11].

[3] NFTs are tokens that may be used to indicate ownership of one-of-a-kind objects.

3.2 Carbon credits and cryptos

Worldwide climate emergency actions to meet the 1.5°C target for global warming, according to the Paris Agreement and following UNO conferences of the parties, imposes severe reductions on fossil based energies and carbon sequestration, and future release avoidances. Carbon markets fulfill a relevant part of these actions towards a global greenhouse gases (GHG) reduction. Carbon credits, purchased voluntarily, enable organizations to compensate or neutralize not yet eliminated emissions by financing the avoidance/reduction of emissions from other sources, or the removal of greenhouse gases from the atmosphere and thus meaningfully contribute to the transition to a global balance of low carbon emissions. The projects generating these carbon credits can be broadly grouped into two categories: (1) GHG avoidance/reduction projects, such as renewable energy or avoided deforestation; and (2) GHG removal/sequestration projects, such as reforestation or technology-based removal [12]. The international carbon market facilitates the exchange of carbon credits. Prices have risen from nearly U$11 per ton in 2018 to U$63, as of May, 2022, or U$26, for the California market. The World Bank has estimated that the price needs to be closer to U$106 per ton by 2030, in order to meet the 1.5°C target. The value of traded global markets for carbon dioxide (CO2) permits grew by 164% to a record U$ 805 billion in 2021. The rise has been followed by voluntary carbon markets, where companies, for instance, trade carbon credits generated from projects to reduce emissions, presently exceeded U$1 billion. These are voluntary, not necessarily compliance-based, markets for carbon credits, expected to reach U$50 billion in volume, by 2030 [13].

Crytocurrencies and Blockchain transaction platforms have rapidly taken the trading and transactions registry opportunities of voluntary carbon markets. This study assessed a sample of coins and blockchains linked to carbon credits, generated by standing forests, tree planting, fossil fuel substitution and many others. Generally, the issue of carbon credits assurance retired in the voluntary market is issued by independent standards (e.g., VCS, Gold Standard, ACR, CAR). If a Coin project linked to a forest conservation concern does not have such assurance, the carbon credit may not be trustable, becoming liable to double accounting effects, where a same object may be “credited” two times from different concerns.

The ideal logic between a cryptocurrency tied to carbon credits should aim at measuring the carbon balance of its object. Take a forest carbon project, for instance. It is observable the concerns of some forest tokens projects informing the issuance of carbon credits from qualfied environmental projects, for instance, in the Amazon Forest (projects that emit, certify and sell credits) and that avoid or capture CO2 emissions to the atmosphere. These credits are traded in a digital platform using a blockchain platform. The observed concept indicates a measured and trustable nature object related to the coin. Nonetheless, a more accurate observation indicates that a forest must be measured by its transitioning balance towards being a carbon sink or a carbon source. A certain financial value can be attributed to the fact that the forest is a sink or a source. For instance, the forest can be a sink with a carbon capture capacity of one million tons, or a source with a capacity of negative one million tons. An effective coin associated to forest carbon storage must be supported by physical observations that indicate that the forest is moving towards being a sink or a source. For instance, negative observations such as traces of fire, tree's cutting and other disturbances indicate that the forest may me decaying, and positive observations such as biomass and forest preservation actions, indicate the forest will maintain its carbon sequestration patterns. Such information is key for a trustable forest carbon project. Overall, most coin based projects linked to forest coverage have not clearly outlined these concerns. They only show a piece of land, registered and bounded by GPS positioning, thus ensuring its existence. That is a fundamental issue to assure the existence and particular allocation of its carbon sink function, however, it does not secure the sink/source flow, thus, its real environmental service and carbon account.

Finally, the issue of double accounting of carbon credits, especially in the voluntary markets, poses a relevant question to all blockchains and coins related to nature projects. The absence of central databases and governance for all voluntary carbon credits does not assure a buyer using digital currencies or not, that his/her credit has not been traded elsewhere. Solutions come with the development of a shared digital data protocol across standards with an open source digital infrastructure to create an operating system for all planetary carbon trades [14]. This data protocol should be tailored to specific project types by defining necessary project data fields and procedures to protect the integrity of the verification process. In this perspective, shared digital data protocols should explore the use of satellite imaging, digital sensors, combined with blockchain distributed-ledger technologies (DLT) to further improve speed, accuracy, and integrity. Implementation of the digital data protocol could be a first step towards broader end-to-end life-cycle and value-chain tracking of all carbon credit data [14].

4. Models for cryptos and nature conservation effectiveness

The observed distances of Blockchains and tokens to their nature objects offers a large room for improvements and proposals to place the nature conservancy objectives clearly outlined and above the blockchain, itself. The major concept here presented is to place the object in first place, leaving the coin and the blockchains to better subsidy funds, provide storage of indicators and resolve governance issues. The issues covered in this part are: fundraising, Fintechs and DeFis, responsible tokenomics, and building a community around the coin.

4.1 Fundraising

The typical sources of funding from cryptocurrencies and blockchains can be summarized in Grant Coins, Initial Coin Offerings (ICOs), Microfinance with capital coins; and Donations through blockchain smart contracts.

Grant Coins, among many applications, are donations from specific purpose coins directly to a project. ICOs are the typical launch of a coin, in this case, with all set of project objectives, responsible tokenomics and clear governance and economic return to investors and the project beneficiaries. Micro finance with capital coins and donations through Blockchain contracts aim at providing funds to communities through coin based donations and micro loans, commonly addressed by Defis and Smart Contracts[4].

Initially, the relationship of blockchain and tokens to a nature project, can be analyzed under a framework containing the project objectives with the blockchain

[4] DeFi is the acromym of Decentralized Finance, in the form of financial applications built on blockchain technologies, typically using smart contracts. Smart contracts are automated enforceable agreements that do not need intermediaries to execute and can be accessed by anyone with an internet connection.

Fundraising from cryptocurrencies	**What Problem will be solved?**	**Funding potential**	**Regulatory complexity**	**Technology Infrastructure**	**Coin Exchange Needs**	**Associate Financial Services**	**Communication with coin investors & donors**
Grant Coins	Raise funds from coin donation sources	medium	low	low	yes, but not for gran-tees	yes	Periodi-cal reports
Initial Coin Offerings—ICOs	Raise funds out of debt & equity and donors' restrictions	high	very high	very high	Yes	yes	constant, chat groups
Microfinance with Capital Coins	Provide microfinance to communities through coin based donations and loans	high	high	high	yes, but not for grantees	yes	moderate
Commited donations through Blockchain smart contracts	Provide donations to communities through coin based funding and loans	medium	low	medium	No	no	moderate

Table 2.
A Blockchain fundraising matrix example for a conservation project.

and coins roles. In this perspective, the major role of a coin and the blockchains are related to funding and financial services for a conservation project including nature preservation and lifequality improvement of locals, representing a group of beneficiaries. Therefore, the project must address questions about the problem to be solved; the funding potential; regulations; technology structure; coin exchange needs; and associate financial needs. An example of fundraising and associate financial needs using blockchain and coins for a nature conservation project involving local populations, is presented in **Table 2**.

4.2 Fintechs and DeFIs

Nature related projetcs can be supported by a specific financial services (fintech) entitity linking coin investors with projects gathering conservation issues and small businesses of sustainable local products. The fintech provides grants and loans, and if applicable, business mentoring support. The ICO must be supported by a thoroughly defined business model with a structured platform under blockchain. In a basic flow of funds, the fintech launches its coin, receiving Bitcoins or fiduciary currency, thus forming an investment account, abiding to all regulatory issues. The raised funds are directed to a specific credit account to perform grants and loans to projects. A DeFi takes the roles to channel and operate the payments, receivables and interest accruals (if applicable) in the scope of supported projects. All projects are selected from a thorough business analysis and once they receive the loans, a clear and continuous communication is established among the projects with the fintech staff and coin investors. The underlying coin is tradeable in a cryptocurrency exchange. This is a summary of an effective financial flow of funds between a coin and supported projects. The methods and systems may vary depending on the scope of projects. The key aspects of this solution is to provide an operating financial solution providing solid links between one coin or token to its supported objectives. A responsible coin will address these isssues, to effectively implement a project together with beneficiares, far beyond simple advertsing and basic web information. A key issue for a successful project fundraing from coins, lies also on a responsible tokenomics.

4.3 Responsible tokenomics

The uses of a token must be well structured. A responsible token project must be supported by a responsible tokenomics for a stable coin with clear outlines about the flow of resources and monitoring of project funds. The relationships of the token with the project must be qualified and measured, as briefly laid out in **Table 2**.

The key point for the creators of a cryptoasset is to understand how the digital currency will be used, in this context, understanding the clear link between the asset, the use of its blockchain, and the value and service offer attached to this asset. A well structured value offers a greater possibility of expanding the services and goods of the asset, thus generating a greater desire for its purchase and use, and therefore a greater demand and increase in the value of the currency. This logic is totally different from the "boom-collapse" effect observed in most currency issues. In this case, the concern with connecting the value of the currency to a clear value construction of the supported object, in this case nature and its beneficiaries, should be above the normal speculative orientation normally observed. In summary, key issues must be addressed, such as:

- How many coins and tokens are in existence and will exist in the future, and when will they be created;
- Who owns the coins? Are there any items reserved to be released in the future for developers?
- Will the coin project beneficiaries be rewarded? How and exactly they will participate? Are safeguards for massive speculation previewed?
- Is there any information to suggest that a large number of coins have been lost, burned, deleted or otherwise made unusable?

Cryptocurrencies and tokens built on the blockchain must have pre-defined issuance schedules created by algorithms for the coin issuers to accurately predict the coin's volume and launching dates. It is relevant to point that yet it is possible for most cryptocurrencies to change this issuance schedule, such an event requires the agreement of stakeholders around the currency, thus offering further difficulties for its implementation. This issue becomes more relevant with the inclusion of the project's beneficiaries in the coin.

Tokenomics should also be a guide to the potential future value of a coin. This prediction should be clearly placed within a relationship of the coins, investors and the projects supported. This measure will, at least, mitigate the speculative nature of the coin's investors. A possible arrangement to improve the relationship between a coin and project beneficiaries is to include beneficiaries as asset holders, with guarantees of stability and tradeability of their assets, as following briefly introduced.

4.4 Build a community around the coin

A nature conservation-based project supported by a responsible token and blockchain should clearly address the areas and beneficiaries to relate, being entities and/or individuals. The scope of the project must be outlined for each purpose, such as nature restoration, fauna protection, income generation, social improvements, education and health support, for instance. The project must have specific, and measurable data to be collected with identified participants and provide veracity and control in DLT (blockchain) for the the guarantee of successfull objectives. The governance structure should join stakehokders, investors and beneficiaries enrolled in the project's activities with clear roles. As seen below, the beneficiaries will have a role to provide key forest information for the project to be effectively measured, thus helping to its better effectiveness. These roles and obligations are better implemented by a Decentralized Autonomous Organization (DAO)[5] linked to the project's token blockchain. The major steps to initialise the funding are:

- Define the project area and beneficiaries with attention to fundraising, financial needs and related issues, as listed in **Table 2**.
- Preview and model an ICO, where buyers acquire tokens and the beneficiaries become providers, and receive tokens for free.

[5] A decentralized autonomous organization (DAO) is an organization constructed by rules encoded as a computer program that is often transparent, controlled by the organization's members and not influenced by a central government, therefore, they are member-owned communities without centralized leadership.

- The value of the token is tied to the providers' obligation as the value of the coin is likely to trend higher if those obligations are fulfilled.

- The incentive for providers to fulfill their obligations is linked to the increase and preservation of the coin value over time.

- The incentive for investors is to hold their coin, buying and staking or bonding their investmets pegged to the currency.

The community around a coin should include the coin's object providers and investors with mutual interest in conserving the forest, because both will prosper with the standing forest. A third group takes part in this DAO in the roles of field implementation, analyses and data verification about the forest and people's social indicators. In a forest carbon project, for instance, the providers are the forest inhabitants or those with access to the forest's dynamics. The coin investors, without access to the land, have a financial interest in preserving the land. The third group is made, possibly, of service partners, such as a nonprofit organizations to implement field projects; auditors; and scientists and researchers providing data verification and technical proposal roles. This last role, may be applicable in changing scenarios, where specific metrics and data may loose value for accounting the forest carbon and environmental preservation, at the same time giving room to new indicators that make better sense for the coin objective of supporting a standing forest, thus a carbon storage.

The tokenomics must address the coin value linked to forest key indicators in a model requiring constant observations and verified contributions. At the ICO, providers are issued with coins, aside from the investors' purchase. The role of providers is to report positive and negative observations, to contribute to modeling the future of the forest; whether it will be in a sink or source state. The better the providers accomplish their role, the more is the certainty over whether the forest will be in a sink or source state. At the same time, the value of the tokens received during the ICO will trend upward, thereby incentivizing them to keep providing observations to the model. Internal data providers contribute to the model by changing the value of the parameters of the model, in accordance of governance principles, pre defined in the DAO agreements. Internal data providers, like researchers or scientists, keep their role to permanently assess and review the intrinsic dynamics of the forest considered in the project objectives as laid out in the digital currency model of contribution to the forest and their inhabitants, the beneficiaries and data providers for the model.

Additional attention to this model must relate to forest protection issues, such as the avoidance of illegal ownership of forest areas under carbon and coins' incentives; the issuing of forest leasing mechanisms, when applicable; and carbon verification to assure a legal framework of compliance with applicable legislations and the avoidance of double accounting of carbon credits. Finally, a responsible token initiative should address their results to nature and social issues according to the Sustainable Development Goals (SDGs) [15] as a measurement for its key project indicators, thus, providing material, measurable and comparable results for all.

5. Conclusions

This study raised questions about cryptocurrencies and blockchains effectiveness to provide nature conservation support. The world has seen the rise and exponential growth, valuation, an extreme devaluation of cryptocurrencies, in short period of time. Most cryptocurrencies have been launched with weak and unclear objectives, leading them to just speculative ends. Among them, many coins have been launched claiming to support forest, fauna and nature related projects; and a relevant group of tokens have been launched to link carbon credits to blockchains providing easy and secured carbon trades. In addition, many other cryptocurrencies have been launched with the promise to help nature related projects. The general observation of these blockchains is that the ones related to carbon projects claims to deliver trustable accounts of carbon credits, especially in voluntary markets; and the non carbon coins present vague information about the nature related objects they claim to support. In the carbon trading projects, specifically for forest conservation, the issue of verified standards and the inclusion of local populations must be improved for a better assurance of the continuous forest coverage and their inhabitants' lifequality.

The general perception about most coins and blockchains related to nature is that they are more relevant than the project itself, thus, placing the solution mechanism on top of the object, presenting few evidences that the nature and local population are really being benefited. Instead, they should place clear and objective nature conservation and social indicators on top of the blockchain solutions. Blockchain and coin solutions shall be framed as intermediary activities because the nature conservation and people benefited by their claimed projects are the "Final/End activities".

This study offered initial ideas for a better use of cryptocurrencies and blockchains for nature conservation projects. The outlined proposals are in the line of a deeper project appraisal, such as defining the fundraising modalities and distribution to beneficiaries, including them in the asset valuation and protecting them from massive devaluations. Such efforts would be better addressed with responsible tokenomics and a fintechc focused in the flows os funds from investors to the projects and its beneficiaries. The safeguards for this responsible approach lies also on clear governance and selection of the projects, after all, placing them above the tokens and ledgers. In a final conclusion digital currencies, cryptocurrencies, tokens, coins and blockchains related to nature conservation and social issues should be tailored to serve a purpose with a clear value proposal to their supported objects, showing a real concern for nature and people, with all possible barriers to speculation.

The ideal framework for a coin and blockchaim relationship with nature conservation projects, should firsthand, include the beneficiaries and stakeholders in the value chain. In this perspective, forest family, landowners become providers of data, supporting value under verified mechanisms. Providers are issued tokens with their value tied to conservation actions. Providers are incentivized to conservation action to either preserve or increase the value of the currency they already own. After all, this is a model based on "Conserve to Earn" principles, where investors do not aim at speculating with a coin, rather, they own, trade, learn and exchange information with stakeholders to effectively provide social and environmental improvements in nature based projects.

Future research and actions on this scope should explore detailed tokenization and project safeguards for effective nature conservation. Contributing technologies must take place, such as web3 platforms, linked to Artificial Intelligence and Internet of Things devices to better monitor the projects' indicators. In addition an advanced DAO governance and operations shall set the path for a more responsible and effective use of cryptocurrencies and blockchains for the environment and social improvements. Finally, sound projects supported by crytocurrencies must include social technologies to better engage all beneficiaries and stakeholders to the assets and exchanges providing a positive perspective of value sharing and nature conservation for all.

Acknowledgements

Robson Rojas Andrade—Instituto Triad Systems; Christian Wiesenthal—OpenEarth Foundation.

Author details

Luiz Cruz Villares
Foundation for Amazon Sustainability, Manaus-AM, Brazil

*Address all correspondence to: luiz.villares@fas-amazonia.org

References

[1] Nakamoto S. Bitcoin: A Peer-to-Peer Electronic Cash System. 2009. Available from: https://bitcoin.org/en/bitcoin-paper

[2] Jake Frankenfield: Shitcoin. Investopedia. Cyrptocurrency. Altcoins. 2022. Available from: https://www.investopedia.com/terms/s/shitcoin.asp

[3] The Economist Newspaper Limited: Why Cryptos's Bruising Comedown Matters. 2022. Available from: https://www.economist.com/leaders/2022/05/18

[4] Springer Nature Limited: Crypto and Digital Currencies–Nine Research Priorities. 2022. Available from: https://www.nature.com/articles/d41586-022-00927-5

[5] Corbet S, Lucey B, Urquhart A, Yarovaya L. Cryptocurrencies as a financial asset: A systematic analysis. International Review of Financial Analysis, Volume. 2019;**62**:182-199

[6] University of Cambridge. Centre for Alternative Finance. Cambridge Bitcoin Electricity Consumption Index. 2022. Available from: https://ccaf.io/cbeci/index

[7] The Bitcoin Mining Council: Bitcoin Energy Consumption from Non Fossil Sources. 2022. Available from: https://bitcoinminingcouncil.com/

[8] CoinMarketCap. 2022. Available from: https://coinmarketcap.com/currencies

[9] Mordor Intelligence: Gaming Market—Growth, Trends, Covid 19 Impact, and Forecasts (2022-2027). 2022. Available from: https://www.mordorintelligence.com/industry-reports/global-gaming-market

[10] Stuit A, Brockington D, Corbera E. Smart, commodified and encoded. Conservation & Society. 2022;**20**(1):12-23. Ashoka Trust for Research in Ecology and the Environment and Wolters Kluwer India Pvt. Ltd

[11] Vatn A. The environment as a commodity. Environmental Values. 2000;**9**(4):493-509

[12] Blaufelder C, Levy C, Manion P, Pinner D. A Blueprint for Scaling Voluntary Carbon Markets to Meet, Challenge. Zurich, Switzerland/London, UK/San Francisco, USA: McKinsey & Company; 2021

[13] Adams T, Winters B, Nazareth A. Task Force on Scaling Voluntary Carbon Markets. TSVCM. Washington, D.C. USA: Institute of International Finance; 2021

[14] Integrated Climate Accounting System. Open Earth Foundation. 2022. Available from: https://www.openearth.org/projects/openclimate

[15] United Nations, Department of Economic and Social Affairs—Sustainable Development. The 17 Goals. 2015. Available from: https://sdgs.un.org/goals

Chapter 5

DAOs: Governance in the Blockchain Era

Joana R. Pereira and Giselle Garcia

Abstract

Blockchain technology promises to revolutionize not only the way we transact among peers but also the way we organize to create socio-economic value. Decentralized autonomous organizations (DAOs) are governed and owned by the community whose members follow a set of blockchain-embedded governance rules that define and control participation. In this chapter, we will clarify what governance decentralization and automation mean, examining DAOs' distinguishing characteristics. We will also discuss the problems that DAOs solve (e.g., lack of extrinsic incentives, censorship, mismanagement and lack of transparency and accountability), as well as the problems they face (e.g., lack of participation, rigidity, voting misbehaviours and legal status).

Keywords: blockchain, organizations, decentralization, automation, censorship

1. Introduction

The rise of Bitcoin and blockchain brought new organizational forms called decentralized autonomous organizations (DAOs) that are blockchain-based organizations, owned and governed by members [1–3]. DAOs push forward the blockchain ideals of decentralization, inclusion and transparency, enabling the communities to own and govern cryptocurrency projects. In DAOs, the community not only owns the organizations but also can propose and vote, having a saying regarding the future of the organization. Therefore, DAOs are distributed instead of hierarchical, power is decentralized instead of centralized and management is autonomous and community-based instead of bureaucratic [2, 3]. DAOs are, thus, an alternative governance form to traditional bureaucratic and hierarchical management models, being a natural governance choice among cryptocurrencies.

One of the first examples of a DAO was Bitcoin itself. Created in 2009, Bitcoin is a cryptocurrency designed to allow people to securely transact and exchange value at a global scale without the need for costly intermediaries [4]. With the purpose to create an independent financial system, Bitcoin is sustained by a community able to validate transactions (miners) and co-create code updates (developers), which translates into new functions and participation rules without the interference or orchestration of a central sponsor. Later on, in 2016, born the first assumed decentralized autonomous organization called the DAO. The DAO raised 150 million dollars in a short period of time, making it the world's largest crowdfunding project at that time. Despite that the

DAO project suffered a massive attack due to security breaches, its governance model came to stay and prosper and thousands of DAOs have been emerging in different areas as decentralized finance (e.g., Uniswap), media (e.g., Global Coin Research and Forefront), gaming (e.g., Decentraland), art and culture (e.g., SuperRare and Rarible) and investment funds (e.g., BitDAO and MetaCartel), among others. While the DAO is a governance model that highly celebrated among cryptocurrency projects, very little is known about its characteristics, as well as the problems they solve (i.e., lack of extrinsic incentives, censorship, mismanagement and lack of transparency and accountability) and face (i.e., lack of participation, rigidity, voting misbehaviours and legal status).

2. Definition and characteristics of DAOs

DAOs are blockchain-based organizations, owned and governed by members. DAOs follow a set of self-executing governance rules that are embedded into a blockchain that run without the interference of a central authority (see **Table 1** for a summary) [1–3].

DAOs are blockchain-based organizations. Blockchain is a distributed ledger technology (DLT), which translates into a public database, where information is written in "blocks" cryptographically linked to the previous blocks, forming a chain of blocks [4]. Blockchain technical features, as distributed data storage, timestamp, consensus algorithm and asymmetric encryption, allow for decentralization, immutability and auditability [5]. The blockchain is, thus, the infrastructure that enables DAOs to be decentralized and autonomous, also ensuring the security requirements and ownership authentication that DAOs require to function [2, 3]. It is important to mention that not all blockchain-based organizations (e.g., DApps and platforms) are DAOs. Indeed, there are several blockchain-based organizations that display a centralized governance system, being owned and governed by private investors, founders, developers' teams or even by a restricted permissioned group. Nevertheless, all DAOs are blockchain-based organizations as blockchain technology is the foundation of their decentralized and autonomous governance form.

DAOs embed a set of self-executing governance rules that are turned into computer code that is embedded on the blockchain [2, 3]. Such governance rules define participation terms, voting and proposing systems as well as rewards and penalties terms. All these rules are open and transparent to all participants. By interacting with this self-executing code, participants can submit proposals that may comprise changes to the functioning of the organization, amends to the governance system or suggestions for future projects. In some DAOs, such proposals go directly for community voting (e.g., community proposal); however, in others, there are a couple of additional steps before a proposal reaches the wider community for voting (e.g., stage proposals and

Blockchain-based	DAOs run on a blockchain infrastructure that allows autonomy and decentralization.
Self-executing rules	DAOs embed a set of self-executing governance rules that are turned into computer code.
Community-owned and governed	Community members own the organization through token holding and govern the organization through proposals and voting systems.

Table 1.
Characteristics of DAOs.

represented team proposals). Once the proposals are up to vote, community members can use their governance tokens to vote on the proposals. There is also a variety of voting systems, such as one person one vote, quadratic voting, weighted votes, holographic consensus, futarchy and liquid democracy, among others. After the voting process finishes, the proposals are implemented according to the parameters defined by the community. In some cases, the execution is automatic or direct (e.g., when the proposal is a code alteration that includes the code) or it might require that developers build the code to implement, leading to a delayed implementation.

DAOs are owned and governed by people without the interference of a central authority. Community members own the organization through token holding and govern the organization through proposals and voting systems that are embedded into the blockchain. Blockchain and smart contracts technologies allow DAOs to operate autonomously without centralized control or third-party intervention [2, 3]. Instead, DAOs run under the regulation rules and collaboration patterns defined by all the stakeholders. The DAO's goal is achieved through bottom-up interaction, coordination and cooperation among members, following the principles of equality, voluntariness, reciprocity and mutual benefit [2, 3].

3. The problems that DAOs solve

3.1 Lack of extrinsic incentives

DAOs resemble online and open-source communities, with the difference that DAOs enact both intrinsic (e.g., self-satisfaction, values and beliefs fulfillment, intellectual stimulation, learning and making a positive difference) [6, 7] and extrinsic benefits (e.g., monetary rewards) such as tokens that can be converted into other cryptocurrencies or fiat currencies [8, 9]. DAOs enact intrinsic motivation as members feel a sense of ownership and excitement to contribute to new projects, but they also guarantee that members have extrinsic benefits through financial rewards for their participation. Members of DAOs have an active stake through holding tokens in the organization, and they can exert that stake, having a word to say and influencing the future of the organization through proposals submission and voting schemes. DAOs solve the problem of a lack of hard incentives shared by open-source software communities and other types of online communities because DAOs are able to offer extrinsic incentives through tokenized business models. The broader nature of DAO incentives that cover both intrinsic and extrinsic rewards allows them to attract and maintain large, engaged communities (see **Table 2** for a summary).

3.2 Censorship

DAOs solve the problem of censorship, a widely discussed issue in nowadays societies. DAOs are censorship resistant as they cannot be shut down by anyone and no one can be censored or expelled from the community. In the limit, even the DAOs' creators cannot plug off the DAO until community of governance token holders approve the shutdown through voting. Nobody can impose their will on a DAO or DAO members regardless of their position or authority. Once members activate the smart contract by meeting some conditions [10] that are predetermined, the transactions are registered in the blockchain without (or with little) human interference and the transaction is registered in the blockchain, being immutable.

Lack of extrinsic incentives	DAO incentives comprise intrinsic and extrinsic rewards allowing them to attract and maintain large, engaged communities.
Censorship	DAOs are censorship resistant as they cannot be shut down by anyone and no one can be censored or expelled from the community
Mismanagement	DAOs' smart contracts enable automated and decentralized decision-making, leading to reduce mismanagement risks.
Lack of transparency and accountability	DAOs' transactions, proposals, votes and even voter collusions are transparent, ensuring that members are accountable for their actions.

Table 2.
Summary of the problems that DAOs solve.

3.3 Mismanagement

Bureaucratic organizations and corporations often face mismanagement problems, in which top-down, hierarchical structures are led by managers that often put their selfish interests in front of the organizations' interests. DAOs avoid mismanagement in two ways. First, operations are autonomously carried by smart contracts without (or with little) human interference, being transparent, auditable and equitable. Second, the DAO operates without the need to have faith on influential individuals of the boardrooms of directors that centralize the decision-making. Instead, the DAO is controlled by the token holders, who can submit proposals for improvement or protocol changes, see each other proposals, and vote on those proposals, defining the future and the success of the organization. DAOs' smart contracts enable automated and decentralized decision-making, leading to reduce mismanagement risks.

3.4 Lack of transparency and accountability

Lack of transparency and accountability are among the biggest issues of our times. From the government to the banking system to big tech corporations, increasing concerns emerge about how governments and corporations are run, how our data and money are being held and used for and, equally important, who is accountable. DAOs' transparency allows for the accountability of every member of the organization. Accountability and transparency are values infused in blockchain technology that are at the very heart of DAOs. Every activity translates into an immutable transaction registered in the blockchain, meaning that it cannot be erased or reversed. In the same fashion, proposals, votes and even voters' collusions are publicly available, ensuring that members are accountable for their actions.

4. The problems that DAOs face

4.1 Lack of participation

While some DAOs display highly engaged communities that often participate in the proposals and voting processes, some face some lack of participation. As vote rights are optional and communities tend to be big, members often think that someone else will make proposals and vote on them, which may translate in low levels of participation [11]. Additionally, sometimes making proposals and voting involves

Lack of participation	In DAOs, voting is optional, which may lead to community members' inertia and lack of participation.
Rigidity	Proposal and voting systems are time-consuming, which can compromise the project's ability to change, adapt or innovate.
Voting misbehaviours	Voting systems can promote power concentration, bribery and collusive behaviors.
Lack of legal status	The absence of a legal framework can promote malicious acts and attacks

Table 3.
Summary of the problems that DAOs face.

time and effort, which discourages members that often participate in many DAOs simultaneously [12]. For these reasons, sometimes the promise of decentralization is not achieved, and most decisions are taken by a small group of members, resembling centralized organizations (see **Table 3** for a summary).

4.2 Rigidity

While DAOs promise to be decentralized, transparent and auditable, such values come at the cost of rigidity. Usually, the protocol and the smart contracts are developed by a creator or a team that launch the project. From the moment that the DAO is active and the governance tokens are distributed across members, the future of the DAO is in the community hands. However, any changes proposed will require a series of activities that encompass a proposal process (that can involve several rounds and checks), voting mechanisms and subsequent implementation. Such a process is time-consuming, which can compromises the project ability to change, adapt or innovate, limiting its growth. Additionally, changing and amending smart contracts increase the likelihood of errors and bugs; therefore, DAOs face a trade-off between flexibility and security [13].

4.3 Voting misbehaviours

The majority of DAOs base their voting rights on governance token that represents ownership, resembling companies' shares. Depending on the voting system adopted by the DAO, it can promote or mitigate power concentration, bribery and collusive behaviors. Some DAOs have employed methods to avoid voting misbehaviours, such as defining shorter/longer periods for voting, limiting/increasing the number of tokens available, controlling voting power, establishing voting thresholds to approve proposals, communicating with all the participants and proposing consensus adaptations. Moreover, the unbalanced power voting can be produced by the technical knowledge required in some decisions giving more opportunities for deciding to developers [13]. Despite all mechanisms employed to avoid voting power asymmetries, it is impossible for DAOs to guarantee that decision-making is not affected by voting misbehaviours [11].

4.4 Lack of legal status

Since DAOs are borderless, it is difficult to define what type of organization they are [2, 3], which codes or regulations their member should follow [14] and which regulations for taxation and management they obey to. Owing to the lack of regulations, it is difficult to determine who will be responsible for liabilities, damage or failures. The absence of a legal framework can promote malicious acts and attacks. Even if the approval relies on voters, there are no clear rules or consequences that protect the ownership and the community from damage.

5. Conclusion

DAOs are owned and governed by people, democratizing ownership and decision-making in the organizational arena. While DAOs bring the promise of transparency, accountability and decentralization, they also face some growing issues that become more evident as new projects emerge and scale. The ability of DAOs to overcome problems, such as lack of participation, rigidity and voting misbehaviours, will determine the future of this new governance form.

While decentralized and autonomous governance modes are increasingly gaining popularity inside and outside the Web 3.0 space, their theoretical and practical implications remain understudied. What are the boundary conditions that enable or prevent the adoption of decentralized autonomous governance modes? How do DApps attract and retain community members to participate in governance decisions? What can other organizations outside the Web 3.0 realm learn from DAOs?

Author details

Joana R. Pereira* and Giselle Garcia
Leeds University Business School, Leeds, England

*Address all correspondence to: j.pereira@leeds.ac.uk

References

[1] Hassan S, De Filippi P. Decentralized autonomous organization. Internet. Policy Review. 2021;**10**(2):1-10

[2] Wang S et al. Decentralized autonomous organizations. IEEE Transactions on Computational Social Systems. 2019;**6**(5):870-878

[3] Wang S, Ding W, Li J, Yuan Y, Ouyang L, Wang FY. Decentralized autonomous organizations: Concept, model, and applications. IEEE Transactions on Computational Social Systems. 2019;**6**(5):870-878

[4] Nakamoto S, Bitcoin A. A peer-to-peer electronic cash system. BIT. 2008;**4**:2

[5] Febrero P, Pereira J. Cryptocurrency constellations across the three-dimensional space: Governance decentralization, security, and scalability. IEEE Transactions on Engineering Management. 2020

[6] Lerner J, Tirole J. Some simple economics of open source. The Journal of Industrial Economics. 2002;**50**(2):197-234

[7] Villarroel Fernandez JA, Tucci CL. Motivating Firm-Sponsored e-Collective Work. 2010

[8] Davidson S, De Filippi P, Potts J. Blockchains and the economic institutions of capitalism. Journal of Institutional Economics. 2018;**14**(4):639-658

[9] Pereira J, Tavalaei MM, Ozalp H. Blockchain-based platforms: Decentralized infrastructures and its boundary conditions. Technological Forecasting and Social Change. 2019;**146**:94-102

[10] IBM. What are smart contracts on blockchain? [Internet]. 2022. Available from: https://www.ibm.com/topics/smart-contracts. [Accessed: May 16, 2022]

[11] Samman G, Freuden, D. DAO: a Decentralized layer for the internet value [Internet]. 2020. Available from: https://static1.squarespace.com/static/564100e0e4b08c9445a5fc5d/t/5ec61d6241af4a208ba87f52/1590041969238/DAO+-+A+Decentralized+Governance+Layer+for+the+Internet+of+Value-1.pdf. [Accessed: May 17, 2022]

[12] Chohan UW. The decentralized autonomous organization and governance issues. SSNR. 2017;**3082055**

[13] Rikken O, Janssen M, Kwee Z. Governance challenges of blockchain and decentralized autonomous organizations. Information Polity. 2019;**24**(4):397-417

[14] Kandova G, Barba R. Governance of decentralized autonomous organizations. Journal of Modern Accounting and Auditing. 2019;**15**(8):406-411

Chapter 6

NFT Legal and Market Challenges in Permissioned Blockchain Networks

Javier Ibáñez Jiménez and Eva María Ibáñez Jiménez

Abstract

This work analyses the distributed-ledger technology as a digital framework for the representation and exercise of tokenized rights, as well as the private-law characterization and EU regimen applicable to non-fungible tokens (NFTs). After a brief introduction to permissioning in blockchain and to permissioned distributed ledgers (PDLs), we reflect on its relevance to NFT contracting and markets, focusing on the legal and economic consequences of NFT trading in PDLs, with particular emphasis on public permissioned networks (PPDLs). We then discuss key challenges about law-compliant trading for NFTs, such as proper legal qualification of their nature, determination of optimal conditions for the legal trading of NFT underlying assets and investing applicable regime, including specific primary-market and essential public surveillance and intellectual-property (IP) issues.

Keywords: permissioned blockchain networks, non-fungible tokens, public-permissioned DLT networks, primary-market obligations, IP protection

1. Introduction

Distributed-ledger techonology (hereinafter, DLT, mostly known in its blockchain version or variant) and tokenization as a device to incorporate and exchange data (in the so called DLT tokens) are jointly considered within a market public- and private-law and also within a contractual legal and economic context, an efficient way to obtain the valid representation of credit rights arisen from contracts. Such credits are enforceable in courts against the issuer of the token or person on behalf of whom such issuer acts. Eventually, DLT tokens can validly represent *in rem*, rights in favor of the token holder.

In both cases, tokenization is used in blockchain DLT markets for fast, safe, and law-compliant:

a. Transmission of the rights represented and possibly incorporated to the token, after the issuance and legal configuration of such rights in accordance with an off-chain contract (ordinarily under initial coin or token offering -ICO/ITO- regulation) between token issuer/offeror and accepting investors, documented in an informative whitepaper similar to the initial public offering prospectus.

b. Exercise of such rights and fulfillment of the corresponding obligations, as the token holder is entitled to claim the contents of the represented rights against the issuer or linked third party.

During the issuance of the tokens, the phenomenon of tokenization technically occurs. It can consist either in the creation of virtual assets (connected or not to a metaverse of investors, as is typical of NFTs) or in the representation of preexisting rights, which have arisen from the celebration of an investment contract, a service or purchase agreement, a specific non-investment contract celebrated to grant tokenized or digitally represented rights between the issuer or creator of such rights and token buyers; or between a generator or creator of tokens other than a financial issuer or financed subject (legal person within the context of the Proposal for a Regulation of the European Parliament and of the Council on Markets in Crypto-assets, and amending Directive (EU) 2019/1937 -Brussels 24.9.2020, COM(2020) 593 final, 2020/0265 (COD); hereinafter, MiCA).

A previous contract between platform manager and issuer is necessary in order to operate under DLT and decentralized financial (DeFi) platform. Such a contract is complementary and additional to the previous contract between the financial issuer and the investing public or interested in receiving tokenized services or products.

In any case, the tokenization or creation of tokens as a mechanism for holding and disposing of virtual assets is an efficient technique for both the creator and the target public of the offer, especially when compared to the classic non-DLT or off-chain (off the DLT network) centralized process. Three practical basic reasons of efficiency linked to the use of DLT support this affirmation:

a. Substantial reduction of issuance and other transaction costs of the issuer (ITO structuring, fees in auxiliary management contracts, global coordination, guarantee, ...), many of them minimal or unnecessary in DLT due to decentralization and suppression of intermediaries. And a parallel reduction of investor costs, particularly commission fees present in the traditional financial intermediation.

b. Significant increase in the speed of market transactions related to the transfer of tokens, with subsequent shortening of contracting terms. This legal and financial characteristic of token transmission entails a huge exceptional competitive advantage both for issuers and specialized operators in a decentralized finance (DeFi) digital environment, namely the marketplaces built on top of a permissioned blockchain, as we will explain later on.

c. Enhanced cybersecurity and privacy of transactions. However, DLT cyber-risks are important and should not be neglected, as organized financial market supervisors continuously warn, without prejudice to the risks associated with deregulation, which the European MiCA Regulation will foreseeably cut down, by disciplining the white paper and the intervention of specialists in tokenized investment services (token platform, crypto exchange, crypto investment advice, market creation, crypto-market making and DLT liquidity, crypto-market orders, and key management and digital wallets, among others; cf. articles 4 to 9 and 67 to 75 MiCA).

Generally speaking, tokens or crypto-assets tradable under DLT conditions can contain rights arising from financing contracts (loan and partnership) entered into or arranged prior to the tokenization, and sometimes even before the creation of the DeFi marketplace or DLT platform habilitating DLT-network token contracting services. On some occasions, tokens arise from a network financial agreement, then being called cryptocurrencies by markets in the sense of payment tokens (serving as a means to extinguish debts). In other cases, tokenization comes from a network service contract, or derives from network developments or the provision of services related or not with the blockchain, but anyway under DLT (as in the case of utility tokens, deserving special rules in accordance with EU MiCA regulation). There are also crypto-assets exclusively used as currencies, thus serving as a mechanism for payment and extinction of credits and debts, or as a means of token exchange, also allowing the disinvestment or profit collection by exiting from a tokenized investing position, as in conventional finance. Finally, some tokens are hybrids, participating to varying degrees in the characteristics of payment, utility, and investment tokens; thus, different types of legal businesses can be configured to exchange virtual or real wealth by representation in a decentralized financial system (DeFi).

2. The private-law characterization and EU regime applicable to non-fungible tokens (NFTs)

The diversity of imaginable contractual configurations and technical standards apt and suitable for the creation of NFTs -on real assets or intellectual rigths, *inter alia*-, does not has uniform and consolidated legal boundaries in many jurisdictions yet.

Thus, the technical characterization of NFTs as traded assets in markets is possible even if the token represents non-replicable or individually predetermined or preconfigured rights derived from the specificity of the token or from the uniqueness of the represented rights or assets. In the US laws, NFTs are conceived as digital certificates representing ownership of a unique asset, or of other rights created or granted on a unique asset or good, or on contracted services or obligations. A massive negotiation of the token or parts of the represented asset is technically possible, though not always desirable from a private-law policy perspective, or even from the viewpoint of public order or macroeconomics.

In order for tokenization to be an efficient way to generate and trade credit rights on the issuer, moreover, it is necessary that the legal regime for tokenized assets adequately protects investors. Otherwise, the NFT token market will lose its still weak legal credibility, on the one hand; and on the other, the decentralized financial (DeFi) market will be affected as a whole to the extent that NFT token markets are linked with the whole DeFi system and all DLT financial markets. Ultimately affecting as well traditional capital markets linked with DeFi by the available mechanisms of digital currencies and associated payment systems (electronic money tokens or EMTs, payments in bitcoin and other digital assets, and mechanisms for exchanging or exchanging digital money for fiat money).

The triple regulation of the European Parliament and Council proposed on 24 of September 2020 (MiCA, operative resilience Act – DORA, and market infrastructure regulation – MIR) composes the so-called European digital finance strategy within the framework of the single digital market. The main regulation is MiCA, concerning crypto-asset markets and token intermediaries to establish efficient market rules and investor-protective new supervisory mechanisms, partly by analogy with respect to the classic protective regulation typical of the MIFIR/MIFID 2 context.

The concern of the legislator in this field is twofold from the point of view of the orientation of the new European regime in the international context of the proliferation of new tokenized markets like NFT markets:

a. On the one hand, financial stability and the defense of European monetary sovereignty, and that of the Member States, mainly translated into the configuration of a special regime for cryptoactives considered stable currencies (stablecoins), either because of their status as tokens equivalent to electronic (e-money tokens, EMTs) or because of their volatility due to their connection with an underlying that is a listed financial or nonfinancial asset or widely traded on the market (asset-reference tokens, ARTs, whose regime of maintenance and deposit and reserve investment or backup by the issuer is very demanding). To this respect, NFT markets are only partially concerned, because NFTs are not conceived as stablecoins since they lack underlying reference, provided that the represented assets are unique in most cases. However, baskets of tokenized assets underlying the NFTs could serve as a reference for specific asset-referenced tokens (ARTs), considered by MiCA as stablecoins, subject to severe specific supervision requirements and governance rules applicable to issuers, including the disposal and control of reserve assets.

b. And on the other hand, the protection of the assets of the issuers themselves and of the investors who operate through specialist market professionals, specifically the new cryptoasset service providers disciplined in the new MiCA regulation from a statutory and contractual point of view. In the case of NFTs, it is to be noted that whenever cryptoassets are "unique" and "non fungible" with respect to other cryptoassets (art. 4.2.c MiCA), rules on whitepaper writing, notification to the national competent market authority, and whitepaper publishing are not applicable to the issuing entity, which could offer them to the public and/or request or apply for quotation or secondary trading in a crypto-asset trading platform with exemption from whitepaper transparency requisites, *ex* arts. 4.1. b, c, d, MiCA, and corresponding arts. 5, 7 and 8 MiCA). Anyway, public offerings as securities or trading in regulated DeFi of such unique NFTs would be subject to the demanding specifications contained in article 13 (cf. 4.1.d, y v. infra, IVC).

Without prejudice to this, NFTs that could be qualified as stablecoins (or fungible tokens derived from NFTs by means of division or fractioning of represented rights on NFTs) would be submitted to severe requisites applicable to ARTs (Title III, Chapter I, arts. 15 et seq MiCA).

In the US, securities laws regime may be applicable if the NFTs represent presales of digital assets intended for use on a platform that is not yet built and the proceeds of the sale are used to build the platform (in the EU MiCA regime these tokens should be considered in general as "utility tokens"). And securities regulation is also applicable in cases of "pooling" or "fractionalization" of digital assets by artists or composers who share revenues with investors owning multiple NFTs representing fractional ownership of an asset, or shares of the revenue obtained from the asset (e. g., sale percentages of represented songs or poems). Amendments of the Proposal by EU Parliament written in 19.10.2021 (Nrs. 50 to 52 from Members of the EU Parliament KAILI and LALUCQ, sub *MiCA Draft Compromises by MEP Berger*, Compromise L) suggest in a new recital (8a) that MiCA Regulation applies only to cryptos "transferable among holders without the issuers' permission." Thus,

NFTs will be excluded from MiCA if they are "unique and not fungible with other crypto-assets, which are not fractionable and are accepted only by the issuer, including merchant's loyalty schemes, IP rights, guarantees, certificate of authenticity of a unique physical asset, or any other right not linked to the ones that financial instruments bear and are not accepted to trading at a crypto-asset exchange ..." Thus, securities laws would apply to fractionable NFTs, or to NFTs accepted by other than the issuer, particularly when NFTs encompass rights linked to those born be financial instruments. In the case that such instruments were negotiable in regulated markets, MiFID 2 Directive (2014/65/EU) would apply. As recital (8a) also states, the fractions of an NTF "should not be considered unique and not fungible," unlike the -multiple and fungible- investment or security tokens ruled by MiFID 2 and related EU Directives (AML, anti-money laundering, 2015/849, and 2014/57) and by Regulations (EU) 2017/1129 on Prospectus and 596/2014 on market abuse. KAILI and LALUCQ believe that "the sole attribution of a unique identifier to a crypto-asset is not sufficient to classify it" as NFT. Anyway, NFTs representing services, digital or physical assets that are unique, indivisible and not fungible (like product guarantees, personalized products or services, or real estate) should not be subject to MiCA, except if the NFT grants to the holder or its issuer "specific rights linked to those of financial instruments, such as profit rights or other entitlements." The reason is clear: in such case, NFTs have been voluntarily (contractually) converted into securities. Indivisible NFTs solely "accepted by the issuer" (new recital 8c proposed), and not accepted to exchange trading, should be subject to a wide bespoke or tailored regime (agreeing, KAILI and LALUCQ, ibid.).

3. A brief introduction to permissioning in blockchain and to public-permissioned distributed ledgers or networks (PPDLs) and its relevance for NFT contracting and markets

3.1 The meaning of permissioning in tokenization and the concept of public-permissioning

From a DLT structure or architectural perspective, the tokenization or creation of cryptos or tokens can be materialized according to various complex technical procedures. We herein consider the options actively present in token markets within the context of permissioned DLT networks (with authorization or permission of certain nodes to record transactions or data exchanges in the blockchain and blocks – onchain transactions), as long as they are publicly accessible and allow anyone to perform operations.

The concept of "permissioned" network or distributed ledger (PDL) is consolidated in international standards since 2019. In general, this adjective opposes to "permission-less" or without permission, in such way that distributed ledgers can be considered as permissioned or permission-less from the double perspective of:

a. Node or nodal requirements for a network member to be approved as capable to validate the recorded transactions, and to be eligible to effectively record them on the ever growing block-chain as a part of the information tied to the ledger, which plays a key role as a practically immutable or indelible and unforgeable data registry. PDLs are governed or administered by permissioned or authorized nodes in accordance with the bylaws or previous contracts governing the network and the competences of nodes ("miners" in permission-less ledgers).

b. Public accessibility to the information recorded in the distributed ledger. Permissioned transactions (corresponding to PDL data traffic) are not publicly accessible or readable without authorization granted by ledger administrators who may require self-identification through certificates or other digital means.

Permissionless ledgers like Bitcoin have received most attention from the markets since 2015, but tokens (including NFT) should be traded only in PDLs, better qualified to address most of the legal requirements posed by token use cases in the DeFi industry from the viewpoint of legislators and governmental financial institutions.

The former affirmation rests on both technical and legal reasons, and also in economic considerations. Total anonymity and decentralization characterizing Bitcoin does not match legal requirements of control and supervision essential to protect DeFi markets and investors. Otherwise, transaction costs and "proof-of-work" classical requirements to tie blocks and computational effort (including electrical consumption) to find out nonce numbers (it takes roughly ten minutes for a miner to sign a Bitcoin block, and fractions of seconds to sign a PDL block) render permissionless costly, environmentally unsustainable and undesirable for token trading and related contract negotiation and execution.

Apart from fastness and computational dramatic cost reduction, other technical aspects favor PDLs as the natural site for optimal token contracting, like the possibility of further control of recorded transaction, the supervision of functioning and cost of the consensus algorithm, and many options to adopt and adapt adequate fairness properties among participants. The legal matters include the support from external legal agreements or the regulatory enforcement in critical financial subsectors like NFT contracting.

The terminology varies from the general perspective of the International Standards Organization (ISO), whose standards recognize the concept of public ledgers in several standards, to the most specific viewpoint of, among others, the specific standards of the International Telecommunications Union (ITU-T) and the European Telecommunications Standards Institute (ETSI). Those institutions consider that PDLs are "trusted" blockchain, whose domain is defined by compliance and interoperability (cf. https://www.etsi.org/deliver/etsi_gr/PDL/001_099/001/01.01.01_60/gr_PDL001v01010 1p.pdf, 1 et seq.) and the corresponding definitions contained in the ISO/DIS 22739 Terminology, ISO/CD 23257.2 Reference architecture set by the Blockchain Technical Committee ISO TC 307. It should be noted that CEN-CENELEC: CEN (European Committee for Standardization) and CENELEC (European Committee for Electrotechnical Standardization) have stable liaisons with ETSI ISG PDL; and a new Technical Committee acts as mirror with ISO/TC 307. This Focus Group decided to continue as a Joint Technical Committee (CEN/CLC JTC19) since 2019.

On the public nature of a blockchain, see the standards of ITU-T FG DLT, Technical Specifications and Technical Reports D1.1, D3.1 and D3.3; cf. ETSI GS PDL-012, developing a layered PDL reference architecture; ETSI GR PDL 003 V1.1.1 (2020-12); Permissioned Distributed Ledger (PDL); Application Scenarios, http://www.etsi.org/deliver/etsi_gr/PDL/ 001_099/003/01.01.01_60 /gr_PDL003v010101p.pdf; ETSI GR PDL 004 V1.1.1 (2021-02); Permissioned Distributed Ledgers (PDL); Smart Contracts; System Architecture and Functional Specification, http://www. etsi.org/deliver/etsi_gr/PDL/001_099/004/ 01.01.01_60/ gr_ PDL 004v010101p.pdf.

The aforementioned standards have become crucial to understand the meaning and scope of permissioning and the possibilities of authorization granted by nodes to DeFi platform managers, intermediaries and NFT investors, as we will show below (section 3.2) since the conditions of legally recognizable markets depend on the technical characterization and structure of permissions granted by and to operating nodes in the network.

It is to be well noted, finally, that in order for a token market to be successfully deployed with full compliance control capacity by market authorities, the optimal solution seems to combine the advantages of permissioning under proof of authority (PoA) protocols and the publicness or public nature of the network, meaning public access to operate and openness for the public control of nodal activity, and therefore, decentralized control of the network, which is not possessed or dominated by a single entity.

The standards of the International Telecommunications Union outline the distinction between public and private blockchains; for instance, in ITU-T Focus Group DLT Technical Specifications and Technical Reports of 2019, namely D1.1, D3.1 ,and D3.3. The European Telecommunications Standards Institute develops a layered architecture in documents like ETSI GS PDL-012 and ETSI GR PDL 003 V1.1.1 (2020-12), and a concept of PDL in several key documents like [1].

The concept of public-permissioned network as a distributed ledger (PPDL) has been introduced in the Ethereum blockchain practice by Consorcio Red Alastria, the generator of the first worldwide essay or alpha version of a PPDL, encompassing ETSI PDL standards and functioning philosophy recently detailed [2], situating the governance and access control in the Platform Service Layer within the PDL network architecture.

3.2 Significance of authority proofing in PDLs for token contracts

In the case of PDLs based on the Ethereum blockchain protocol, transactions are developed applying IBFT or Byzantine-type consensus protocols to grant permission to record transactions on blocks, under a DLT network system configuration where different alternative tests or validation protocols can coexist (e.g., specific test of permission or authorization by validator nodes, known as Proof of Authority or PoA [3]). For the proper execution of such protocols and the deployment of smart contracts to automate the fulfilment of programmed transactions related to payment and transmission of tokens, and other effects linked to their normalized trading on exchange platforms (and their corresponding asset and benefit transference among node addresses), the developers of Ethereum have devised different standards, called Ethereum Request for Comments, ERCs, which entail substantially different characteristics, serving to carry out transactions a variety of tokens.

Derived from Proof of Stake (PoS) and in the Ethereum 2.0 blockchain (today ready for the mining of more than 100k transactions per second, thanks to new shards or channels for massive-data distribution, which guarantee the scalability of the network), the PoA protocol aims to keep the strengths of PoS (financial incentive for validators, limited computational effort and hardware sophistication; sharding scalability) overcoming its drawbacks, mainly the assumption that 'staked-tokens' owners have incentive to act in the network's interest (they actually do not, since dominant holders 20% accrue more investment in the network's success than other having 1% less holdings staked, regardless of the actual stake size). PoA algorithm, instead of tokens as stake, takes participants' identity as stake, thus implying the disclosure of validators as known entities risking their reputations for the right to validate the blocks. Thus, unlike in PoS, monetary discrepancies between validators are irrelevant, ensuring equal motivation to work for network success.

Regarding DeFi token trades, the specification of ERCs depends, among other relevant factors, on the content of the rights associated or incorporated to the cryptos. And such rights granted to the token holder or acquirer previously depend on the mechanism or mode of representation of digital assets involved in the standard.

In the case of NFTs, one of the key characteristics of the rights associated with the token is non-fungibility which implies the full identification of the token, its individuation, plus the limitation to massive trade whenever the token is unique and there is no prevision for massive token trade-off on a DLT market platform. In a PDL-Ethereum context, different protocols generating different types of ERCs constitute an ideal and flexible way for the orderly management of the corresponding transactions on tokens and associated smart contracts to efficiently execute trades. The structure of such trading varies considering the NFT intrinsic structure and tradability as an investment token.

It is to be noted that the first characteristic of the NFT to consider to this extent is the market tradability of the crypto. Some NFTs could be traded despite they are technically (apart from its referred or underlying asset) non-fungible or unique, since massive tradability could be obtained by means of fractioning or by means of contractual linking or connections to tradable assets or rights constituted on such massively tradable assets in regular or specific DeFi markets, which should be regulated. Deputy members of EU Parliament (MEPs) and EU INATBA (International Association for Trusted Blockchain Application), representing DLT industries, have been suggesting according considerations within the MiCA Task Force Working Groups created to feedback MEPs with relevant criteria from DLT industry and PDLs. Spain's DLT Alastria Consortium (cofounded by Prof. J. IBÁÑEZ) is a founding member of INATBA, thus participating since 2021 in finance WGs operating in INATBA MiCA Task Force (cf. https://inatba.org/mica-task-force/).

Proof of Authority (PoA) blockchain protocols, within this context, may play a key role to determine the consent of public supervisors in order to authorize the legal trading of NFTs in the scope and within the boundaries predetermined by the securities markets applicable rules. The reputation of validators is held by market authorities, which makes PoA impractical for public blockchains like Bitcoin and Ethereum, wherein hundreds or even thousands of validating nodes operative, and the identity requirement becomes a clear advantage to comply with regulations in the context of public-permissioned blockchains subject to legal specific requirements. In a PoA-network PPDL, a relatively small number of validating nodes re-centralize or make the market less decentralized, facilitating supervision, with enough degree of decentralization but gaining in investor and market confidence (like in the cases of Hyperledger Besu networks, an implementation of Enterprise Ethereum offering two PoA alternatives; or in the Alastria Consortium Telsius and Besu networks), with much higher throughput capacity than in permissionless blockchains, minimal computational effort and no specialized equipment (like in PoS). The acceptation of established financial institutions (as investment banks, financial intermediaries, or crypto-asset service new providers) as validators for crypto-transactions should be a legal previous condition to the settlement of DeFi market platforms for cryptos.

Previous authorization of intermediaries in traditional financial markets is essential to the market confidence and investor protection. Similarly, previous authorization of validating nodes in crypto-transactions will probably be a legal and also economic-efficient precondition for the development of token transactions in general, and in particular for high-risk token investments like NFT trading. Although PoA-based algorithms are unlikely to power big market platforms with thousands or millions of users, they are already optimal to build networks tailored to the needs of a limited number of known stakeholders (agreeing, [4]), as in the case of certain limited-client oriented NFT platforms and markets.

4. Approach to the legal and economic consequences of NFT trading in PDLs. Particular reference to public-permissioned networks (PPDLs)

DLT, in its blockchain version, facilitates the creation of markets built on decentralized applications (DApps) and platforms wherein token marketplaces can be installed and open to the public for the online confluence of investors seeking high returns operating from any blockchain-associated (operating or regular) node, directly or through representatives—cryptoasset or cryptotoken brokers or market specialists.

As mentioned supra (section 2), in PPDL blockchain networks like Ethereum, token transactions use to be executed using specific protocols of permissioning like PoA. In the case of NFTs, each token is autonomous, identifiable in the ledger and separable to be examined for efficient traceability and proper exercise of holders' rights. Its API interface, available on the upper layer of a PDL network architecture, facilitates traffic and exchange of tokens in accordance with market legal requirements and economic constraints of the business model settled by the issuer, with no quantitative boundaries within the issuing process, as a consequence of the PDL structure of protocols.

Frequently, the ERC 20 protocol is used in PDL Ethereum networks to represent and transmit fungible assets (typically, of a financial nature, such as fixed or variable income products, and derivatives on these); while others such as ERC 721, the first in use chronologically, and also others of later construction, such as ERC 1155, are used to represent and assign or transfer rights over infungible cryptoassets like NFTs.

The substantial content of the ERC 721 protocol is specified in a standard smart contract of Solidity, on whose original code the creators of open or free software (developers) can write lines or scripts of new smart contracts (Smart Contracts, SCs) or clauses compatible with the ERC 721 standard itself in order to adapt it to new uses, by means of transcription and import from a virtual library or code library called OpenZeppelin. The ERC 721 SC standard itself implements application programming interfaces (APIs) also designated as ERC 721 or, under certain conditions, APIs under another compatible protocol designated as ERC 165, which facilitate specific interaction between API and SC for traceability and transmission of NFTs on the PDL.

Regardless of the PDL ledger registration of NFT transactions on blocks, each NFT is virtually maintained, possessed and (under private-law full compliance conditions present in each competent jurisdiction) owned virtually in digital wallets. The NFT may also be transferred individually to a singular purchaser, or eventually be traded by speculative investors or their authorized representatives from nodes, directly or through brokers or auctioneers specialized in operating with these wallets.

Each NFT token can represent new born virtual or digital non-fungible individualized assets such as artistic or ludic representations (like the famous recreational "cryptokitties" from Axiom ZEN, coded in 2017 to ERC 721 and identified by their respective unique set of "genes" of appearance, passable onto offspring by "breeding" with one another; or the 2021 Bored Ape Yatch Club –BAYC- pics and drawings); pre-existing goods or physical assets (art works or linked intellectual property of authors – IP); specific commodities; individuated or isolated credit rights on any asset, e.g., music fragments; or even offline services.

The referred or underlying asset, financial (credits) or nonfinancial, will normally be non-fungible, and such non-fungibility justifies the issuance of such kind of token. But since the token individuation is independent from the degree of fungibility of the physical underlying (property, work of art, collectibles, ...), NFT can be issued on groups of fungible assets functioning unitarily or characterized by convention with unicity, or on identified parts of such fungible assets, parts, or fractions of them. Thus, a PDL platform manager or DApp programmer (NFT issuer or not) can also create and involve the trading of rights over prefigured new financial assets derived from the NFT, or on the NFT itself, or create guarantee contracts or other kind of guarantees or akin securities also recordable on the ledger, and offchain as well.

PDLs permit the negotiation of an unlimited number of NFTs, as was already the case of the ERC 20 fungible token standard created in 2016, from which NFT-suitable ERC 721 itself derives, unlike the Bitcoin system where the cryptocurrency resulting from the mining or composition of transactions is quantitatively limited.

Ethereum PDLs have also enlighten the factual NFT-trading utility of other ERC protocols like the ERC 1155 standard, generated in 2017 and officially deployed in permissioned systems since 2019. This much more evolved ERC, wherein a record number of developers have participated, allows the issuance of both fungible tokens and NFTs, all of them under a more flexible smart contract modality (SC), unlike the precedent ERC 20 and 721, which require a new SC per token. Under ERC 1155, the investor or broker uses a secure interface and the SC allows to adapt a menu of unique features, such as, in the case of war games, different weapons, defense modes, or battle scenarios; if they are connected to a metaverse, it is possible to introduce unique parameters and conditions by NFT, and similarly individualized replicas from the real world. On the other hand, the ERC 1155 standard makes it easy to send multiple tokens in a single block transaction (even more than a thousand), shortening the wait for block closures and therefore increasing the speed of DeFi operations. Likewise, this standard allows the exchange of tokens in a safe and disintermediated way, enhancing the construction of atomized financial products (e.g., "atomic" derivatives like swaps and other).

In summary, each NFT token can relvolve around fungible or non-fungible goods or rights, irrespective of its own non-fungibility or full individuation as a standalone asset, that prevents its massive or impersonal exchangeability that exempts its issuance from primary-market public or administrative control applicable to MiFID, MiCA DLT, or DeFi regulated trading (cf. again art. 4 MiCA).

Regardless of the concurrence and level of legal control of token issuances, it should be borne in mind that, in any virtual space of a DeFi platform on the upper DApp (decentralized application) layer of a PDL network, a permissioned (previously authorized) mechanism for representing the rights of the parties contracting NFTs is always created. Such authorization is made under private-law boundaries. Smart contracts (SCs) required to execute NFT contracts are also previously authorized and deployed on the service layer of the PDL. And before SC authorization, PDL PoA or similar protocol to execute transactions has been implemented.

Such chain of permissions is prior to the provision of investment or exchange services for NFTs, within a PDL context. The NFT contract for the transfer of the underlying property or connected services or investment is concluded autonomously, prior to the technical (and legal) birth of the token. In the case of NFT tokenization or minting of nonexpendable property or rights (for example, real estate, art, or copyright), the ERC standard allows for the subdivision, not of the tokenized sole or indivisible property, but of a representation of property (*in rem*) rights or credit rights of investors in a public offer made by an offeror, against the issuer of the tokens, which may or may not be that offeror. In case of initial token offering (ITO), the rights have been "fungibilized," which does not imply a reconversion of the non-fungible asset into a fungible one, but rather a substitution or a subdivision or fractioning of the object of investment to allow massive or market negotiation. In the case of real estate tokenization, for example, virtual fungible goods (real estate tokens) are created to represent credit rights on properties that can be registered using Ethereum ERC 20 tokens, or to provide an alternative for the transmission of full or limited real rights.

We have emphasized [5] [132, 218–221] that the token fungibility does not impede the existence in the real world of the tokenized represented underlying assets (services or goods) nor the possibility of contracting on it also in the real world. The link between an NFT and the real asset through annotations in a public registry privileges and prioritizes the legal position of the token holder to the point of "establishing it as a means of representation deserving of a regime analogous to that of negotiable securities (including nonfinancial assets such as listed merchandise), for the purposes of recognition of ownership transferred in a market" – ibidem.

Summarizing, PDL networks, in particular public-permissioned DLT networks like Alastria's, facilitate the legal compliance with private-law contractual issues related to the identification of NFTs and their underlying asset or property, and also with public-law issues concerning the supervision of node protocols and activity in contract registration and authorization of DeFi platform managers and issuers. Hereinafter, we deal with particular conditions of compliance that challenge the NFT trading activity in such DLT permissioned spaces.

5. Specific law-compliant trading challenges for NFTs

Without intention to explain all legal problems related with NFT commerce, we hereinafter show some of the key issues posed by NFT market practice in PDL DeFi environments considering basic differences among common law and civil law jurisdictions, and between negotiable (massively traded in market places) and non-traded tokens.

5.1 The nature of the token and its underlying good and the legal regime applicable to NFT trading

First of all, it is to be outlined that tokenization has different legal consequences and scope in the case of NFT trading and in the one of fungible virtual assets.

But the bunch of law-compliance specialties widens when taking into account the nature of the goods, services, or assets underlying the NFT. In the case of listed deliverable goods, like agricultural products quoted in commodity exchanges, NFTs can confer rights to the possession, transfer, disposal, withdrawal, or delivery of goods previously identified (e.g., selected wine bottles, or containers, with trademark and specifications of quality and provenance from a denomination of origin guaranteeing such characteristics) in favor of the token holder. The same structure would be applicable to collectable or individualized artworks or cataloged antiquities. The nature and extension of faculties granted by the token can differ substantially in the case of NFT unlisted goods, as the corresponding administrative or public-law regimes applicable to their registration vary depending on the civil-law regimes or common law principles for the delivery and efficacy of the transmission of property, and according to the nature of the tokenized object and specialties of its delivery and transmission. The NFT negotiable tokens, or alternative fungible tradable tokens on which "derivative" rights on NFTs or underlying physical assets are created, require the implementation of regimes such as the one provided for in MiCA for their legal circulation as investment tokens, with specific supervision and provision of services efficiently supplied by professional intermediaries of blockchain services (virtual asset service providers – VASPs or crypto asset service providers – CASPs).

In civil law countries, legislators usually adopts Roman law antecedents to consider that the cession of any property, including rights annotated or recorded in a public registry of properties (mostly for real state but also for mobile financial property), requires, for legal efficacy of the transmission in favor of the acquiring party or purchaser, the compliance of registration standard principles like legitimacy. In most countries, transferor shall be deemed to remain the holder of the token until the name of the transferee is entered in the ledger or the token is received in the digital wallet of the purchaser.

Otherwise, most EU jurisdictions under MiFID domain set regimes on mobile fungible assets (namely listed assets) establishing specific rules of control and recognition of fungibility in order to effectively protect the position of both the registry (in terms of public trust) and the registered holders of traded assets. Thus, concepts like "tradability of securities" as noted in art. 18 of Spain's Decree 878/2015 developing national securities law (art. 6.4 Ley del Mercado de Valores) defines the "fungibility" of the book-entry securities associating it with the equality or identity of "characteristics" of the book-entry. The grouping of the property of the fungible securities in "balances" or "determined amounts" identified in the account of the holder, as in the case of classes and series of securities issued by the same company, demonstrates the capability of fungible assets for massive trading, without prejudice to "the needs of specification or breakdown of registered securities derived from special situations" (art. 18.3, ibid.) such as embargo, execution of sentences or usufruct contracts, modification of encumbrances, or other relevant circumstances that do not affect the identity of the characteristics among the fungible values but can be registered (as in the case of the pledge, whose registration is equivalent to possessory displacement of the recorded value, *ex* art 14.1 Decree 878/2015). Similarly, special regimes could consider NFT as eligible negotiable assets grouped in such amounts within segregated accounts to protect the rights of holders at the same level of fungible securities under the aforementioned circumstances, applying similar rules considering the singularities of NFT identity or individuation as a key characteristic for the autonomous recognition of the virtual asset as a category of DeFi tradable (and ledger-traceable) security.

The non-fungible characterization of NFTs, in a PDL network, and more specifically in the realm of the PPDL, can become preconfigured and programmed by the platform managers and by DApp/SC developers (with the acquiescence and participation of the issuer in the process) from the very beginning of the preparation of the platform in accordance with the PoA consensus protocol, and in any case they must be set to prepare the transfer of the cryptos no later than the start-up of the SCs layer and the contemporary provision and insertion of the tokens in the platform. In particular, in accordance with the specifications of the ERC requests for comments for the case of Ethereum network protocols and derivatives (for example, those of the Quorum-Ethereum kind).

5.2 Investment and betting with NFTs

All investments entail a bet in the sense of probability of nonreturn, partial, or total loss of committed funds, or materialization of the fundamental economic risks of price, market, and solvency of the issuer, in addition to operational, network resilience, and regulatory risks. On the other hand, not every rational investment involves a bet in the classic sense of "game," limited or even prohibited according to civil law private regimes (e.g., Spanish civil code of 1889, arts. 1790 and 1801). Both ethical and social reasons, and the reasonableness of NFT trading control from a public order perspective and from the viewpoint of investor protection, should prevail in future specific NFT regulations. The magnitude of the interests potentially affected, also in the strict legal-private sphere, is exponentially growing in the case of the NFTs market. Thus, investments could require betting or gambling legal boundaries in the purest sense of market speculation. Surveillance should foresee the measurement of NFT returns and the consistency and soundness of their value and valuation methods, still in infancy and precarious, since the bul of the NFT investment value depends on unknown factors such as the issuer credit risk, the fulfilment of contractual promises of profitability, or the effective avoidance of cyber risks incorporated into the architecture of the PDL network wherein the transfer of tokens takes place.

Even the risks associated with the potential misuse of protocols or abuse of technical trading platform should be considered when legally qualifying NFT trading as gambling in this particular context. However, this does not exclude their generic character of "investments," and therefore, to the extent that investment contracts are made in the market, they must be subject to a specific regime, which many issuers seek to avoid by selecting unregulated DeFi trading forums and movements of headquarters or establishment of the platform and the underlying business (forum shopping), triggering overall risks borne by unaware investors, as already denounced by the best mercantile-law doctrine regarding speculative stock market positions.

Transactions made by virtual characters (avatars) proliferate exponentially (purchase of virtual yachts or virtual plot in a metaverse like Sandbox, created by Republic Realm Co.). The association between a metaverse and NFTs representing unrepeatable unusable objects (because they are unusable) is a symbol of elitism and selection of smart investment today. Decentraland, a "virtual local" of 560 m2 meters, was recently sold for 2.4 M USD in Mana cryptocurrency, located in the exclusive virtual street of Fashion Street, hosts fashion shows. Burberry, Gucci, Louis Vuitton, Adidas, and Nike (developing Nikeland metaverse) already operate in these virtual environments. Microsoft Teams video conferencing app, widely used in the professional and educational world, will soon allow Mesh to operate in the metaverse trading NFTs in virtual markets, creative form of a parallel financial system, but also of a mechanism of social separation, where operators develop systems possibly widening the digital social breach. This is why investment valuation must be redefined considering the difficulty of access to the market and use of sophisticated DeFi payment tools and NFT contracts by which a few investors gain stratospheric profits in the purest classic stock-market speculation on extra-volatile instruments.

Aside from ethical-philosophical considerations, these economic facts should lead to legislative reflection on the creative value mechanisms of NFTs, even with a possible redesign of the legal concepts of "utility" in the Paretian sense, and "value," since the emergence of new asset valuation models determining reactive mechanisms to protect the integrity of the market and the health of the financial system itself, aligned with the single European digital market and within the conceptual framework of the European digital finance strategy, led by the MiCA-IM-DORA proposed regulations. Measuring the value of NFTs on works of art in new NFT DeFi marketplaces connected on metaverses requires imagination and prompt reaction to volatility. Investment caprice, artistic novelty, or simply the sudden mobilization of large investments, in the absence of institutional market creation systems or control mechanisms such as the DeFi AMM, restrict a rational control of volatility. Classic CFROI-type valuation systems, and even other more accurately adjusted to market valuation like value-at-risk (VAR) or real-option methods, are practically useless in these virtual contexts, considering their geographic offshoring, their lack of jurisdictional control (and even competent jurisdiction), and the dilution of responsibilities associated with the use of DLT and linked digital technologies, whose scope does not seem difficult to intuit, despite their yet unknown full dimension. The role of national and regional regulators on NFT volatility control mechanisms is crucial for financial stability even when such tokens represent unique creations and cannot be legally reputed ARTs or present extreme valuation complexity according to traditional parameters.

An acceptable solution to these issues is the attribution of specific role to legislator in the control of NFT volatility when traded (directly or by means of rights on fractions of the underlying asset, virtual or not) in relatively profound and liquid marketplaces, including OTC and nonregulated or deregulated specific contracting spaces (IP e-markets, artwork virtual auctions, DeFi metaverse-linked platforms, ...). Legal measures to set volatility boundaries or prohibit certain NFT trades poses the eternal dilemma about the efficiency of law intervention and regulatory gatekeeping. Anyhow, we believe that the use of stablecoins as an anchor to limit the volatility of the value of the NFT requires three constructive assumptions:

a. The existence of a stable reference underlying currency or asset to establish a correspondence or value relationship, as in the case of electronic money tokens or EMTs and those referenced to assets or ARTs, as contemplated by the MiCA regulation.

b. Setting up legal support, like endorsement guarantee or reserve system and mandatory coefficients specific for NFT risk hedging or coverage, associated with the value relationship or correspondence system, including collateral execution and reserve governance mechanisms, among others, to limit credit and insolvency risk of issuers and ensure the execution of investment contracts on the tokens.

c. Minimal state prudential supervision ensuring DeFi NFT market compliance and regulatory updating, including proper monitoring of the actions of operators for the sake of market integrity, financial stability, and investor protection, especially in cases of high pre-contractual informative asymmetry or disinformation of holders.

5.3 Considerations related to specific primary-market obligations of NFT direct or derivative

As outlined before, MiCA does not regulate NFTs, apart from the exemption rule excluding them from the MiCA whitepaper regime. The EU regulator assumes that NFTs are "unique" (individuated) and also presumes that the token is not related to other cryptos. But in the case of existing a predetermined contractual relationship between the NFT and the represented object, either in form of economic indexing or stabilization arrangement, or in form of contractual commitment (i.e., servicing, lending, or fund depositing) with respect to underlying assets (like in asset-referenced tokens – ARTs), the MiCA primary-market whitepaper transparency rules on construction, notification, and legal publication would be applicable to issuers. Even if that were not the case, NFT issuing entities would be anyway constrained, under MiCA EU provisions, by the general obligations applicable to crypto-asset issuers or petitioners of admission to trading in authorized DeFi platforms, when the tokens are not stablecoins (e-money—EMT—or asset-referenced—ART—cryptos; when issuing ARTs or EMTs, issuance obligations intensify to protect investor, market stability, and monetary local or national policy). But all MiCA-ruled tokens are cryptoassets requiring adequate investor protection standards in all jurisdictions.

Some authors do not qualify tokens other stablecoins as crypto-assets in the sense of MiCA regime—*per omnia*, [6], arguing that EU regulation is designed to prioritize the supervision of stablecoins (AMTs/EMTs), thus excluding other cryptos. It is true that stablecoins are monitored with particular intensity, but encompassing most crypto-tokens present in markets (including NFTs under particular circumstances) is a key aim of the MiCA legislator as well.

As ruled in article 13 MiCA, such common minimal burdens involving all crypto-issuers refer to:

a. Acting honestly, impartially, and with professionally in the best interest of token holders, treating them fairly except in case of due preferential treatment as shown in whitepaper provisions and publicity or commercial communications.

b. Impartial, prompt, relevant, complete, and not misleading or deceitful communication with token holders.

c. Detection, prevention, and management of all conflict of interest arisen, which should be communicated to interested parties.

d. Safekeeping and maintenance of all access systems and protocols in accordance with EU pertinent standards specified by the European Securities Market Authority (ESMA) in cooperation with the European Banking Authority (ABE) *ex* art. 16 of UE Regulation 1095/2010.

e. Timely reimbursement of funds deposited by token holders or possible investors; in cases of cancellation of an initial public offering of tokens (ITO) for any reason (cf. par. 3 of article 13 MiCA), ensuring due immediate payment by themselves or by fund custodians, depositary banks, or escrow agents.

5.4 Essential intellectual property (IP) protection issues

NFTs can represent, in the case of author artistic or scientific underlying creation, IP rights. Such intellectual work can adopt the form of virtually accessible videos, pics, graphics, lines, drawings, or artistic elements contained in the token or associated to it. The NFT can be created *ad hoc* to be linked to the PDL transaction containing and thereafter, recording the NFT data. The NFT is virtually possessed in the wallet of token holders who pay for it in the DeFi market.

In other cases, NFTs do not represent virtual artworks but preexisting assets (e.g., songs, movie images or paintings on canvas) on which authors or coauthors have also moral and economic IP rights. In many cases, authors retain copyright and token buyers just acquire rights of use as licensees, having faculties of loading, visualizing, or sharing images or objects accessible by possessing the token in a digital wallet. Such noncommercial rights should be clearly communicated in marketing channels and in clear licensing terms and conditions as well in order to prevent future claims derived from insufficient or inadequate pre-contractual information, as recent US doctrine spotlights [7].

NFT trading facilitates the sale of works, either selling the whole value of IP economic rights or parts of it in case of high-value artistic unique pieces. But it is to be well noted that these tokens representing IP rights do no generate themselves a proof of authorship. To this extent, it is necessary to reflect on the circumstance of the previous generation of the authorship, always preceding the constitution of the token as a virtual support on the PDL.

DeFi NFT marketplaces usually present and show determined cryptographic elements containing or representing IP rights. These elements are sold by means of DLT transactions but were previously created by authors. Buying and selling of IP, from a legal perspective, can be executed by means of DLT transactions meaning the effective purchase and sale of those works. The sale means the virtual traditio (delivery for the cession of property) of the IP rights, from the moment in which the transaction is recorded in the ledger.

The generation of proofs of authorship needs, therefore, to be prior to the sale, or circulation, of any artwork. But the risk of distributed (DLT) use of digital work (or work in general) by unauthorized persons (misappropriation, IP usurpation) is permanent. Professional scammers or IP usurpers utilize virtually supported or mounted artwork and upload artworks to the blockchain as their own, creating the representation (NFTs minting) from previous art represented that they did not compose or fabricate. This is one of the main perils around NFTs and even around blockchain PDLs in general. Unscrupulous NFT creators pretend to be art composers of works they do not have even permission to distribute or exhibit, creating derivative works of original works and selling them without the author's consent or even knowledge. Additional phishing, impersonation, identity fraud or theft, and unfulfilled promises of NFT distribution or delivery are common disloyal additional practices tied to these shadow markets.

The activity of NFT minting of digital assets including IP (artwork, music, video clips, ...) or industrial property (trademarks, patents) by persons other than authors, owners, or valid licensees are liable for infringing third-party IP, and cannot grant to the token purchasers such rights either since they do not have them (nemo plus iura transferre potest quam ipse haberet). Misstating the rights conveyed in connection with the NFT sale entails possible additional claims against token minters and DeFi platforms involved in such sales.

Some legal remedies (still not ruled in most EU countries) entail:

a. Specific legal recognition of NFT transactions in PDLs, including virtual IP transmission, as cession of IP rights as regulated in IP laws currently in force in each competent jurisdiction.

b. Specific legal recognition and regulation of platforms wherein such rights are granted or transmitted, in order to avoid civil liability of platform and DApp managers where NFTs infringing IP are being traded, even without acquaintance or even knowledge of those administrators. Since most NFTs have a single owner, they are not likely to be securities but, as shown before, in some cases they may; in the US, if they meet the Howey test, for the token buyer invests with a reasonable expectation of profits to be derived from the efforts of others.

And in the EU under MiFID rules, when NFTs are listed in regular markets as financial instruments (e.g., in case of pooling or division of credit rights, securitized in liquid markets).

c. The creation of specific public registries encompassing DLT IP represented by NFTs concerning virtual or material artworks or creations. Alternatively, the constitution of sections for virtual properties in the existing registries, in order to properly cope with the peculiarities of NFT-related IP conflicts.

d. Collective social reflection led by market authorities and IP competent public institutions and registries, in order to instruct virtual artwork professional authors, particularly on the economic relevance of legal artwork registration, in order to foment the generation of early proof of authorship, and thus effectively defend both original underlying IP rights, and token-associated IP as well.

e. Facilitation by PDL administrators, in cases of fraud, of a ledger proof of fraudulent transactions or scams on tokens, in favor of prejudiced authors. If an NFT contains robbed, plagiarized, or forged IP, transactions also reflect such infraction, and PDL managers can help locate and consult the concerned transaction in the network.

f. Ruling on market disclaimers warning potential NFT buyers on the fact that (in most cases) the only property they are acquiring is just a unique trace recorded in the ledger, containing a link to images of an artwork. NFTs detail transaction data (selling date, price, contracting parties' public-key addresses, wallet identifiers, ...). but they do not specify what has been bought or sold.

g. Ruling on the minimum mandatory contents of the NFT regime, with detail of the exploitation rights associated to the holding of the crypto, and the way to exercise such rights. Like in securities legal theory, NFT can lack, like traditional negotiable titles of credit, literality, and not refer to the full content of rights to be exerted by the holder. Anyway, the DeFi platform or NFT market managers should document such and register such contents (off the NFT, off-chain at least; e.g., by recording in Interplanetary File System IPFS with link to the crypto) to fully inform investors or purchasers of tokens. Otherwise, these might soon become empty properties.

Immutable and inseparable linking between IPFS info and NFT impedes a new creation of NFT containing its copyright data previously registered in an authorized or private DeFi DApp or platform, as practitioners resourcefully indicate [8].

5.5 Miscellaneous public-law compliance issues

Market illegal practices have been detected in recent years concerning the misuse and abuse of crypto-assets, particularly utility tokens and high value NFTs, as money-laundering vehicles. European authority (ESMA) and the U.S. Department of the Treasury have published recent studies on the facilitation of money laundering and terrorist financing through the NFT and stablecoin trade in ARTs. In the case of NFTs, digital art is involved as a relatively easy instrument for capital laundering, considering the volatility of art assets, which is extreme in the case of high-value art pieces, vulnerable to financial crimes.

In the realm of the administrative control of gambling, players pay for a chance to win betting on an NFT position traded on a secondary market. Such position may be deemed a "thing of value" potentially implicating gambling issues. The increased use of chance-based mechanics in games (e.g., lootboxes, casino games) has led to enhanced scrutiny under gambling mandatory laws and to class-action lawsuits. Many game publishers have prevailed those lawsuits filed against them when terms of service grant only a license to use an in-game currency or items in the game and prohibit their transfer or exchange. Such currencies and items have no value for gambling law purposes, but NFTs issued or promoted by gambling companies create true tradable cryptoasset in DeFi markets. That is why DLT-based games use few chance-based mechanics and more play-to-earn or user-generated content models.

Finally, other public-law related issues deal with market abuse. Within this scope, NFT insider trading policies should be completed and updated by NFT massive professional issuers and by DeFi marketplaces and brokers trading with NFT. Recent incidents of directors, top-level employees, and executives at NFT companies and marketplaces outline the need of engaging in specific preventive self-regulatory activity. NFT insider trading policies should restrict, condition, or even prohibit the intermediation, purchase, or sale of NFTs based on publicly reserved or undisclosed information, also prohibiting various types of token market manipulation (including underlying markets) by whistleblowing or manipulating the optimal prices by means of company NFT trades designed to distort the perceived price level or the trading volumes.

6. Conclusion

NFTs pose huge market and legal challenges, which should be guided by legislators with full respect to technological neutrality and entrepreneurial innovation initiative in PPDL emerging networks.

Fungibility is a legal-economic concept concerning tokens and their underlying assets. Forthcoming NFT regimes still dismiss this dichotomy and its relevance for the right insertion of tokens in legal taxonomies entailing the applicability of securities-law requirements (like EU MiFID2 primary-market rules) or crypto-asset specific laws (like MiCA Regulation), and their further developments concerning NFT ownership and transferability of rights tied to their exchange in DeFi platforms.

In the field of IP, effective protection of NFT purchasers demands additional efforts to balance the interests of DLT-platform managers and DApp industry with adequate protection of IP rights and authors of digital art. Such balance is necessary too for effective protection of investors and purchasers of rights linked to NFTs in the cases of gambling, money-laundering and macro-financial instability involving in DeFi NFT illicit trading. PPDL architecture surveillance is an excellent vehicle to this extent, requiring bespoke intervention in node governance, both on and off chain.

Author details

Javier Ibáñez Jiménez[1] and Eva María Ibáñez Jiménez[2]*

1 Universidad Pontificia Comillas, Madrid, Spain

2 Universidad Nacional de Educación a Distancia (UNED), Madrid, Spain

*Address all correspondence to: eibanez@cee.uned.es

References

[1] European Telecommunications Standards Institute (ETSI). Permissioned Distributed Ledger (PDL), Application Scenarios. Available from: http://www.etsi.org/deliver/etsi_gr/PDL/001_099/003/01.01.01_60/gr_PDL003v010101p.pdf [Accessed: March 27, 2022]

[2] Wang C et al. An Introduction of Permissioned Distributed Ledger (PDL). ETSI White Paper No. #48. Sophia Antipolis CEDEX; 2022. pp. 1-12

[3] Ibáñez J. Consorcio Red Alastria. Madrid: Reus; 2020. p. 280

[4] Bogdanov D. Proof of Authority Explained. LimeChain blog. Available from: https://limechain.tech/blog/. [Accessed: February 21, 2022]

[5] Ibáñez J. Tokens valor (Security tokens). Madrid: Reus; 2021. p. 402

[6] Muñoz Pérez AF. Algunas consideraciones sobre la taxonomía de criptoactivos en el regimen de transparencia. In: Ibáñez J, editor. Transformación digital y nueva regulación de personas, instituciones y mercados –Estudios Jurídicos en Homenaje al Profesor Prades Cutillas. Madrid: Reus; 2022

[7] Gatto J, Parsafar Y, Koury G, Almasi P. Tokenization and the Law: Legal Issues with NFTs, The National Law Review, vol. XII, Nr. 131. Available from: https://www.natlawreview.com/article/tokenization-and-law-legal-issues-nfts#:~:text=AnNFTcanbesubjectfromtheeffortsofothers. [Accessed: April 30, 2022]

[8] Iglesias A. NFT y propiedad intelectual: cómo se articula la protección de los derechos de los autores en la era del 'blockchain'. Invertia-El Español, blog; 2021. Available from: https://www.elespanol.com/invertia/disruptores-innovadores/politica-digital/espana/20211108/nft-propiedad-intelectual-articula-proteccion-derechos-blockchain/624938006_0.html. [Accessed: May 11, 2022]

Chapter 7

Perspective Chapter: Audit Digitalization – Key Impacts on the Audit Profession

André Schreuder and Hanlie Smuts

Abstract

The rate of technological advancement has influenced existing digitalization programmes of audit processes. The adoption and use of emerging technologies in audit digitalization programmes present numerous opportunities for innovation and improvement. These opportunities come with many threats to the current ways of working, which impacts the audit profession, audit process, and auditor. This study is based on two qualitative data sets to identify the impact of digitalization on the audit industry and audit profession, specifically big data analytics, blockchain technology and artificial intelligence. Existing literature was used to identify areas where the biggest impacts were experienced. Based on the themes identified, semi-structured interviews were conducted in the financial services sector in South Africa to gain insights from industry lived experiences. The interviews focused on the impacts experienced and the realized and potential benefits. Data collected was synthesized to address the research questions of how the adoption of emerging technology will impact the audit industry and shape the audit process, audit profession, the role of an auditor, and aims to define the auditor of the future.

Keywords: audit automation, digitalization, audit, emerging technologies, blockchain technology

1. Introduction

In a world where every facet can be digitally enabled through emerging technologies, enterprises in all industries will need to adapt to keeping pace and prosper or will ultimately risk unplanned obsoletion [1]. Advances in emerging technologies have the potential to impact and disrupt the audit industry, most notably in big data analytics [2–5], artificial intelligence (AI) [6, 7] and blockchain technology [3, 8, 9]. Audit firms cannot simply be familiar with emerging technologies any longer [10], as these principles and concepts have the potential to integrate with their existing processes and skills or replace them [11, 12]. As manual labor was replaced by machines during the industrial revolution [13, 14], from big data [15, 16], audit procedures can be codified in machine learning (ML) algorithms [3, 17], and ultimately the use of AI for cognitive tasks [7, 18], such as audit opinions and judgments. The introduction and use of blockchain

IntechOpen

technologies in financial services, specifically accounting, has led to the rise of triple-entry accounting that would negate the need for transaction level auditing [19, 20].

These technological advances can be regarded as disruptors to the audit industry, or they can be harnessed by audit firms and used as opportunities [7]. Opportunities that can change the audit industry, the audit profession, and the role of the auditor [3] to better assure stakeholders while increasing profitability and maintaining or creating a competitive advantage in the marketplace [21]. Therefore, the purpose of this study is to identify the impacts of digitalization on the audit industry and was guided by the research question "*What is the impact of digitalization on the audit industry considering the new advances in technology, specifically focusing on big data analytics, AI, and blockchain technology*?" The impacts identified were themed and tested in the South African financial services industry to uncover how these impacts have materialized in the market. Furthermore, to gain insights on the lived experiences of the changes to the audit process, audit profession and the skills required by auditors. Organizations can use the results of this study to plan and prioritize their digitalization strategies and programmes and obtain a view of the skills required by auditors in a digitalized audit environment.

The rest of the chapter is structured as follows: Section 2 describes the background to the study and the research methodology is presented in Section 3. Section 4 describes the data analysis and findings and Section 5 the contribution. Section 6 concludes the chapter.

2. Background

The initial studies that focused on the impact of technology, specifically automated auditing, were completed by Vasarhelyi [22] over 35 years ago. Although this was a theoretical paper focusing on future impacts of audit automation, it did conclude on three anticipated impacts that have materialized today. These are the physical location of the auditor, audit timing, and the audit's dependence on technology [22]. More recent studies have indicated that audit automation leads to significant reductions in time and productivity gains [2, 23]. The key drivers for audit automation were identified as the ongoing changes in the regulatory landscape [24, 25] and the increased complexities in the financial system and financial products [23]. In addition, further studies have revealed that audit automation will impact the entire audit industry and that the impacts are not limited to internal or external audits or any specific type of review [23]. Since these pioneer studies, there have been dramatic advancements in technology enabling adoption and changes in all industries, including audit. The adoption of these technologies has impacted the audit industry [3]. Different ways of working were required in the audit industry to leverage the benefits posed by digitalizing old audit processes [3].

2.1 Audit and the audit process

According to the ISO 19011:2018 standards, an audit can be defined as a "systematic, independent and documented process for obtaining audit evidence (records, statements of fact or other information which are relevant and verifiable) and evaluating it objectively to determine the extent to which the audit criteria (a set of policies, procedures or requirements) are fulfilled". An audit involves an independent evaluation of information with a view to express an opinion on whether the information being represented is reliable, understandable, and relevant [26]. There are many types of audits, mainly financial audits, operational audits, statutory audits, or compliance audits. Audits are generally carried out by either internal audit or external audit teams [27]. The standard audit process can be followed for all audit types and consists of three main steps, namely: audit planning, audit execution and audit reporting. The audit follow-up process takes place after remediation of issues and exceptions noted have been remediated by the audit client. This assures that the risks identified during the audit have been addressed [26].

Internal audit has a broad scope related to the stringent control over all company business practices and risks, including assurance of the authenticity of financial records and the efficiency of the operations [28]. An external auditor is responsible for providing an independent opinion on an organization's annual financial statements. Therefore, providing reasonable assurance that the published financials are a true representation of the financial position reported [29, 30]. Any misstatements in the financial statements identified or breaks identified in the control environment that might lead to a material misstatement of the financial statements are reported to the shareholders and the public [31].

2.2 The application of technology in auditing

Digitalization of audit processes started as continuous auditing [2, 32], audit automation [2, 3, 23] or computer-assisted audit techniques [33–35] with tools or toolsets. These studies highlighted the following advantages and benefits of adopting these techniques as faster execution times, wide scope coverage, consistent execution of audit tests, full population testing rather than sample testing [2, 33, 34, 36]. The above studies were completed in industries, including telecommunications, mining, and the financial sector, specifically banking [2].

Research completed specifically focusing on the impact of audit automation of accounting practices and financial auditing indicated significant productivity gains and a reduction in labor cost [37]. The automation of repetitive tasks has made teams more efficient, allowing them to focus on the outcomes rather than the execution itself [2, 37]. There was a significant spike in the repetitive nature and the need for these tests with introducing the Sarbanes-Oxley Act of 2002 (SOX) [38]. It increased the defined scope of internal audit functions and the responsibility and accountability of external audit firms leading to staff and skills shortages in the market and increased audit costs [39]. Introducing SOX regulation served as a catalyst for organizations to adopt automated audit techniques [40]. In recent years, the continued introduction of new acts focusing on specific topical areas such as data protection, information security, open banking services and payment security added pressure on organizations to comply and audit functions, both internal and external, to assure compliance statuses to the relevant regulatory bodies.

2.3 Emerging technologies and audit

Many emerging technologies have revolutionized industries. As this study is focused on the impact of digitalization on auditing, the study was limited to key technologies that have impacted the general automation of processes and digitalization of processes. The most notable emerging technologies considered are big data, AI and blockchain technology. Due to massive advances in storage technology, costs have tumbled, and in line with increased computing power, the amount of data generated has seen an exponential rise [3]. The term big data refers to datasets that are too large to manage and process with standard tools. In many cases, the data is from multiple sources, in differing formats and fulfills different objectives; therefore, it is not generally stored in a structured/controlled way [41].

The four-V paradigm is used to describe big data, and these are volume, velocity variety and value [41]. Of the four metrics, the most pertinent is value, and the biggest concern, finding value in these large datasets and not wasting resources and time is important [3, 41]. Big data analytics is the utilization of techniques and technologies to gain insight into these large, unstructured data sets. Audit traditionally will focus on financial data, which is well-formed and more easily understood and interrogated; this can make analytics on big data potentially counter-intuitive for the audit industry [42]. The techniques and technologies, however, can be applied to the more structured financial data, and testing of all data is possible rather than the more routine sample data approach [3, 42]. Links between financial data and external non-financial sources from a big data source can be found; potential sentiment analysis might be an option [42]. Linking what is said on social media with an increase in transactions or trying to determine a value of reputational risk following a negative article being published are options that did not exist before [43]. The risk lies in not understanding these large unstructured datasets. An auditor must make sure that they understand the content and value of the dataset [3]. Another risk lies in trying to link disparate datasets (structured and unstructured) in a simple and reproducible way. Domain knowledge will have increased importance, and the skillset of the auditor will need to include several technologies in the future [3].

AI denotes that a computer can be utilized to accomplish more complicated tasks that require a human operator and skill set [44]. AI, and ML, a branch of AI, is used for the automation of tasks that require decision making rather than a linear progression that was the focus of the industrial revolution [18]. Most AI solutions are focused on language recognition, logical problem solving through iterations and visual pattern recognition [45]. For auditing purposes, the most common use for AI is for the identification of irregularities in accounting data [3, 18, 45]. As financial markets and products become more complicated, so does the process for identifying fraud and financial crime; this can be achieved with ML. It is noted that some companies are using AI for data collection and validation during the audit execution phase [45]. While the use cases of AI are still being identified, the full utility of AI in auditing is not codified. The World Economic Forum, however, predicts that 30% of audits will be completed with AI by the year 2025 [46].

A blockchain is a decentralized database that chronologically stores information about transactions of any kind [47]. As it is decentralized, a public blockchain does not have the weakness of a normal database that it can be modified or, without proper disaster recovery, data can be lost. Every member that has access to the network has an identical copy of the database, and every new transaction is validated by each member of the network [47–49]. As it is available to all parties on the network, there is an exponential increase in transparency, and this itself helps illuminate or prevent financial misdeeds.

A potential use for a public blockchain has been floated as a new way to transact and eliminate the current system where a financial institution (most commonly a bank) is the central authority in verifying and overseeing financial transactions [49]. Since auditors validate the accuracy of financial transactions and financial reporting, their role in this system would be greatly reduced [3]. It is highly doubtful that all enterprises will prescribe using a public blockchain and will prefer the security and safety of having a private blockchain, then allowing for the skills required, auditors will still be needed to validate the accuracy of the transactions [3, 49]. The regulations are currently unclear regarding regulating and the use of blockchain technology in financial systems. Hence, full-scale adoption has not occurred, and risks and benefits of using blockchain in auditing have yet to be discovered [48, 49].

3. Research methodology

The research philosophy used to identify the impacts of emerging technology on the audit process, audit profession, and the role of an auditor is interpretive in nature [50]. The researcher aims to understand, analyze, and interpret the participants' everyday experiences in specific situations [51]. A qualitative approach was followed, which was based on two qualitative datasets. This allowed the researcher to gain a rich understanding of how impacts on audit were experienced by participants and how these experiences changed over time to support the chosen research philosophy [50]. Data collection for this study was done with a structured literature review (SLR) and semi-structured interviews. An SLR is a process whereby literature is identified and selected based on predefined search criteria designed to address specific topics. It is a comprehensive and transparent search that can be replicated to provide a balanced and impartial view of existing literature coverage based on the search criteria [51]. The SLR was used to identify the impacts of digitalization on auditing from existing academic literature. This is a "comprehensive pre-planned strategy" to identify and review existing literature concluding on what is known and not known in a specific field or covering a specific topic [51]. Thematic analysis was used to identify themes from the outcome of the SLR. These themed impacts identified were used as focus areas, which were further synthesized into detailed questions during the formulation of the interview guide. The semi-structured interview guide was used during the semi-structured interviews for the collection of the second qualitative dataset used for this study.

Google Scholar was used for the identification of academic literature for this study. The search included "audit AND blockchain", "audit AND artificial intelligence", and "audit AND big data" as keywords. The search was refined by limiting the search to return only articles where keywords appeared in the title, which reduced the search results to 403 matches (145, 83 and 175, respectively) that were selected to be read first. After an initial investigation, the search was further refined to exclude citations and limit considered literature to English articles, excluding popular press and non-peer-reviewed publications. The total number of matches was 242 (80, 54 and 108 respectively). A screening was completed of the 242 articles, after which 107 studies were identified as eligible for this research. An abstract review was completed on the 107 articles. Preference was given to papers that focused on the impact of digitalization on auditing after or during the implementation and use of these technological capabilities, specifically blockchain, AI and big data analysis. Of the 107 eligible articles, 52 were selected and included in the full article review. The papers were analyzed, impacts identified and themes were created by grouping related impacts. Appendix A depicts and extract of the detailed analysis dataset.

The themes identified as an output of the SLR, were then considered and included in the interview guide created for the semi-structured interviews. Each of the themes was supported by 2-4 questions. Questions were derived from the detailed impact identification from the SLR. Semi-structured interviews were conducted to obtain the second qualitative dataset to assess if and how the themes identified in the SLR have materialized in the South African context. The interview guide consisted of three sections, the respondent's professional and demographic information, followed by baseline questions, with the last section focusing on the themed impacts of digitalization on auditing as identified in the SLR. The interview guide was used for each interview, and all themes identified were covered with each participant. Flexibility was allowed in the conversation to identify the lived experiences of participants [51].

Purposive sampling was used for the participation selection. Purposive sampling allows the researcher to apply judgment when selecting participants for a study [51]. All participants that participated in this study were selected based on their industry sector, job level, specialty, and experience. Only candidates that work in the financial services sector in South Africa, with a job level of Executive or Senior Management, working in either Internal Audit, External Audit or Risk Management with at least 10-years' experience, were eligible for this study. Prospective participants were screened based on the information available on LinkedIn. Each prospective participant was contacted with a telephone call to confirm the accuracy of the information obtained. After a prospective participant was confirmed to be eligible for the study, the purpose of the research was shared, and the prospective participants had the opportunity to inform the researcher if they would like to participate in the study. Participants that took part in the study were contacted via email, and interviews were scheduled. A total of 12 interviews were conducted [52]. Informed consent was obtained and all interviews were voice recorded and transcribed. Amidst the COVID-19 pandemic, interviews were conducted via collaboration software, Microsoft Teams.

4. Data analysis and findings

Fifty-two papers were identified as relevant for this study. Each paper was reviewed to identify the impacts that the adoption of emerging technologies might have on the audit process. In papers where the impact was not specifically stated but the benefits of adopting emerging technologies, a thematic analysis was performed to identify the changes required within the audit process and profession.

Table 1 shows the frequency count of impacts identified from the SLR (detailed analysis extract in Appendix A). Note that in most cases, more than one theme was identified in an article due to the relationships that exist between elements in each theme.

Table 2 shows a summary of participants' professional and demographic information referring to each participant number. The participant number serves as a reference point and is used in the next section as part of the interview analysis.

A detailed analysis of the transcribed interviews are presented in the next section.

4.1 Audit process

All participants indicated that the automation of the audit execution step had had a significant impact on their teams, with the main benefits as less time required from auditors for mundane, repetitive tasks during the audit execution phase. Time is spent on exception investigation, opinion-forming and audit judgment. This has improved the overall quality of audit deliverables and client relationships. Typical audits consist of three main phases: audit planning, audit execution or testing phase, and the audit reporting phase. Audit opinions and judgments are formed during the reporting phase based on the outcome of the testing performed during the execution phase [27].

Themes	Description	Frequency Count	%
Audit Process	Change required or impact expected in the end-to-end audit process, specifically annual planning, audit planning, audit execution (manual versus automated) and audit reporting. Literature echoed the increased frequency of audits, reduced audit execution times and ultimately better assurance as to the main outcomes and benefits in this category.	51/52	98%
Auditor's Professional Profile	The adoption of emerging technologies requires audit teams to ensure their skills remain relevant. Skills to implement automated procedures to utilize these new technologies and techniques and skills to interpret the output of automated audit procedures. This was specifically evident in literature covering big data analysis (BDA), which contributes 45% of the frequency count. Literature indicates that a shift is required from the traditional audit skills to a multi- or cross-skilled individual to allow audit teams to embrace and benefit from the adoption of disruptive technologies.	33/52	63%
Audit User Perceptions	The readers of financial accounts and audit reports are the customers of the audit profession. These include the shareholders, stakeholders, and regulators (not an exhaustive list). With the move to automation, these customers will have to rely on and trust the outcome of automated procedures and automated reports. This trust relationship between customer and auditor, specifically the trust of judgments by auditors on automated procedures, will be impacted.	22/52	42%
Auditor-Client Relationship	Audit clients are the individuals or teams being audited or auditees. The shorter execution times and the potential to increase the audit frequency will impact the current audit costing models in use.	11/52	21%

Table 1. *Themed impacts identified.*

Participant 7 described his experience as:

> *"This is where we have experienced the biggest impact on our traditional audit testing approach whilst on the audit automation journey. The biggest impact which has led to the biggest benefits. The relieve of specialist auditors to perform mundane tasks and allow them more time for analysis of the output produced to form audit opinions."*

There was a consensus from all participants that automation of the audit execution phase is where they have experienced the biggest benefit through time saving and the speed of execution. Participant 8 highlighted a unique issue that emerged in their organization as they progressed on their digitalization journey in that auditors would require guidance on how to spend the time saved by automated audit procedures to best serve the end-to-end audit process. **Table 3** provides a summarized breakdown of responses received per participant.

All 12 participants indicated that automating the audit execution phase resulted in the biggest impact with the most benefits. This allows auditors more time to apply their audit skills on opinion-forming and judgment. Seven participants indicated an enhanced overall assurance service to the clients.

The interviewees also reflected on the anticipated or lived changes of the auditor's role in the audit team and the organization, specifically focusing on the role of the auditor potentially shifting from classic auditing to a consulting role. The feedback from participants was not consistent for this question. The inconsistencies noted

Participant number	Job level	Function	Specialty	Internal / external audit	Years' experience
Participant 1	Senior Management	Audit	IT Audit	Internal Audit	16 or more
Participant 2	Executive	Audit	Financial Audit	Internal Audit	16 or more
Participant 3	Senior Management	Audit	Financial Audit	Internal Audit	16 or more
Participant 4	Executive	Audit	Financial Audit	Internal Audit	16 or more
Participant 5	Senior Management	Group Risk	Risk Management	N/A	16 or more
Participant 6	Executive	Audit	IT Audit	External Audit	16 or more
Participant 7	Senior Management	Audit	IT Audit	Internal Audit	16 or more
Participant 8	Senior Management	Audit	IT Audit	External Audit	9–12
Participant 9	Executive	Audit	Financial Audit	Internal Audit	13–15
Participant 10	Executive	Audit	Financial Audit	External Audit	16 or more
Participant 11	Senior Management	Audit	IT Audit	External Audit	9–12
Participant 12	Executive	Group Risk	Risk Management	N/A	20 or more

Table 2.
Participant summary table.

Audit execution automation—common themes per respondent	Response count
Automating of the audit execution phase leads to significant time saving and shorter lead times between audits.	12
Allow auditors to spend more time on exception investigation, root cause analysis, opinion-forming, judgment and client engagement.	12
The results experienced are richer, better quality audit reports and improved client relationships.	7

Table 3.
Audit execution automation—Summary.

were mainly due to the different definitions of what a consultant is and the different interpretations of a consultant's role. However, there was a consistent element in all responses from all participants in that the role of the traditional auditor will not be replaced as the need for opinion-forming and judgment will remain a human action.

Participant 1 indicated that the time saved during audit execution might be applied in a consultant capacity. However, this remains a gray area due to the need

Auditor lived experience impact – common themes per respondent	Response count
The role of a classic auditor will remain due to the need for opinion-forming and judgment.	12
Performing consultant services might be possible but not advisable due to the need for the auditor to remain independent.	6

Table 4.
Auditor lived experience impact – Summary.

for auditors to remain independent and objective. These consultant services provided might lead to "[...] *auditors marking their own homework down the line* [...]" if they are involved in the design and operation of certain controls and required to provide assurance at a later stage. **Table 4** provides a summarized breakdown of responses received per participant.

Diverse opinions were obtained for this question due to the participants' perception and definition for a "consultant"; however, all participants indicated that the role of an auditor would remain due to the need for opinion-forming and judgment, and six highlighted the potential lack of independence should auditors be assuming the role of a consultant in an organization.

4.2 Auditor's professional profile

Overall, all participants agreed that there would be a shift in the skillset required of the auditor. In summary, audit teams will consist of cross-skilled individuals that are tech-savvy to complement their existing audit skillset and understanding of risks and controls in the financial services market. All auditors would not need to be IT experts but should understand what capability emerging technologies provide and how these technologies can be used in the automation process.

The lived and anticipated changes were described by participants based on where they are on their individual digitalization journeys, with Participant 1 describing the anticipated changes as *"[...] updated tech-savvy skillset and an inquisitive mind would define the auditor of the future oppose to the traditional box ticker or a cookie-cutter mindset."* Participant 9, which had the least exposure to automated auditing, shared her upskilling journey. She specifically highlighted the need to analyze large data sets, including trends analysis and pattern identification. These skills were obtained through completing online Udemy courses, specifically focusing on data analytics. She emphasized the need for all new auditors joining their firm "[...] *to have a basic understanding of data analytics and data visualization."*

Participants that have made the most progress on their respective audit digitalization journey are provided insight into their current recruitment practices. Such as Participant 4, who responded:

> *"Being in a digital organization, the auditors we have recruited over the last two years all have an IT background or moderate to strong IT knowledge and understanding of the capabilities and use of emerging technologies."*

Follow-up questions revealed that the ask is not for all new hires to be IT specialists, but a moderate understanding, use and applying emerging technologies (specifically data analytics) in the audit process have become a standard requirement. These skills were described in the context of auditors still having their

Skillset required of the auditor—common themes per respondent	Response count
The auditor of the future will need to understand technology, specifically, how the technology can be used in the audit process.	12
Data analytics will be a required skill for all auditors irrespective of their field of specialty in an audit.	10
All auditors need not be IT specialists and understand the underlying technologies, just the use thereof.	6
Existing recruiting processes require auditors to have data analytical skills, and strong IT background is a plus.	6

Table 5.
Skillset required of the auditor—summary.

specialized audit knowledge and experience. **Table 5** provides a summarized breakdown of responses received per participant.

All participants expect a shift in the skillset required of an auditor with 10 participants highlighting the need for data analytical skills as a key requirement. Although the participants were asked their expectations for the future, six participants indicated that data analytics is already a required skill for all new joiners in their respective audit teams. In terms of adoption of emerging technologies, most participants are expecting an impact on the balance within audit teams with the ongoing adoption of emerging technologies. Eight participants expect the balance in audit teams to shift leaning towards more IT auditors than financial auditors. However, follow-up questions led to diverse views from some participants.

Participant 1 does not expect a change in the balance in audit teams, as the skills required from both financial auditors and IT auditors will remain. He describes the expected change as "[...] *the forming of more rounded auditors with a balanced skillset between auditing and IT skills*." He highlighted that in his view, "[...] *auditors with stronger business acumen to provide correct contextual judgements based on automated audit activities* [...]" will be required. The increase of data points consumed by automated auditing will provide "[...] *much greater insights to the control estates and the risks to the business*," which will require auditors with institutional knowledge of the organization to understand and correctly interpret the outputs for opinion-forming and judgments.

Participant 8 agreed with the views of Participant 1 that the existing members of audit teams will need to upskill and obtain a more balanced skill set and that the need for the original team compositions will remain. He, however, describes the required shift in team composition as:

> *"[...] current audit teams are made up of financial auditors and IT auditors, where the latter is focused on the auditing of IT. The need for these skills will remain and should be supplemented/supported by data analytical skills. The introduction of a new capability is required, audit with IT. This is a team that focuses on digitalization and automation of audit procedures."*

This is a team of emerging technology specialists focused on "[...] *codifying existing audit procedures and institutional knowledge* [...]" and driving the digitalization journey. He describes this as the key contributor to all the progress they have made to date. Therefore, the team balance in his organization has seen a dramatic

Auditor skills profile—common themes per respondent	Response count
The existing balance between financial auditors and IT auditors that make up audit teams will be impacted.	11
The team balance will be impacted as the need for IT auditors will increase.	8
The team balance will not be impacted as both skillsets are required; the continued adoption of emerging technologies will lead to cross-skilled individuals. No shifts expected in the team.	1
The need for IT auditors will reduce as automated auditing will provide indirect assurance over ITGCs.	1
The current teams will remain, new specialized teams will be formed to drive automation.	1

Table 6.
Auditor skills profile—summary.

shift with the restructure. **Table 6** provides a summarized breakdown of responses received per participant.

The collective response from other participants highlights the increased need for data analytics skills in the teams to support their respective initiatives. Adding these specialized skills will impact the team balances leaning towards the IT auditor being more dominant in the composition of the audit team.

The consensus among participants is that the professional profile of the audit profession will be more attractive for young professionals. The audit profession will become "[...] *funky and sexy* [...]" as described by Participant 2 and "[...] *glamorous and exciting* [...]" as described by Participant 9. All participants agreed that the impact would be positive and that the audit profession would be more attractive for young professionals. Participants 8 and 12 believed the profile has started to change and will continue to change as the adoption of emerging technologies in auditing matures in South Africa.

The discussions that flowed from this question varied between participants. The researcher allowed for the flexibility to gain insights covering topics, such as what the updated auditor profile will mean for the audit profession, the type of individual that the profession will attract, and the additional skills these potential new auditors will require.

Participant 2 suggested that the title of the auditor will change to better reflect the changing audit environment. He explains that "[...] *the title auditor suggests that you, as the person, is the process, and you go in and audit by looking at transactions or working through files* [...]" however, this will not be the case. The role of an auditor will be to focus on the output provided by an automated procedure or work through exceptions noted. Therefore, as the expectation of an auditor will change to focus on "[...] *root cause identification, risk assessment, reporting, client engagements* [...]," the title of the profession should change. He suggested a "[...] *Strategic Assurance Analyst* [...]" would be a better fit which will lead to an improved and more attractive profile in the job market.

Participants 6, 9, 10 and 11 all noted their concerns around the slow rate of change in university auditing degrees and diplomas to include more comprehensive coverage of technology principles and emerging technologies in current curriculums. Exposure to and using data analytics in the audit process was raised by all four participants as a shortcoming they have noted in newly graduated professionals. Participants 6 and 9 indicated that organizational training paths had been defined, focusing on data analytics short courses, for the upskilling of new graduates joining their organizations. **Table 7** provides a summarized breakdown of responses received per participant.

Auditor's professional profile—common themes per respondent	Response count
The adoption of emerging technology will have a positive impact on the professional profile of audit in the job market.	12
Concerns raised that the rate of change to current curriculums of university degrees and diplomas are not keeping up with the rate of change in the IT environment and the rate of adoption of emerging technologies in the audit industry.	4
The skillset of newly graduated audit professionals does not meet the requirements demanded by organizations.	2
The job title of the auditor is not appropriate for the function served by the auditor of the future. The robot auditor will perform the audit; the human will interpret the output; therefore, the suggested future title is Strategic Assurance Analyst.	1

Table 7.
Auditor's professional profile—summary.

All participants agreed that the adoption of emerging technologies would make the professional profile of an auditor more attractive in the job market. The field of auditing will require individuals skilled in technology principles, a general understanding of emerging technologies and using data analytics. This raised the question of whether the current curriculums at the university level have changed at the same pace with the changes experienced in the audit industry.

4.3 Audit user perceptions

All participants agree that audit users will trust automated audit procedures more than manual procedures; however, trust comes over time (all participants). Most participants indicated that the initial trust levels were low. Participants 2, 4 and 7 shared that the first audits supported by automated audit procedures took significantly longer than similar audits supported by manual audit procedures. This was due to the lack of trust in the automation capability within the audit team, the automated testing procedure itself, and, as noted by Participant 2, "[...] *the lack of human interaction with the audit client made them feel left out of the process and not involved in the execution* [...]."

Participant 3 noted that "[...] *population testing by means of automated auditing* [...]" rather than "[...] *sample testing during manual audit procedures* [...]" eliminated sample risk. This was noted as one of the key contributors to increasing the trust levels in automated auditing and "[...] *realization of value* [...]" by the audit clients. Like Participant 2, Participant 3 highlighted the need for client engagement as:

> *"There is still the need for an auditor to play the role of a trusted adviser to interpret results and maintain the auditor-client relationship. Maintaining the relationship is a key ingredient to build and maintain trust during these engagements."*

Participant 7 added that automated audit procedures remove the subjectivity from the audit procedure and audit bias during the audit procedure execution. The removal of the human element during audit execution improves the overall quality of the audit deliverable. This improved quality and value provided are the ingredients for building trust. **Table 8** provides a summarized breakdown of responses received per participant.

All participants believe that more trust will be placed in automated audit procedures. At first, trust levels will be low, but as the audit department progress on their digitalization journeys, trust levels improve. It was noted that the inclusion of the

Audit user perceptions – common themes per respondent	Count
All participants believe that more trust will be placed in automated audit procedures.	12
Most participants experienced low trust levels in automated audit procedures during the initial phases of their automated audit journeys.	9
Some participants indicated that the first iterations of automated audits took longer than the manual procedures.	3
Participants noted the importance of the human element in auditing, the client interaction and client engagements during the audit process.	3
Some participants indicated that the trust must not be placed in the tool or the technology used but rather in the application thereof.	3
Client inclusion and consultation are important during implementing automated audit procedures as this helps the initial building of trust in the automated audit process.	2
Reducing sample risk and auditor bias contribute to the trust in the automated audit process.	2

Table 8.
Audit user perceptions—summary.

audit client in the audit automation journey promote trust levels and assist audit teams in reaching a state of digital maturity quicker.

All participants indicated that the impact would be the ability to form better audit opinions and judgments. This was mainly supported by the "[...] *move from sample testing to full population testing during audit execution* [...]" (all participants) also, "[...] *the fact that more data points are consumed by automated audit procedures, specifically in case where principals of big data analytics are applied and used* [...]" (Participants 1, 4 and 8). Therefore, the coverage of multiple full population datasets provides the opportunity for the forming of better audit opinions and judgments.

Sample risk (Participant 3), subjectivity and auditor bias during audit execution (Participant 7), human error and oversight during manual testing (Participant 9) are all factors that affect the quality of the testing outcome, which audit judgments are based on. The removal of these factors improves the testing outcomes and better informs the audit opinion and judgment. This was best described by Participant 4 as:

> *"[...] the major change is how the opinion or judgement gets informed. In our current environment, most of the fieldwork in the audit process is a data-enabled body of evidence. Gone are the days to manually understand controls and sampling. The auditor needs to understand the digital process and step back and interpret the output and ask, "what does that mean?" Therefore, audit judgement and opinion will be better-informed."*

Table 9 provides a summarized breakdown of responses received per participant.

The adoption of emerging technologies supporting audit automation projects and programmes will enhance auditors' ability to provide better, informed audit opinions and judgments. This is supported by the increased coverage of automated audit procedures. This is not limited to the automated audit procedure that can test the full population but also the fact that multiple datasets are included in the design of these tests. This allows for exception and theme identification that is not possible with sample testing. Furthermore, with using automated audit techniques, human errors and oversight during audit execution are removed, and sample risk, audit subjectivity and auditor bias.

Audit user perceptions of automation – common themes per respondent	Count
Audit automation using emerging technologies will lead to better audit opinions and judgments.	12
Rationale for better audit opinions and judgments: Audit opinions and judgments are formed on full populations of multiple datasets.	12
Rationale for better audit opinions and judgments: The removal of the human element during the execution phase of the audit. No human errors and oversight, elimination of sample risk, reducing of subjectivity and auditor bias.	4
Using automated audits allows auditors to spend more time on opinion-forming and judgment resulting in better quality audit reports and audit outcomes.	3

Table 9.
Audit user perceptions of automation—summary.

4.4 Auditor-client relationship

In general, most participants agreed that continued adoption of emerging technologies in the digitalization of auditing would have an impact on the current pricing models but not necessarily the cost of the audit opinion and judgment.

Participant 11 started his response by highlighting the sensitivity around "[...] *pricing of assurance services* [...]" and highlighted the "[...] *legal liability that comes with audit opinions* [...]." He pointed out that the "[...] *legal liability on both the auditor and audit firm* [...]," and the "[...] *auditor responsibility towards the director, shareholders and stakeholders* [...]" remain unchanged. Therefore, the cost of the audit opinion or judgment should remain the same "[...] *irrespective of the approach followed to conclude* [...]." He indicated that "[...] *the current pricing requires a revamp as it based on hourly charge-out rates, which will not be as relevant going forward due to the reduction in audit execution time* [...]." Similar views were shared by Participants 4, 6, 8 and 9, with Participant 4 explaining:

> *"The cost will not change; the quality of the assurance will improve. This is evident with the automation we have already implemented move from sample to population testing. So, the quality of your view, opinion and judgement is exponentially better. Therefore, for the same price, a client will get a much better quality of audit."*

Participant 2 shared similar views and agreed that the current pricing models should be adjusted to consider the quality of assurance. However, he suggested that a hybrid model should be considered. He suggested that audit departments should "[...] *start with calculating the cost of automation of a single audit test. This should be extrapolated across all data-driven audit procedures that can be automated to get a view of the total future investment. This should consider all costs such as research and development, licencing cost, procedure maintenance, staff upskilling* [...]." These are the elements that should make up the calculation for the cost of quality. **Table 10** provides a summarized breakdown of responses received per participant.

The consensus is that pricing models must be adjusted in time, but there will not be an immediate cost saving for the audit client and audit user. Clients will receive better assurance deliverables, better quality audit reports, better-informed audit opinions and judgments on a more frequent basis at a similar or higher cost. The expectation from some participants is that there might be reduced costs in the future

Auditor-client relationship – common themes per respondent	Count
Most participants indicated that audit pricing models for professional fees would need to be changed or revamped.	11
Some believe the cost of assurance services will decrease with the increased adoption of emerging technologies in auditing.	4
Some believe the cost of assurance service will increase with the increased adoption of emerging technologies in auditing.	3
Some believe the cost will remain the same, the quality of the assurance will increase.	5
Some believe the cost will increase as the quality of the assurance will increase.	2

Table 10.
Auditor-client relationship—summary.

after the initial investment costs have been depreciated. However, as the technology and risk land space change there will be a need for different assurance that will introduce new costs.

During the interview introductory questions, it was established that none were currently using or are planning to use blockchain technology as part of their audit digitalization journeys. Participants 4, 6, 9, 11 and 12 indicated that they know little about the technology other than the relationship with cryptocurrencies such as Bitcoin. They chose not to provide an opinion. Both Participants 2 and 3 indicated that they know little of the technology itself and the different applications thereof, but they both shared the view *"[...] as with the introduction and use of all new technologies, there will be a shift in the risk landscape or the introduction of new risks, therefore, the impact on audit and the audit process will be the need for assurance over this risk resulting in a higher demand for audits [...]."* Participants 7 and 10 agreed that the need for audit would increase, and at no point will audit not be necessary. The expected change is only a shift in focus of the auditor.

Participants 5 and 8 shared their understanding of triple-entry accounting made possible by blockchain technology. They noted the advantages and efficiencies this approach has over the double entry accounting principles currently in use as "[...] *it will fundamentally improve accounting and address the current transparency and trust issues that exist.*" Both participants concluded by stating that if accounting is enabled by blockchain technology, the need for audit will just increase, and it will focus on different controls and risks. Five of the participants chose not to share a view or opinion due to their limited knowledge of blockchain technology. The other eight participants highlighted that using new technology such as blockchain will change the risk landscape and introduce a new set of risks and controls, therefore, different focus areas for audit. Two participants indicated that the need for audit might even increase. None of the participants aim to use blockchain technology as part of the audit digitalization journey.

5. Discussion and contribution

The aim of this study was to investigate the impacts of audit digitalization and in this section the findings are discussed.

5.1 Current state of audit digitalization in audit in South Africa

Although the focus of the initial data collection for this study was focused on the impacts big data, AI and blockchain technology had on the audit industry, none of these technologies is actively used for auditing services by South African audit teams and departments that formed part of this study. The data indicates that 100% of the sampled participants are using technology as part of their audit processes, and 84% of these organizations have formally initiated their audit digitalization journeys. Of the 84% of organizations, 50% have made significant progress. These participants have successfully implemented end-to-end automated audits. These are referred to as 'click-a-button' audits, 'on-demand' audits, 'audit-as-a-service' engagements, and real-time auditing.

All participating organizations are currently using some form of data analytics during the execution of the audit process. This is achieved with a wide range of tools and techniques implemented to best suit this audit test. These analytical capabilities in use ranged from basic data analysis in Microsoft Excel to advanced functions built-in Microsoft Azure. The 84% of organizations that have moved past mere data analytics have implemented RPA to codify all audit steps required for a specific audit test. This combination allows for end-to-end audit automation with the robot logging onto multiple systems, extracting all required information for a specific audit test, and executing the audit test based on the predefined audit procedure. This is possible for all data-driven audit tests. The organizations that have made the most progress on their digitalization journeys (16%) were in a proof-of-concept state with an AI solution to address a wide range of audit tests that were of a predictive nature.

Most participants (75%) indicated that they expect that AI, specifically expert systems and neural networks, and blockchain technology, will fundamentally impact the audit industry.

5.2 Digitalization and the audit process

The standard audit process can be followed for all different audit types and consists of three main steps, namely: audit planning, audit execution and audit reporting. All participants that have moved towards audit automation indicated that the biggest impact was experienced in the audit execution phase, which also resulted in the most benefits gained. The impacts were described as a complete redesign of the process as the execution of the audit tests were no longer completed by human auditors. This led to significant time saving, shorter lead times between audits, faster audit execution and wider audit coverage. These outcomes allow auditors to spend more time on exception investigation, root cause analysis, opinion-forming, judgment and client engagement, which results in richer and better quality audit reports and improved client relationships.

An industry view was obtained regarding the potential shift of an auditor's role within the audit team from being a classic auditor to a consultant within an organization. Although diverse opinions were obtained, all participants indicated that the role of an auditor would remain due to the need for opinion-forming and judgment, and 50% of the participants highlighted the potential lack of independence, and objectivity should auditors assume a consultative position in an organization.

With the advances in audit automation already evidenced in the South African market, industry views were obtained on the impact and need for interim- and year-end financial reviews. All participants agreed that the need for these reviews would remain. However, the nature of these audit engagements will change. Instances already exist where these reviews have been automated, where the execution of these reviews is no longer completed by human auditors. The change in the approach has impacted the audit frequency from bi-annual to monthly, it changed the auditor from human to electronic (or robot), and this can impact the audit department responsible for the review between internal audit and external audit depending on who developed the robot.

5.3 Digitalization and the audit's professional profile

All auditors will require a good understanding of technology, specifically, using data analysis in the audit process. The expectation by organizations is not for all auditors to be specialists in the field of data analytics; however, the need to analyze large data sets, interpret the outcomes of audit tests based on data analytics and the ability to visualize data in an understandable and comprehensible way is required. This is supported by the already changing recruitment processes implemented in South African organizations that stipulate the need for 'a strong data analytical background' as a requirement for new joiners. This requirement has been added to all role profiles, irrespective of an auditor's field of specialty, and it is regarded as a core competency. A general information technology (IT) background was listed as a plus. The organizations at the forefront of audit digitalization indicated that their recruitment processes have been updated and have been aligned to their digitalization strategies. All new joiners must have a moderate to strong IT background, data analytics is key, and understanding of new emerging technologies and the application thereof is a plus.

There is an expectation that the existing balance in audit teams between financial auditors and IT auditors will be impacted as organization's progress on their respective digitalization journeys with the increased need for IT skills. The expected shift is not definitive as different variants exist to achieve and deliver digitalization strategies. The two extreme cases identified are a major shift from the traditional balance with more IT auditors than financial auditors or there will be no shift to existing teams, and the digitalization of audit processes is driven by a specialized team outside the audit department. In both these scenarios, the required skillsets of the audit team should be revisited with the increased need for cross-skilled individuals in audit departments to benefit from audit digitalization.

There is an expectation that the digitalization of audits will have a positive impact on the role profile for auditors in the South African job market. The role profile of a cross-skilled auditor will attract younger professionals, and it will also attract more individuals at a university level to move into the audit profession. The need for specialized technical IT skills in emerging technologies will also open the door for non-auditors to join audit teams to deliver audit digitalization strategies. These technical specialists are currently being used in the South African audit industry to codify the knowledge and held by existing audit team members.

Although market expectations for future graduates look bright, concerns were raised over the level of IT coverage and IT exposure current university degrees and diplomas include. It was noted that newly graduated individuals do not meet the minimum requirements to join audit departments due to a lack of IT knowledge and a general understanding of technology.

5.4 Digitalization and the audit user perception

More trust will be placed in the outcome of automated audit procedures than in the traditional, manual audit procedures. The increased trust levels will be based on introducing end-to-end process testing covering multiple data points, eliminating sample risk and human error during the audit testing process and reducing subjectivity and auditor biased. These benefits will lead to better, more informed audit opinions and judgments and better richer, more informed audit reports and audit insights. These were the combined views shared by all participants based on their lived experiences.

5.5 Digitalization and the auditor-client relationship

One of the key outcomes of the study was the expected, and proposed changes to the current costing and pricing models in use for assurance services and the impact digitalization of audit will have on these models.

At the time of the research, all participants in this study were still employing an hourly-based pricing model for assurance services that makes up the total cost of the audit opinion. Introducing emerging technologies and the ongoing digitalization journeys seen in the South African audit industry have led to shorter audit execution times. Audit procedures that previously took days or weeks to complete can now be completed in a few minutes, depending on the computing power available to the codified audit procedure (robot auditor). The assurance provided by the robot auditor can be completed monthly or even daily, whereas the manual procedures are generally scheduled twice a year or on an annual rotation basis based on risk. The coverage and assurance provided by the robot auditor are not based on sample testing but full population testing, leading to better assurance outcomes, as discussed in this paper.

Most participants (91%) indicated that the current pricing model used in South Africa is not fit for the purpose of a digitalized audit function, leveraging emerging technologies in their automated auditing capabilities. Participants proposed that a value-based costing model would be more fitting in a digitalized audit world as clients will receive better assurance deliverables, better quality audit reports, better-informed audit opinions and judgments on a more frequent basis. The expectation from some participants is that there might be reduced costs after the initial investment costs have been depreciated. However, as the technology and risk landscape change, there will be a need for different assurance that will introduce new costs.

The expected impact of blockchain technology on the audit industry is that the need for assurance will be different, and that it might increase. Although it is expected that blockchain technology will have a major impact on the financial accounting processes and principles as described by Penkin and Pehrsson [19], participants in this study believe that the assurance over a typical triple-entry accounting system will be different, and the need for assurance might increase.

5.6 The auditor of the future

Based on the research completed and the industry views and opinions obtained, the auditor of the future can be best described as a digitally fluent individual, cross-skilled to use a combination of IT and audit skills to deliver audits. Considering the four themes covered in this study, various aspects of the auditors' ways of working will be impacted, including the audit process, the auditor's professional profile, the audit user perception, and the auditor-client relationships.

Looking at the audit process, the audit execution will be the most impacted. The traditional audit steps of inquiring, observance and performance by a human auditor are falling away as audit processes are digitalized. Auditors will rely on audit steps executed by a robot auditor, and an auditor's audit skills will be applied for opinion-forming, judgment, and client engagements. Based on the progress made in the South African context, evidence suggests that digitalization will result in richer and better quality audit reports and improved client relationships.

The need for cross-skilled individuals will impact the profile of the auditor, which suggests making it more attractive in the job market and at the university level for new entrants. The study highlights the role universities must play to ensure that degrees and diplomas remain fit for purpose, and in line with the skills required by the audit industry to equip young professionals. These changing requirements will also cast the net wider and allow for IT specialists and technologists to join audit departments as their specialized skills are required for the digitalization efforts within the audit industry to deliver organizations' digitalization strategies.

Auditors will still provide assurance over the accuracy and completeness of financial statements and provide assurance over current and emerging risks. However, the way this assurance is derived will change, including the quality, frequency and focus area. This will result in a different value proposition to clients who will need a different costing model. The current hourly-based model in use in South Africa does not fit this new emerging value proposition, and this study suggests a value-based costing model for audit opinions derived from automated auditing or audits completed by robot auditors.

6. Conclusion

The study set out to identify the impacts of digitalization on auditing focusing on the audit process, the auditor's professional profile, the audit user perception, and the auditor-client relationships. Data was collected in South Africa through an SLR and 12 semi-structured interviews. The biggest impact was noted in the audit process with implementing automated auditing levering emerging technologies such as Robotic Process Automation, which resulted in the execution phase of an audit being completed by a robot auditor shifting the focus of the human auditor to exception investigation opinion-forming, judgment, and client engagement. Digitalization in auditing has led to the need for moderate to strong technical IT knowledge by existing auditors, specifically data analytical skills. This had a positive impact on the professional profile of an auditor as it has opened the door for technical specialists to join audit departments to meet the skills demand required for delivering organizations' digitalization strategies. In addition, there is an expectation that the professional profile of an auditor will be more attractive for young professionals and university graduates. The biggest anticipated impact identified is the need to define appropriate costing and pricing models for audit opinions in a digitalized audit environment. The study suggests a value-based costing model would be better suited as it will incorporate the benefits gained from digitalization. This would require further studies to define and validate.

By incorporating all the outcomes for each theme covered in this study, the auditor of the future can be best described as a digitally fluent individual, cross-skilled to use a combination of IT and audit skills to deliver audits in a digitalized audit environment.

Future research could include a focused study on implementing automated auditing using emerging technologies to confirm the impacts on auditing as identified in this study. Research can also be expanded to include the skills required by an auditor working in a digitalized audit environment compared to the skills new market entrants are equipped with after the completion of their auditing degrees or diplomas. This will address the concern highlighted in this study to provide a view of the alignment or potential misalignment between the demand and supply for IT and technology skills in the audit industry. As data was collected in the South African audit industry only, the data collection may be extended in further studies in order to generalize findings.

Conflict of interest

The authors declare no conflict of interest.

A1. (SLR output-extract only)

Impact(s) identified	Theme allocated	Title and Theme/Impact extract	Ref.
Change in the audit process required. New set of risk and controls will emerge leading to new skills required to provide assurance.	Audit Process Auditor's professional profile	A Primer for Information Technology General Control Considerations on a Private and Permissioned Blockchain Audit.	[53]
Change in audit process required. Changes required in all assurance services including Audit, Governance and Risk Management disciplines.	Audit Process	Technology Innovation Management Review: Is Internal Audit ready for Blockchain?	[54]

Author details

André Schreuder and Hanlie Smuts*
University of Pretoria, Pretoria, South Africa

*Address all correspondence to: hanlie.smuts@up.ac.za

References

[1] Loebbecke C, Picot A. Reflections on societal and business model transformation arising from digitization and big data analytics: A research agenda. The Journal of Strategic Information Systems. 2015;**24**(3):149-157

[2] Alles M et al. Continuous monitoring of business process controls: A pilot implementation of a continuous auditing system at Siemens. International Journal of Accounting Information Systems. 2006;**7**(2):137-161

[3] Tiberius V, Hirth S. Impacts of digitization on auditing: A Delphi study for Germany. Journal of International Accounting, Auditing and Taxation. 2019;**37**:100288

[4] Hou B et al. Research on unstructured data processing technology in executing audit based on big data budget. Journal of Physics: Conference Series. 2020;**1650**(3):032100

[5] Rakipi RA. Correlates of the internal audit function's use of data analytics in the big data era: Global evidence. Journal of International Accounting, Auditing and Taxation. 2021;**42**:10035

[6] Kirkos E, Spathis C, Manolopoulos Y. Audit-firm group appointment: An artificial intelligence approach. Intelligent Systems in Accounting, Finance and Management. 2010;**17**(1):1-17

[7] Couceiro B, Pedrosa I, Marini A. State of the Art of Artificial Intelligence in Internal Audit context. Iberian Conference on Information Systems and Technologies, Seville, Spain: CISTI; 2020. p. 2020 -June

[8] Kokina J, Mancha R, Pachamanova D. Blockchain: Emergent industry adoption and implications for accounting. Journal of Emerging Technologies in Accounting. 2017;**14**(2):91-100

[9] Palestini AR. Blockchain Technology and Audit, Next Steps and Future Opportunities for Corporate Managers. 2020

[10] Tang X, Kostic N. The Future of Audit: Examining the Opportunities and Challenges Stemming from the Use of Big Data Analytics and Blockchain Technology in Audit Practice. Sweden: Lund University; 2017

[11] Alina CM, Spătariu EC, Gheorghiu G. Internal audit role in artificial intelligence. Ovidius University Annals. Economic Sciences Series. 2018;**18**(1):441-445

[12] Chu MK et al. Big data analytics for business intelligence in accounting and audit. Open Journal of Social Sciences. 2021;**9**(9):42-52

[13] Brynjolfsson E, McAfee A. The Second Machine Age. New York: W.W. Norton and Company Inc.; 2014

[14] Gultom JB et al. Reciprocal use of artificial intelligence in audit assignments. Journal of Accounting, Business and Finance Research. 2021;**11**(1):9-20

[15] Li W. Analysis on application of big data technology in audit practice. Advance Intellegent System Computer. 2020;**1303**:1042-1048

[16] Stensjö G. The Changing Nature of the Audit Profession-Opportunities and Challenges with Digital Transformation and the Use of Audit Support Systems, Big Data and Data Analytics. Sweden: University of Gothenburg; 2020

[17] Tallqvist V. The Role of (Big) Data Analytics in the Audit Process Through

the Position-Practice Perspective - An Evaluation of the Necessary Skillset of an Auditor. Turku, Finland: Åbo Akademi University; 2021

[18] Smith SS. Blockchain, Smart Contracts and Financial Audit Implications. Journal of Accounting Research & Audit Practices. 2020;**19**(1):7-17

[19] Penkin I, Pehrsson P. Future of Financial Audit: Impact of Blockchain Technology. Metropolia University of Applied Sciences, Sweden. 2019

[20] Cai CW. Triple-entry accounting with blockchain: How far have we come? Accounting and Finance. 2021;**61**(1):71-93

[21] Rohrbeck R, Kum ME. Corporate foresight and its impact on firm performance: A longitudinal analysis. Technological Forecasting and Social Change. 2018;**129**:105-116

[22] Vasarhelyi MA. Automation and changes in the audit process. Auditing: A Journal of Practice and Theory. 1984;**4**(1):100-106

[23] Alles MG, Kogan A, Vasarhelyi MA. Audit automation for implementing continuous auditing: Principles and problems. Rutgers Business School 1. 2008. pp. 1-24

[24] Bierstaker JL, Burnaby P, Thibodeau J. The impact of information technology on the audit process: An assessment of the state of the art and implications for the future. Managerial Auditing Journal. 2001;**16**(3):159-164

[25] Janvrin D, Bierstaker J, Lowe DJ. An examination of audit information technology use and perceived importance. Accounting Horizons. 2008;**22**(1):1-21

[26] Huerta J, Salazar P. Audit process framework for data protection and privacy compliance using artificial intelligence and cognitive services in smart cities. In: 2018 IEEE International Smart Cities Conference, ISC2. Vol. Missouri, USA; 2018

[27] Bedard JC et al. Risk monitoring and control in audit firms: A research synthesis. Auditing. 2008;**27**(1):187-218

[28] Soh DSB, Martinov-Bennie N. The internal audit function: Perceptions of internal audit roles, effectiveness and evaluation. Managerial Auditing Journal. 2011;**26**(7):605-622

[29] Pong C, Fraser I. The future of the external audit function. Managerial Auditing Journal. 2009;**24**(2):104-113

[30] Bratten B, Causholli M, Sulcaj V. Overseeing the external audit function: Evidence from audit committees' reported activities. A Journal of Practice & Theory. 2020;**41**(4):1-31

[31] Archambeault DS, DeZoort FT, Holt TP. The need for an internal auditor report to external stakeholders to improve governance transparency. Accounting Horizons. 2008;**22**(4):375-388

[32] Appelbaum D. Kogan A, Vasarhelyi M. Moving Towards Continuous Audit and Big Data With Audit Analytics: Implications for Research and Practice. Symposium State University of New Jersey. 2015

[33] Braun RL, Davis HE. Computer-assisted audit tools and techniques: Analysis and perspectives. Managerial Auditing Journal. 2003;**18**(9):725-731

[34] Alles MG, Datar S. How do you stop the books being cooked? A management-control perspective on financial accounting standard setting and the section 404 requirements of the Sarbanes-Oxley Act. International

Journal of Disclosure and Governance. 2004;**1**(2):119-137

[35] Mohamed IS, Muhammad NH, Rozzani N. Auditing and data analytics via computer assisted audit techniques (CAATs): Determinants of adoption intention among auditors in Malaysia. Proceedings of the 3rd International Conference on Big Data and Internet of Things, ACM Digital Library. 2019. pp. 35-40

[36] Appelbaum D, Kogan A, Vasarhelyi MA. Big data and analytics in the modern audit engagement: Research needs. Auditing: A Journal of Practice and Theory. 2017;**36**(4):1-27

[37] Banker RD, Chang H, Kao Y. Impact of information technology on public accounting firm productivity. Journal of information systems. 2002;**16**(2):209-222

[38] Klamm BK, Kobelsky KW, Watson MW. Determinants of the persistence of internal control weaknesses. Accounting Horizons. 2012;**26**(2):307-333

[39] Pizzini M, Lin S, Ziegenfuss DE. The impact of internal audit function quality and contribution on audit delay. Auditing: A Journal of Practice and Theory. 2015;**34**(1):25-58

[40] Šindelář M, Dlask L. The future of audit: Literature review of possibilities of automation and blockchain technology. In: Annual Conference on Finance and Accounting. Prague, Czech Republic: Springer; 2019

[41] Gantz J, Reinsel D. Extracting Value from Chaos, IDC iview 1142, no. 2011 (2011). pp. 1-12. 2011

[42] Breur T. Statistical power analysis and the contemporary "crisis" in social sciences. Journal of Marketing Analytics. 2016;**4**(2-3):61-65

[43] Cuquet M, Fensel A. The societal impact of big data: A research roadmap for Europe. Technology in Society. 2016;**54**:74-86

[44] Smith SS. Blockchain Augmented Audit-Benefits and Challenges for Accounting Professionals. The Journal of Theoretical Accounting Research. 2019;**14**(1):117-137

[45] Gershman SJ, Horvitz EJ, Tenenbaum JB. Computational rationality: A converging paradigm for intelligence in brains, minds, and machines. Science (New York, N.Y.). 2015;**349**(6245):273-278

[46] Schwab K. Global Competitiveness Report, World Economic Forum. 2015

[47] Christidis K, Devetsikiotis M. Blockchains and smart contracts for the internet of things. IEEE Access. 2016;**4**:2292-2303

[48] Dai J, Vasarhelyi MA. Toward blockchain-based accounting and assurance. Journal of Information Systems. 2017;**31**(3):5-21

[49] White GRT. Future applications of blockchain in business and management: A Delphi study. Strategic Change. 2017;**26**(5):439-451

[50] Oates BJ. Researching Information Systems and Computing. London, UK: SAGE Publications Ltd; 2006

[51] Saunders M, Lewis P, Adrian T. Research Methods for Business Students. 2012;**3**:215-218

[52] Onwuegbuzie AJ, Leech NL. Sampling designs in qualitative research: making the sampling process more public. The Qualitative Report. 2007;**12**(2):19-20

[53] Sheldon MD. A primer for information technology general control considerations on a private and permissioned blockchain audit. Current Issues in Auditing. 2019;**13**(1):A15-A29

[54] Rooney H, Aiken B, Rooney M. Technology Innovation Management Review. 2017;**7**(10):41-44

Section 3

Blockchain Adoption in Government

Chapter 8

Perspective Chapter: Actor-Network Theory as an Organising Structure for Blockchain Adoption in Government

Reyan M. Zein and Hossana Twinomurinzi

Abstract

Blockchain technology (BT) is a promising technology with compelling distributive and security capabilities for digitalising organisations and social systems. It is, however, often approached from a deterministic and technical perspective yet requires social, cultural and institutional changes as part of the process of adopting new technology in the context of the digital government sectors. This study uses actor-network theory (ANT) for its closely related hybrid middle, translation features, token passing through to network stability and interessement, as a lens to shape and understand the complexities surrounding the adoption and use of BT, particularly in the public sector organisations. Using the land registration system in Sudan as a case study, the findings show that ANT provides an adequate lens through which to examine the role of emergent distributive technologies such as BT in shaping social and organisational processes. ANT further contributes to a more holistic adoption of BT in public sector organisations. In the case study, the proposed blockchain guided by ANT simplified the complexity of land processes for registration, selling, buying and ownership, eventually replacing multiple processes with single transactions while at the same time embedding security and transparency.

Keywords: actor-network theory, blockchain technology, translation, obligatory passage point, inscriptions

1. Introduction

Blockchain technology (BT) is an emergent technology that is being adopted and considered for its key features of decentralisation, peer-to-peer authentication, persistency, anonymity and auditability [1]. Other attractive features include its openness, extensibility and anonymity in transactions [2, 3]. Some authors have even suggested that blockchain is comparable to a legal institution in its own right [4].

These BT features are value-laden as each feature presupposes certain belief systems, meaning that BT adoption is accompanied by the social meanings embedded in the technology [5]. There is, therefore, bound to be a clash of social structures when BTs are implemented [6–8].

This study illustrates BT adoption through the lens of actor-network theory (ANT) [9, 10]. ANT represents the complex socio-technical world [11] in which new technologies such as blockchains are introduced based on the interests of different actors – human and non-human [12, 13]. More specifically in the public sector, the adoption of BT as part of digital government efforts, though seemingly ideal, has mainly focused on the technological aspects at the expense of the transformatory potential inherent in the technology [14]. Institutionalising such technologies as part of digital government efforts implicitly involves introducing extrinsic social structures that are embedded in the digital technologies [15, 16]. For example, BT is recommended by the United Nations as an essential low-cost approach to digital government in low-income countries without considering the associated social transformatory effects and the resistance to such changes [14, 17].

The primary objective of this chapter, therefore, was to identify the suitability of ANT as a tool to holistically adopt BT in the context of digital government. The secondary objective was to assess the influence of ANT as an organising theory for the adoption of blockchain in digital government efforts. Specifically, the study seeks to answer the question, 'How can ANT inform the implementation of BT in the public sector?'

The remainder of the chapter is structured as follows: The next section presents the literature on ANT and BT. It is followed by a section that maps BT to ANT terminology and then offers an illustration using a case study from Sudan's land registration. Analysis and discussion of the illustration are then presented. The final section provides conclusions and recommendations on the efficacy of ANT as a lens through which to introduce BT into digital government.

2. Literature review

2.1 Actor-network theory

The boundaries between the technical and the social, between human and non-human (machine), are frequently contested and negotiated. ANT is concerned with the interaction between the social and the technical, and the creation and maintenance of stable coextensive networks of humans and non-humans [18]. In the case of digital technologies, this includes people, organisations, software, computers and communications hardware and infrastructure standards.

ANT symmetrically treats the social and the technical as inseparable, arguing that humans and artefacts should be analysed with the same conceptual apparatus. Latour [19] illustrates the rationale for the symmetric treatment by identifying that it is no longer clear if digital technologies are a limited form of an organisation or if the organisation is an expanded form of digital technology [11]. ANT is often described as a systematic approach exploring the infrastructure that supports the 'scientific and technological achievements' within a network, making it a more profound approach to researching and understanding service networks [20]. ANT is, therefore, a useful theoretical lens for understanding socio-political phenomena, especially where digital technology plays a critical role. One of the most distinctive ANT features is that actors can also be non-human – a text, a machine, an institution. It values the interaction between humans themselves or between human and non-human actors [21].

ANT provides added explanatory power over existing socio-technical theories that are either deterministic such as structuration theory and diffusion of innovations, or are theoretically too narrow such as the social construction of technology theory. ANT does not exclude a priori non-human actors from the analysis, allowing for a more explicit examination of the enabling or the restricting role of digital technology in a socio-technical process; second, ANT does not distinguish a priori between micro (e.g., individuals) and macro actors (e.g., organisations), but acknowledges the inherently unstable nature of actors [22–24]. This allows for a flexible consideration of a socio-technical collective as a single actor or as a group of individual actors, depending on the level of analysis desirable [25].

Concept	Description
Actor	Both humans and non-human actors such as technological artefacts. Any element, which bands space around itself, makes other elements dependent upon itself and translates their will into a language of its own.
Actor-network	A heterogeneous network of aligned interests, including people, organisations and standards.
Punctualisation	The treating of a heterogeneous network as an individual actor to reduce network complexity (encapsulation).
Focal actor	The key actor behind gathering other actors' support for a change initiative.
Translation	The successful alignment of the interests of a diverse set of actors with the interests of the focal actor by encouraging one another towards the pursuit of self-interest and collective objectives [20]. Translation is a process that creates 'a temporary social order, or movement from one order to another, through changes in the alignment of interests in a network' [29].
Problematisation	It is the first moment of translation that defines the nature of the problems or opportunities to be solved in the network. The common problems precipitate the actors to align their interests with the network.
Obligatory passage points (OPP)	An essential step that has to occur for all of the actors to be able to achieve their interests, as defined by the focal actor.
Interessement	It is the second moment of translation where actors are convinced they share the interests of the focal factor [30]
Enrolment	The third moment of translation, wherein all actors in the network accept (or get aligned to) interests defined for them by the focal actor.
Mobilisation	The fourth moment of translation that is achieved when the actors are successfully enrolled.
Inscription	It is a process of artefact creation that would ensure the protection of certain interests in order to indicate how the network should operate.
Irreversibility	The point where it is impossible to go back to a point where alternative possibilities exist.
Block box	A frozen network element, often with properties of irreversibility. The way scientific and technical work is made invisible by its own success.
Immutable mobile	Network elements with strong properties of irreversibility, and effects that transcend time and place.

Table 1.
Actor-network theory terms and descriptions.

The most important ANTs characteristic is providing an analytical lens to understand the socio-technical components of hybrid contexts [26]. An essential ontology of ANT when executed is to track and clarify the translation moments by which networks of aligned interests are formed and preserved, or to inspect why those networks are unable to be created [24, 27]. Successful networks are established through the enrolment of a set of allies that successfully translate its interests, therefore, becoming well prepared to participate in specific methods of thinking and acting, in order to preserve the network [11]. Specifically, ANT supports the goal of this study because of the following network features [28]: ANT networking, which means the strength of the involved actors in the networks over communities and individuals who are not part of these networks; the second feature is ANT network, which is the strength of the standards needed to control the network social interaction through the imposition of the inclusion laws; the third feature is ANT networked, that is the strength of some social actors over other social actors in the network; and finally ANT network making, the strength to program alliance networks based on the dominant actors' interests and values [28]. **Table 1** illustrates ANT terminology and concepts.

2.2 Blockchain technology

BT revolves around the smart networks theory [29] in which value is replicated in a network using a sophisticated protocol that validates, confirms and controls transactions through the network. These protocols offer a peer-to-peer transfer of value using algorithmic trust compared with the classical typology of trust between human agents [30–32].

Blockchain is an append-only database technology because the moment the data is stored in the database, it cannot be changed or deleted [33], since the blocks are added in chronological order using timestamps and hashes to form an incorruptible chain. This chain is shared and distributed to all the participating entities [34, 35] in instances where blockchain transactions are public. Due to this transparent behaviour, security features were introduced by categorising blockchains as either public or private blockchains [34, 36, 37].

Public blockchains (permissionless), where the actors are anonymous, have more security challenges as each blockchain is able to be part or quit at any time [38]. On the other hand, private blockchains (permissioned) are predefined groups of specific actors who are authorised, authenticated and allowed to be part of the blockchain network in order to decrease the existence of malicious actors inside the network [31, 33, 39]

An essential part of blockchain development is its distributed intermediary governance where no single entity has full control, but a consensus between the different groups has to be reached using consensus algorithms [35]. Consequently, blockchain-based services are not maintained by a central authority, but by a community of miners and developers [24]. Miners are powerful and important actors in the network since the continuation of the blockchain depends on them. They collectively validate and bundle batches of transactions into blocks and add them into a chronological chain through a 'consensus' process that uses multiple consensus algorithms depending on the types and applications [40]. Consensus is used to ensure enhanced security and privacy for various applications in many domains using different mechanisms such as proof-of-work (PoW), which is a mathematical challenge that ensures the security of peer-to-peer transfer by maintaining a digital ledger of transactions that is considered to be unalterable [38]. As a consequence, a novel approach of 'algorithmic trust' is established, which is completely different from the classic typology of trust in human actors [32].

BT uses cryptography to provide the security, immutability and rightful ownership of the transactions being stored on the block using the hash function [40]. Further, cryptography helps the receiver to verify the authenticity and integrity of the transactions on the network [31]. It uses a changeable public key (PK) to record the users' identity that provides an extra layer of privacy [41].

The process of safe transfer of value is undertaken using smart contracts, which can simplify the process by automating verified transactions [42]. Smart contracts are computer programs that run automatically when certain criteria are met within the system, which are used to transfer value of any kind between the peers in a blockchain without the service of a trusted third party [31].

The next section presents an example of blockchain in the land registration context that was technically superior yet excluded critical social processes and therefore failed.

2.3 Bitland case in Ghana

Bitland Ghana is a sophisticated permissionless open blockchain model implemented for informal land ownership in Ghana using the open-ledger platform [43–45]. A proof-of-stake was used as the consensus mechanism utilising cadastral-tokens formerly issued by the Danish Cryptocurrency Exchange (CCEDK). It also used GPS and open-map API to map lands [46–48]. Rather than save the identity data on the block, Bitland uses a unique value to link the title to the owner [46, 49–51].

Bitland focused on the customary and local authorities outside the major cities to establish a land ledger for farmers and local communities [45, 52] without any official government institutions [45, 50]. The exclusion of the official land registration authority in government meant that Bitland was not accepted beyond the local community [52] because customary courts do not have any legislative power. This example illustrates the challenge of focusing on BT's technological supremacy at the expense of important wider social and network processes.

3. Methodological approach

ANT lends itself to interpretivism in its fundamental thesis which seeks to understand and give a 'voice' to technology artefacts in a social context [27]. Therefore, the interpretive case study methodology is used as an analytical framework to guide and analyse a case study. This study, consistent with ANT, does not adopt an ontology of natural realism. In other words, data is not viewed as objective evidence supporting or falsifying an assertion but as texts and text analogues, whose meanings, when read hermeneutically, can go beyond the original intentions and meanings attributed by their sources. This study illustrates the mapping of blockchain to ANT with an interpretive case study from the Sudan land registration authority.

3.1 Blockchain technology through the lens of actor-network theory

ANT has previously been used to describe groups of actors and how those groups sought to define and inscribe particular codes and standards into particular electronic record technologies. Once the rules have become part of the network, they are hard to reverse [53]. ANT has immense potential for understanding the complex social interactions associated with digital technology in various contexts [24]. Modern digital technologies represent the values of the policy-makers frozen in complex digital representations, such as algorithms, codes, electronic thresholds and applications. The idea of software as frozen discourse is an example of an inscription that resists change and irreversibility and leads to a frozen organisational discourse [11].

ANT's emphasis on the dynamic and relational aspects of a problem is a useful lens for studying non-linear change and the unintended outcomes of technology projects, including the stochastic events that are known to characterise BT projects and

Concept	Blockchain equivalent
Actor	Human: Participants, Top management, Miners, Developers. Non-human: Infrastructure hardware, System framework, Software platform, Organisation rules and standards, Documents, Distributed ledger
Actor-network	Distributed ledger + Human Actors + Non-human Actors
Punctualisation	The defining of blockchains as sets of different actors such as nodes, miners, developers, consensus mechanism, smart contracts, hash functions. All are encapsulated together to be seen as one actor.
Focal actor	The professionals who lead the change process (blockchain developer, senior manager, executive)
Translation	Alignment of the actors' interests with the interests of the blockchain through organisational process rules reformulation. In a temporary order, each actor addresses the effect of introducing the blockchain in order to adjust some of the organisational rules that correspond to decentralisation, transparency and information sharing.
Problematisation	The moment of defining the need to transfer the organisational model from centralised with limited access to decentralised peer to peer in order to share information safely between the actors. The type of blockchain is defined in this stage. Public blockchains are permissionless, and anyone can join the network. They also are decentralised: no central authority or administration has control over the network. Private or consortium blockchains, such as Hyperledger, are permissioned and therefore impose restrictions on who is allowed to participate and make transactions.
Obligatory passage points (OPP)	Blockchain developer (technical) and top manager (social) both participate to create the socio-technical standards that are used to complete the accepted actor's registration.
Interessement	Actors commit to the focal actors and adapt their interests with those of the blockchain developer and the top manager. Sometimes encouragement and promises are offered to attract actors to accept.
Enrolment	The blockchain to be created and the consensus mechanism are determined. Human actors accept their roles and commit by signing up to join the blockchain using agreements that are registered as inscriptions. The needed non-human actors are included and involved in the network.
Inscription	The definition of the BT rules that guarantee each interest using smart contracts or software as frozen organisational discourse.
Irreversibility	When smooth, an integration of organisational workflow process occurs between the blockchain parties, leading to societal adoption. The BT then becomes indispensable.
Black box	The process of creating, accepting, mining, and adding new blocks successfully to the blockchain, could be considered as the black box when it runs efficiently. The focus moves to its inputs and outputs and not on its internal complexity.
Immutable mobile	The whole system could always lead to the same results of efficiency regardless of place and time.

Table 2.
Blockchain representation using ANT concepts.

programmes [53]. The purpose of this theory is to discover and describe the processes for making patterns, social regulation and resistance [54].

Table 2 illustrates BT through an ANT lens, followed by a discussion using a case study taken from Sudan's Land Registration system.

4. Case study background (land registration)

Land registration refers to a system whereby a government entity records ownership and land-related rights. These records provide evidence of title, facilitate transactions and prevent fraud. It means that there is an official record (the land register) of rights on land or of deeds concerning changes in the legal situation of defined units of land [55]. Out-dated land registry systems introduce delays in ownership verification, slow down legitimate transactions, and in the worst-case scenario, could enable land misappropriation [56]. Given their significance to individuals and society as a whole, it is important to guarantee that land transaction registers are created and stored in a way that allows their availability, accuracy and management according to the law. The transparency, public accountability, financial stability and human rights are peril in case of no suitable treatment and care of land transaction records [57, 58].

Recently there has been a global movement to add more accountability to land registry systems and particularly increase the validity of land titles. It is vital for present-day governments in terms of curbing corruption, reducing red tape, enhancing transparency, improving the speed of the stated public service and eradicating risks of possible disputes [36].

The potential of blockchains to authenticate ownership can legitimately be transferred without needing third-party verification [45]. Once the registrar affirms the land title transfer, smart contracts are triggered to update ownership data to the buyer and accordingly the corresponding transaction is saved on the blockchain, as a result, all ownership records history could be traced [45]. By introducing this technology in the land registration systems, blockchains could introduce an era of network computing where private value transfer [37] of money, assets and contractual preparations can be conducted in an automated and dependable mode through computational systems [30, 36]. Such capabilities reduce forgery and the risks associated with the transfer and distribution of land data [59].

Most of the literature that has been published is theoretical analysis of the potential of using BT in the field of land registration, theoretical proposals and conceptual models using different types of mechanisms and tools. Other research included empirical studies of this technology in several countries [14]. Fortunately, early blockchain land registry adopters, like Sweden, Dubai, Ghana and others, report excellent results, such as cost savings, a better quality of service and elimination of fraud and corruption [60]. On the other hand, there is a lack of the studies that organise BT adoption in land registration. This chapter, therefore, seeks to present a process of introducing BT as a social and technical implementation.

4.1 Land registration in Sudan

Sudan is a land-locked country whose e-government development index (EDGI) has fallen from middle- to low-EGDI level due to adverse political, socio-economic and natural conditions [61]. It is considered a low-income country.

The land policies have undergone transitions according to social standards from the pre-colonial era, during colonial times and post-colonial times [62]. Socially, land has a unique position in most communities in Sudan, particularly in the rural areas. It is a means of livelihood and a source of wealth, tribal identity and social peace [63, 64].

The land has also been a source of deep conflict. From the political perspective, the civil war in Sudan between 1983 and 2005 was often seen as associated with land tenure, which was both a cause of conflict and a factor in ensuring peace and social stability [62]. On the other hand, the ownership and use of land, especially among traditional producers in most low-income countries, is not just a source of livelihood but a symbol of identity, dignity, solidarity and peace. The World Bank devoted some of its resources to the interrelated issues of access to secure property rights, exchange and distribution of land in markets and the role of governments in land management [64].

This chapter seeks to theoretically represent and explore the networks and associations that underpin the situated land practices through the lens of ANT [15]. The next section uses the Sudan land registration systems for illustration.

4.2 Sudanese land registration current actors

In ANT, anything that modifies a state of things making a difference is an actor [21]. The current practices of land registration in Sudan involve several actors (human and non-human) who contribute to the flow of the process. They are distributed among different stakeholders to achieve various tasks (**Table 3**).

4.3 Sudanese land registration current practices

4.3.1 The registration process

The registration is the process of documenting the citizen's land ownership at the official authorities to ensure the rights when needed. Registration is conducted in case of land purchasing, granted land by the government, inherited land or gifted land or through a mortgage in the case of banks, large companies or legal entities.

This is the process of granting land by the government to a citizen.

1. The citizen accesses land's office – planning authority.

2. The employee checks the identity and the land's number.

3. The employee contacts the survey office and confirms all the land's data.

4. The employee asks the citizen to pay the fees.

5. The land entitlement decision has been issued and signed.

6. The contract is issued between the lands' office – planning authority and the citizen as an owner of this land.

7. Three copies of the contract are issued, one for the use of the lands' office – planning authority, the second is for the owner, and the third is sent to the registration office.

Actor	Definition
The land Actor	It is the land with the boundaries that identifies it as a property that is owned by a person.
The Registration Employee Actor	An employee at one of the land registration offices who helps citizens to register or check a land.
The Head of Registration Actor	A person who is responsible for confirming the process of the land registration. In addition, the head decides any procedures that must be conducted according to the case.
The Registration Payment Employee Actor	The person responsible to collect money and issue an invoice accordingly.
The Head of Engineering Affairs Actor	The person who grants the permission to check the land details.
The Survey office Employee Actor.	The person responsible to check the land location, area and coordinates.
The Engineering Affairs Payment Actor	The person responsible for the payment at the engineering affairs.
The Owner Actor	The person who owns land registered under their name or has some issues about land.
The Buyer Actor	The person who wants to have an agreement with a land owner through a trusted lawyer and wants to transfer the land ownership under their name.
The ownership document actor	Any document that proves that ownership of a land belongs to an owner, such as titles, certificate and/or statement.
The lawyer actor	The person who has academic experience in law and passed successfully all the conditions and standards of the justice's authority.
Electronic System Actor	The electronic program that stores some of the land's records. This is designed by the information technology department and distributed to some registration branches.
The General Ledger Actor	The program that stores some of the land's records. Designed by the information technology department and distributed to some registration branches.
Lands Prosecution	The party where complaints are initiated due to forgery or frauds or any land ownership problems.
The judiciary authority and courts actor	The party that is responsible for resolving disputes to decide the property ownership.
The Bank Actor	The party responsible for issuing mortgages.
Civil Registry	The governmental body that issues or verifies the national IDs for the citizens according to birth, immigration or displacement. It also issues death certificates.

Table 3.
The main actors of land registration in Sudan.

8. At the registration office, the owner shows his/her copy to the employee to compare with the copy of the registration office.

9. The employee asks the owner to pay the fees.

10. The receipt details are appended to the land's record.

11. The land's record is revised and confirmed by the head of the registration.

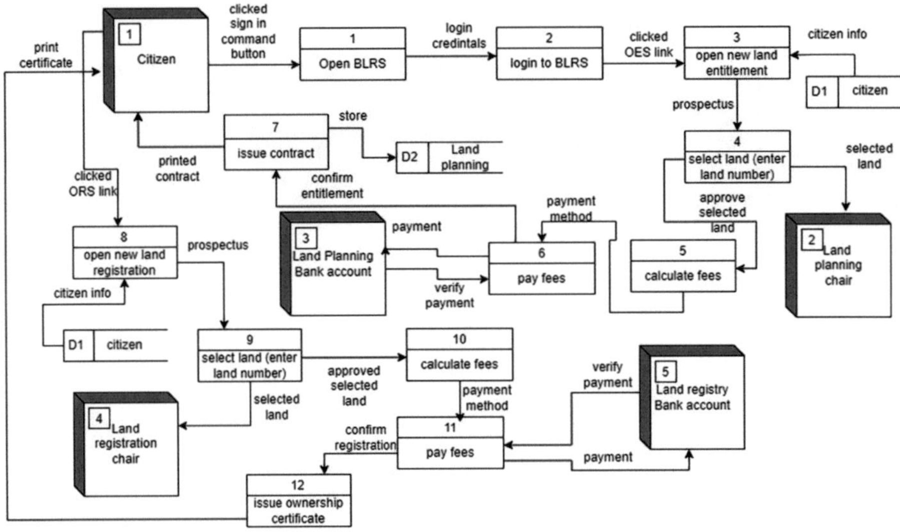

Figure 1.
Land registration process.

12. The land's record is signed by the head of the land registration, the head of the branch.

13. The land's record is updated in the paper ledger and the computer (**Figure 1**).

4.3.2 The land ownership certificate for selling purpose

Once the owner has delivered the ownership certificate, he/she must access the land registration to request a land selling certificate in order to be able to sell his/her land to the buyer through the selling process.

This is the description of land ownership certificate issuance.

1. The owner accesses the land registration office taking a valid identity or passport and a recent photo.

2. The employee at the window of the head of the registration office checks the identity and asks the owner to fill a form and then sign and put a fingerprint.

3. The owner then is directed to the accountant to pay the computer fetching fees, then the documents are passed to the computer fetching employee.

4. The fetching employee finds the land data and compares it to the form's name to make sure they are the same.

5. The fetching employee sends the land's number to the employee who is responsible for fetching the papery records and asks him/her to bring the folder of the land.

6. The fetching employee inserts the form into the land's folder and returns to the employee and asks him/her to issue a selling certificate.

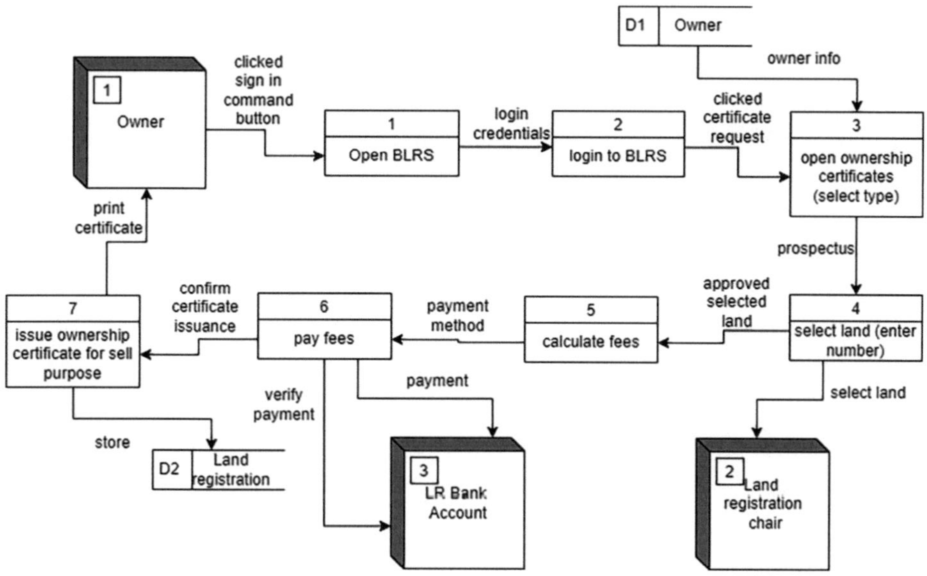

Figure 2.
Issuance of ownership certificate for selling purpose.

7. The employee issues the land selling certificate and sends it to the head of the registration office for the final signature.

8. The certificate is then sent to the employee who is responsible for delivering the certificate to the owner after checking the identity, the payment receipt and ensuring that he/she is the same person who requested the certificate and asking the owner to put a fingerprint again for certificate delivery (**Figure 2**).

4.3.3 The selling process

The lawyer is the main actor who controls the process of selling the land. He/she is responsible for documenting and contracting to ensure that the agreement flows in a proper and legal manner.

This is the description of selling a land.

1. The buyer assigns a lawyer and holds his/her fees after ensuring that the lawyer stamp is valid.

2. The buyer asks to see the land title (ownership verification certificate).

3. The lawyer asks for a land selling certificate.

4. The lawyer ensures that the land selling certificate is valid as it works only for 1 week from the issuance date. If it is more than a week it should be returned to the registration office to be cancelled and to issue another one.

5. The lawyer checks the header of the land selling certificate, if it is ('Hikr' contract: Hikr is a property right that grants usufruct in return for a specific fee.) then he/she must ask for Ornic 3A.

6. The owner goes to the land's office – planning authority asking for Ornic 3A.

7. The employee makes sure that the land selling certificate is valid to issue Ornic 3A.

8. The documents are ready and the lawyer proves all of them and then begins preparing the selling contract.

9. The contract must be signed by four witnesses.

10. The lawyer fills his/her part on Ornic 3A and asks the owner for a signature and a fingerprint to be used with the land title in transferring the ownership at the registration office.

11. He/she also prepares a petition letter headed by the lawyer, asking the land registration authority to transfer the ownership from the owner to the buyer according to the attached documents (the selling contract, the land selling certificate, Ornic 3A).

12. The seller takes the money with thanks (**Figure 3**).

4.3.4 The ownership transfer process

By the end of the selling process, the owner takes his/her money and leaves, and the buyer continues the process of transferring the ownership until delivering his/her ownership certificate.

This is the description of transferring land ownership to the buyer.

1. The buyer or (the lawyer) takes (the selling contract, the land selling certificate, Ornic 3A and the petition letter) to the land registration office.

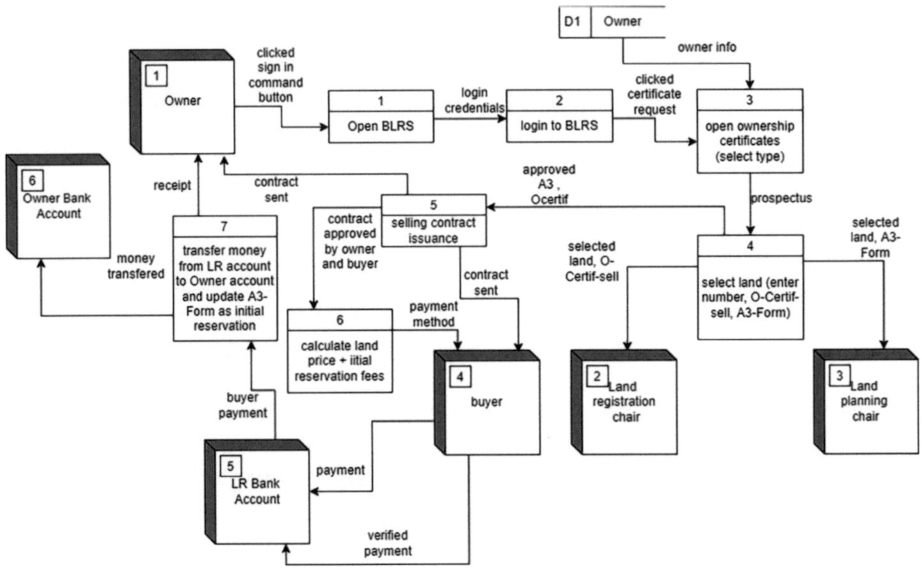

Figure 3.
Land selling process.

2. The buyer pays an initial reservation fee for the buyer's name (for a period of 3 months).

3. The employee stamps Ornic 3A by reservation.

4. All the documents then are taken to the land's office – planning authority at the researcher's employee's window.

5. The researcher employee then checks the reservation period and asks the buyer to fill a form.

6. The researcher employee prepares a letter to the survey office asking for the land's details.

7. The buyer is directed to the survey office to pay the fees and bring an affidavit that contains the land's details and the estimation of the mitre price.

8. The documents now are complete, and a photocopy of all the documents and a computer update is done.

9. The documents are sent to the head of the land's office – planning authority who will sign by the acceptance of ownership transferring to the buyer and send the documents to the land's registration authority.

10. The head of the registration checks the identity and the documents and sends the documents to the registration specialised employee.

11. The registration specialised employee inserts the documents into the land's file and calculates the fees of the last reservation then takes the calculations to the accountant.

12. At the accountant, the buyer should pay 3% if the land is (Hikr), and 4% if the land is (Ain).

13. Once the buyer pays the fees, the ownership is then transferred to the buyer and he/she is able to ask for any land's certificate (**Figure 4**).

4.4 The proposed blockchain land registration

BT could guide the private value transfer of assets, and contractual preparations in an automated and dependable mode [30]. BT in Ref. to the land registry system can be described as a distributed ledger with functions, such as storing all the transaction records, land owners for a certain period and transaction times [48]. Therefore, digitising land transactions via BT could enable secure confirmation by all parties in a land transaction before a transaction is completed, possibly leading to credible and reliable land transactions. The blockchain register is also tamper-proof and not subject to discretionary or inadvertent modifications by those who have access to it (**Figure 5**) [44].

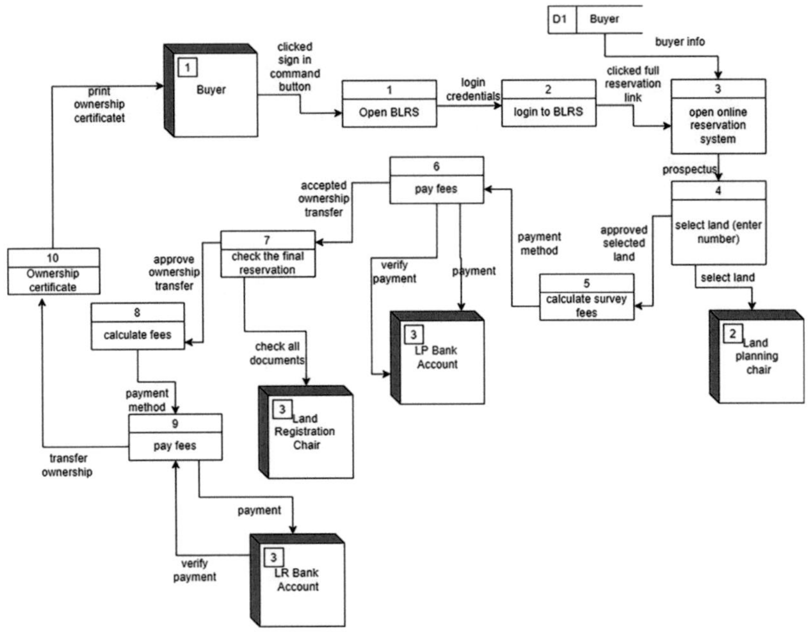

Figure 4.
Land ownership transfer.

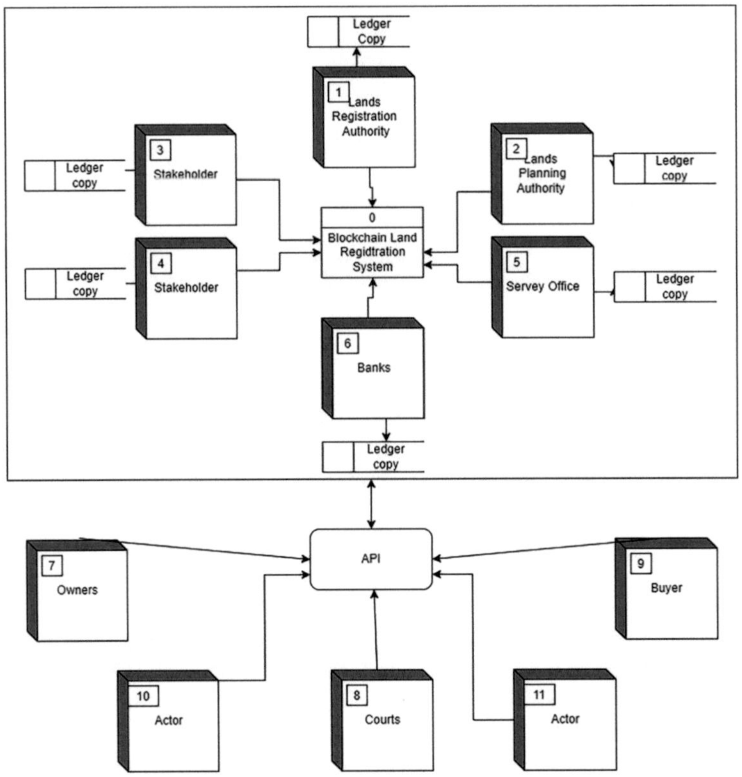

Figure 5.
The proposed land registration system using blockchain technology.

5. Mapping the land registration system using ANT and blockchain

5.1 Actors of land registration

The BT equivalent of actors is the nodes of the blockchain. There is always a need to define the actors [65] within the network and consider their culture level, interests, views and ability to use the technology. An actor is not just a 'point object' but an association of heterogeneous elements themselves constituting a network, so each actor is also a simplified network, which means that any changes affect this actor and the networks it simplifies [18]. Technology artefacts can transform networks by influencing relationships between human actors as the active role of the land registry record in mediating social relationships between land registration staff.

The list of actors (**Figure 6**) includes the legal, technical, professional staff, managers and the general registrar as human actors; the courts, ministry of finance, ministry of engineering affairs, and the information system are the non-human actors. The system user interface such as a mobile application or the internet browser could be considered a non-human actor granted access with view-only privileges. Such an interface is needed to avail open data that connects sellers and buyers without the need for a third party.

5.2 Actor networks

ANT is formed of several actants that join an alliance to achieve their different goals. Every actant tries to enrol new actants and attempt to convince them to support

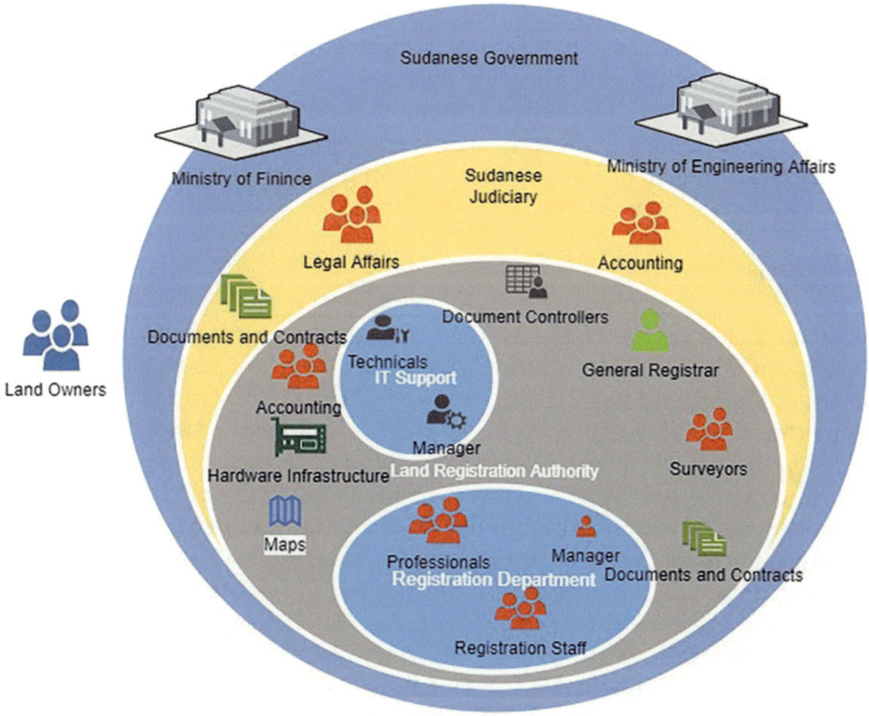

Figure 6.
An ANT view of the actors (institutional stakeholders).

their own goals. Those networks get stronger and more durable as long as more actants are being enrolled [19]. Each of the actants can be enrolled as a blockchain node with different types of access and authorisation privileges. This can be done using the consensus mechanism according to the private blockchain using one of the consensus algorithms that assists to share the exact same copy of the data on the network [31].

The general format of a research question from an ANT perspective – "What is the network, and what phenomena are emerging from it?" – offers broad and flexible scope for mapping the relevant terrain [53]. The Sudanese land registration could be considered a network in a larger network of politics and other organisations. Actor-networks are concerned with the information flow between actors without changes inside domains specified by the networks and can change dynamically [8]. For example, the availability of internet services throughout the country changes the relationship [66] between the urban and rural areas by enabling illiterate farmers to use land registration systems to track their land records [7]. The adoption of BT comes as a consequence of the actions of everyone in the chain of actors who has anything to do with it, and in the same way, each of these actors shapes the blockchain to their own ends [18].

The next section illustrates land registration through the four moments of translation: problematisation, interessement, enrolment and mobilisation [24].

6. Translation process

ANT suggests that successful social networks of aligned interests are created by enrolling actors and translating their interests so that they are willing to participate in particular ways of thinking and committed acting that will maintain the network [26]. This alignment is achieved through 'translation', which includes four phases of a problem definition for that a specific technology could solve, convince the others to consent to this solution, determine the main roles and practices in the network and at last distribute others to perform the roles, commence the practices and connecting with others in the network [53].

6.1 Problematisation

Problematisation defines the problem or opportunity with which an actor proposes a solution. The proposed solution acts as the obligatory passage point [20]. In the context of Sudan land registration, the process of land registration, or the procedures for extracting the testimonies, or the stages of selling the land, all involve many repeated steps for scrutiny, review, verification and confirmation within each party and in the interaction between the parties with each other. Despite all that, there are losses, fraud, forgery and other manipulation cases that lead to conflicts. Therefore, the top management of the authority would seek to introduce BT to guarantee transparency, immutability, persistency and auditability through decentralisation peer-to-peer and information sharing. Therefore, the challenges of corruption would be dealt with, accelerating the process, reducing fraud cases, bringing transparency with smart contracts, eliminate forgery and allow openness of the land registration data. According to the above, the top management would be responsible for pulling other actors to adopt this solution.

6.1.1 Obligatory passage point

The obligatory passage point (OPP) can be initiated by facilitating communications with the stakeholders. Each stakeholder would have to pass through the OPP to gain their desired interests. However, in passing through the OPP, they may have to experience some changes and need to adjust their situations to suit the new network [15]. For example, the Sudanese land registration would need to initiate the BT development as OPP to imply the concepts of decentralisation and information sharing that support the process of BT adoption. Therefore, a radical change of the key organisational processes in the organisation would be necessary [67].

6.1.2 Focal actor

The focal actor determines the problem, related actors and clarifies the influence of this problem to those actors, therefore figures out strategies to handle the problem. This focal actor sets itself as an OPP among the other actors and the network to make itself 'indispensable' [68]. This could be achieved by understanding the organisational process of land registration and figuring out all its stakeholders. Different focal actors are identified in the land registration network but mainly the top management decision-makers who are the most aware of the organisational rules. The other focal actors are the BT developers who are familiar with the BT properties and behaviours. The enrolment of the top management can be seen as the first focal factor that represents the emerging network and a change process, carrying the responsibility of identifying and enrolling other key actors [25]. The second focal actor would be the land registration blockchain, which defines the set of rules through algorithms. Accordingly, the decentralisation, information sharing and anonymity concepts are then also reflected.

6.1.3 Actors

ANT focuses on the actors within the socio-technical network and their contributions. Each human and non-human actor has interests and features that must be met (Latour, 2017) to obtain smooth adoption within a network. The current socio-technical network in Sudanese land registration includes the index technology; other technical systems which could be interoperable; the individuals (registration employees, citizens, administrators, surveyors) who provide and use the registration services; a specific alignment of policy makers, technical professionals and technical elements whose infrastructure and its data models were established; civil servants, monitoring and evaluation top management and decisions makers (general registrar, head of branches).

By introducing BT, a set of actors under different categories is also proposed and divided into four divisions as follows:

- Acceptors: Human (owners, buyers, participants)
- Providers: Human (blockchain miners, practitioners, professionals, employees, surveyors) and objects (registration branches, court, information system, banks)
- Supporters: Human (blockchain developers, administrators, legal professionals, technologists, accountants) and objects (maps, records, contracts, software tools)

- Controllers: Human (managers) and objects (government, land registration authority, land record blockchain)

6.2 Interessement

Interessement is the process of getting others to accept this problem-solution [53]; it is when the primary actor recruits other actors to assume roles in the network, roles that recognise the centrality of the primary actor's role [65, 68]. The process involves convincing other actors to have interests that are aligned with the focal actor [25]. This would allow each actor to identify all the actors and resources needed to accomplish their tasks and achieve their goals [18, 25, 67]. Moreover, incentives are created for actors such that they are willing to take a detour from their earlier charted paths and pass through the OPP defined by the focal actor [25]. In other words, Interessement aims to attract other actors in the proposed solution to favour a new opportunity that confirms the problematisation phase [20]. Managers could be enrolled and empowered to achieve the aims of the changes. Those managers, in turn, would have to enrol individual groups within the organisation so that established networks can be reorganised. Those separate groups may also utilise software or tools to enhance the tasks executions in different procedures. As a result, power relationships would shift from 'top-down' implementation strategies led by the government to horizontal and distributed. The focal factor could do this through devices that seek to lock that commitment in place, blocking the actors from alternative courses of action. The signing of the formal agreement between the Sudanese lands' registration, the judiciary authority and the ministry of urban planning is one part of this. It commits them to the blockchain framework project as the only technological choice and the only course of action.

At a general level, interessement involves 'actions by which an entity attempts to impose and stabilise the identity of other actors it defines through its problematisation'. It includes locking new allies into place and cornering entities not yet enrolled. Successful interessement 'confirms (more or less completely) the validity of the problematisation and the alliances it implies' [69].

BT consensus mechanism could be considered as OPP because the data authentication processes are being verified directly before acceptance. It would be necessary to amend some procedures, dispense others; and in some cases, replace a group of tasks with only one procedure. For example, any number of service requests could be accepted at any time using a web or mobile application.

6.2.1 Alignment

Alignment involves all the strategies through which an actor identifies other actors and arranges them in relation to each other [18]. In ANT the actors' interests could differ in a way that they may support or constrain the technology. Therefore, the technology needs the alignment of actor interests in order to stabilise the network [7]. Translation involves negotiations among human actors and representatives of material actants. Negotiations establish common sets of definitions, conditions and meanings for understanding the network's phenomena. The outcome of successful negotiations is an ANT characterised by aligned interests. The degree of alignment is the degree of convergence of an actor-network [19]. Translation implies that an actor reinterprets or appropriates the interests of other human actors and the interests embedded in non-human actors according to one's own and has these interests represented in the inscription. Actors must identify the other actors who may react to it differently. They may modify it, deflect it, betray it, add to it, appropriate it or let it drop [18]. As for the land registration case, each actor has to define the chains of tasks and resources that determine the processes and the other participants who are needed to achieve this task. If conflicts arise, they must be addressed by adjusting some process.

The actors' interests are flexible and can be translated, enabling the interest alignment and the maintenance of an actor network [70]. For instance, the procedures that the owner should do to transfer land ownership are a task of multiple actors and different resources. Each of them has various processes that are needed to be done in a particular order starting with the owner's request until the buyer delivers. For example, the owner's interest may focus on reducing the time of the overall process, while the buyer's interest may focus on the accuracy of these procedures. Due to these contrasts between the interests of actors' groups, actors' networks in the Sudanese land registration authority should be updated to include the concepts of information sharing, decentralisation and transparency. The alignment of all interests could lead to a clear visualisation of the proper data flow and necessary changes in the organisation's business process.

6.3 Enrolment

Enrolment is the moment when roles are defined and actors formally accept and take on these roles [68]. It is a negotiation process to exhibit how the interessement meets the actors' interests and needs and persuades them to accept the new actor-network [20]. This process involves defining the accepted roles of each actor in the new actor-network. As a part of the enrolment process, the enrolment commitments can be recorded in a shared memory through inscription. This is considered as the foundation of a settled actor-network and needs suitable enrolment strategies that could handle attitudes, power and politics. The goal is to boost lock-in, in which digital land registration becomes socially acceptable, has a positive social texture and socially solidify as a safe feasible interaction between the actors [7]. The process involves defining the accepted roles of each actor in the new actor-network. As a part of the enrolment process, the commitments of enrolment can be recorded in a shared memory through inscription [71]. The actors' interest's alignment within the actor network happens by enrolling others into the network by the actors who are known by cooptation [19]. However, it should be noted that enrolment is temporary and a betrayal by enrolled actors is a possibility. Actors enlist other actors into their world and they bestow qualities, desires, visions and motivations on these actors [18]. For instance, an owner may encourage another to use the land registration blockchain system and, if successful, enrols them into the actor-network. In doing this, the owner may also call on other actors to support his case [11]. The owner may call on texts (land registration authority literature), stories (of successful use and benefits gain) and the technology itself (through demonstrations).

6.3.1 Inscriptions

'An inscription is the result of the translation of one's interest into material form' [24]. They are common procedures such as managerial practice, employee contracts, standards, regulations or software requirements documentation [20]. Inscription devices (for example, pull-down menus in a piece of software) may help to stabilise the network and thus shape and constrain land registration work [53]. All the interest within the network is translated using inscriptions attached to the technology [7, 70]. These inscriptions may involve maps, programs, user requirements, regulations, documents and even the messages and marketing related to the technology and the technology services, which typically impact actors' roles.

A Sudanese owner would be interested in increasing the price of land and reducing the intermediaries' roles in selling land. The government is interested in increasing the registered lands through land registration authority with owners of unregistered lands. Inscriptions are typically provided with more concrete content to record actors' interests within a material that varies in their flexibility, for example, policy and regulations. Therefore, the strength of the inscription may be determined by the possibility of irreversibility [20].

Smart contracts are one of the BT features that could be considered as inscriptions as they are scripts that reside on the blockchain that allows for the automation of multi-step processes [72]. Smart contracts are therefore an essential feature that supports utilising BT in land registrations. They offer a third way to perform contracts, a new paradigm, wherein legally binding agreements are backed up by real-world agreements, and can be built to run within a network of computers without any single party sabotaging parts of the agreement [45].

For instance, the lawyer actor could be disappearing in the new network. Only his/her actions could be digitised in a smart contract that checks all the conditions at the different stakeholders to conduct the selling process and automatically executed to transfer the ownership to the buyer. This can eliminate several steps that may cost the owner and the buyer much time, effort and money.

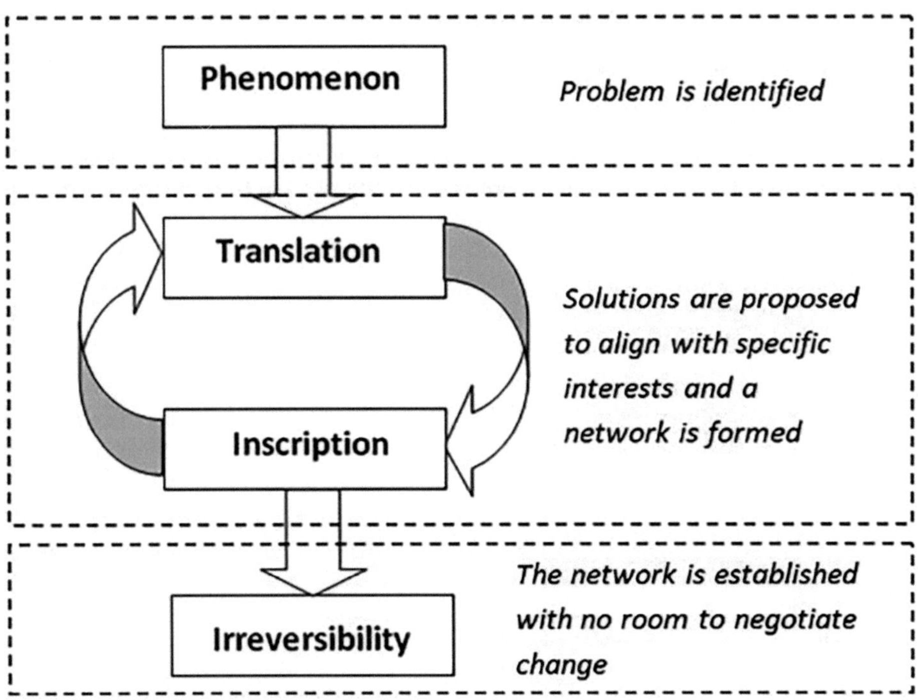

Figure 7.
Relationship between translation and inscription [20].

Figure 7 depicts the relationship between translation and inscription to address a phenomenon (the formation of an actor-network) through various interests and to establish an irreversible network [20].

6.4 Mobilisation

Mobilisation is necessary to ensure that actors represent other actors' interests [20]. Mobilisation describes network elements that display strong properties of irreversibility and are mobile across time and space; various software standards provide illustrations of immutable mobiles elements [11]. If actors enrolled in the network adequately represent the masses, enrolment manifests as active support and mobilisation occurs [24]. Primary actors assume a spokesperson role for passive network actors (agents) and seek to mobilise them to action [68]. Mobilisation could include the process of migrating records from the traditional system to the blockchain system, which leads to the same expected results.

6.4.1 Black boxes

Actor-networks could be seen as heterogeneous, and developed open systems, however, the stability of an actor-network is a kind of pact, fulfilled via 'black boxes' – settings of actors (human and non-human) that are being taken for granted and therefore they are no longer enquired [53]. A black box is a technical term for a device, system or object when it is viewed in terms of its input, output and transfer characteristics without any knowledge required of its internal workings. Latour [19] defines a black box as a term 'used by cyberneticians whenever a piece of machinery or a set of commands is too complex'. When the actor-network determines and accepts the standards, it would be challenging to inverse them, and the actors become locked-in under these restrictions. The standards related to technology and communication are locked into a black box, and there is no need to consider the contents and the processes in this black box [15]. When the digital land registration is admissible, there is no need to inquire about how it works or seek if it follows the best method to do it. Black boxes are not necessarily restricted to physical artefacts or technologies; instead, black boxes consist of knowledge that is accepted and used regularly as a matter of fact [65]. Making a black box does not require consensus.

The actors join or quit, or even alliances changes may lead to the 'black boxes' of the network to be opened in order to revise and reconsider their contents [18]. Fortunately, the black boxes can be opened [7] to check, revise, and alter their contents, because ANT allows for such dynamism [19]. For example, land ownership transfer under the inheritance clause could have been taken for granted. However, if legal or social issues are raised according to different owner's religions, new texts issued based on the Personal Status Law based on the owner's religion may become part of actor-networks as different potential standards. At that point, the land transfer process has to be queried, and the black boxes should be opened. The reality of institutionalisation status of digital government system functions can be represented and measured via the ANT construct of black-box behaviour [15].

6.4.2 Irreversibility

A network becomes durable when actors feel no need to spend time opening and looking inside black boxes, but just accept these as given [18]. The optimum state of the actor-network should be to become stable which is known as irreversibility state. If this is not achieved, then it can die out as fast as it began which leads to the failure of technology adoption. Digital land registration has to be an essential part of society to reach the irreversibility state and become indispensable in the context of the land registration actors and the government [11].

Below is the discussion of the potential contributions of ANT and the practical applicability to BT adoption in land registration in Sudan.

7. Discussion

7.1 Changed relationships

BT, guided by ANT, offers an approach to technology adoption that organises information and human action in a decentralised as well as offering more credibility between organisations and their stakeholders. One of the most interesting challenges of BTs is the transition of trust from institutions and social interactions to semi-automatic and semi-autonomous technological systems. ANT offers a smooth approach to transitioning through the concept of interest alignment, which includes a comprehensive discussion between the actors to figure out how to identify the motivations to adopt BT to guarantee easy adaptation.

ANT illustrates how such a network may, for example, look like in relation to the introduction of a digital land registration system using BT. The integration of the new land registration system requires the formation of new norms and other established network components that reorganise around a new actor. ANT offers deeper insights into the change processes involved in the new relationships. This can then result in recommendations on how to make the new network more stable and in so doing facilitate the effective integration of the technology into the land registration environment.

7.2 Transformative effects

The ANT attempts impartiality towards all actors, whether human or non-human and makes no distinction in approach between the social, the natural and technological. Introducing such innovative approaches using BT technologies that change the current social and political system will probably be resisted initially. This resistance includes the beneficent, the owners and buyers in the case of the Sudanese land registration, the intermediaries and the critics who would reject the new. The resistance could be attributable to different reasons such as legal, social, cultural or even educational reasons. Offering some awareness upfront through the ANT lens to the actors regarding the BT implementations could reduce the expected resistance (**Figure 8**).

Some digital government practices can be accurately captured via the ANT construct of obligatory points of passage. The associated strategies of the stakeholders who attempt to reconfigure networks of artefacts in order to institutionalise digital government systems are usefully analysed via ANT translation stages. However, achieving full enrolment in the blockchain network could require more time and effort from the actors to have full interests' alignment for all the participants. The parallel implementation of the blockchain network with the current network could make the adoption smoother in the future as the existence of reassurance factor regarding the potential of incidence in the current work which boosts confidence with the new artefact with continuous progress.

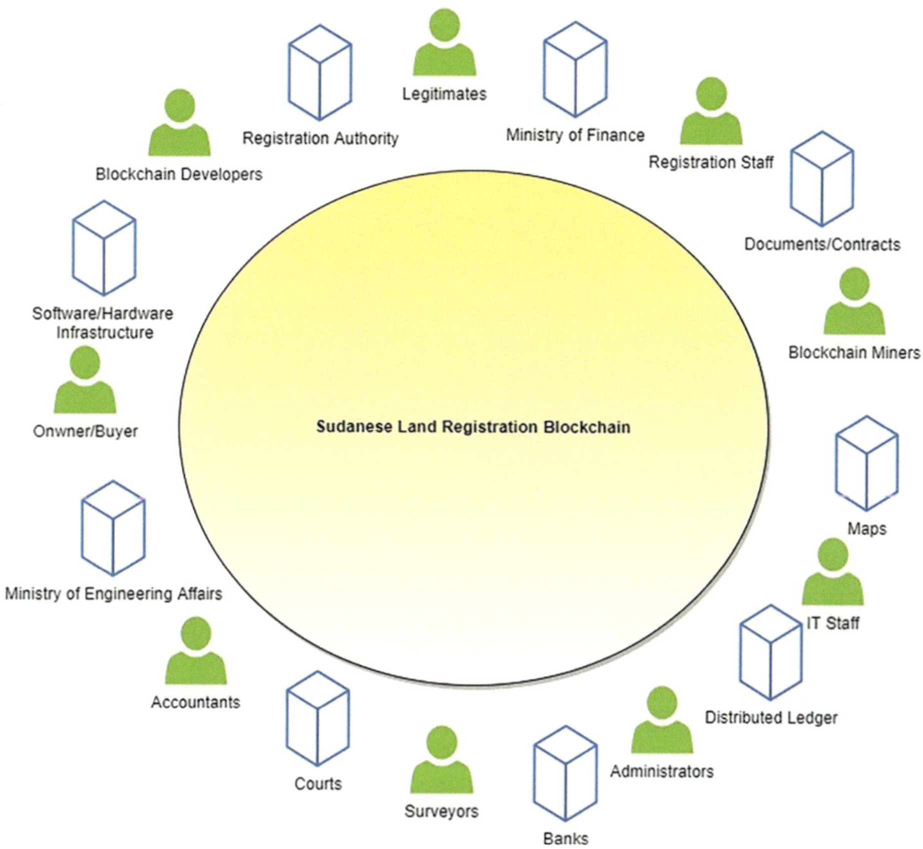

Figure 8.
An ANT view of the intended land registration network after the blockchain adoption.

7.3 Flexibility for social context

The approach presented attempts to organise the new technology's structure using ANT basis to facilitate its mapping process with the real-world actor-network through negotiation and dialogue sessions, leading to compatibility. ANT illustrates BT's flexibility according to actor social desires through negotiated smart contracts. For example, the ability to dynamically open black boxes to take into account local and situated contexts in BT means that actors can gradually make transactions that are mutually acceptable and move towards a shared view of land tenure. This shared view is important for social and financial stability in reducing land challenges that often arise from over- or under-valuing, and as collateral for needed finances.

7.4 Persistent records and transparency

ANT illustrates the importance of having a full and transparent history of records about land using BT, and the reasons for land transfers. This transparency and persistence of records further contribute to a stable social system around the land. The appeal of BT revolves around the smart networks theory which posits that value replication is conducted by the network itself where smartness is created within the network's tasks using a sophisticated protocol that validates, confirms and controls transactions through the network [30].

7.5 Dismantling technocentric assumptions and power relations

ANT brings to the surface the technocentric assumption that transparency is always desirable by all actors. Power changes that are often not easily given up without resistance occur naturally. ANT, therefore, allows for a balanced implementation approach that considers the important social, political and technical negotiations that accompany the introduction of transformative digital technologies. The challenge of studying the introduction of new technologies in a highly institutionalised field of land registration, and especially in emerging economies such as Sudan, brings to the surface strong power, cultural and social structures that enable and constrain human action.

7.6 The potential of using ANT in Bitland

ANT sees that broad, multi-stakeholder engagement is key; therefore, networks, meet-ups, building expertise etc. are essential to guarantee the implementation and promulgation of the technology [43]. Bitland records can be disputed if the official institutes do not accept the legality of these records that is why Bitland was not supported and preserved through law. Regardless of how property is registered, a dispute over land titles must always be facilitated by some form of the court system. Moreover, the engagement of governmental actors is essential to get the accuracy and legitimacy of the asset because land registration utilises the concrete assets that should be confirmed by professionals. The government is considered as an organisational body that has the supervising and setting rules and regulation role for all actors like transactions costs reduction and increase information symmetry.

The government cannot be ignored as it is one of the necessary actors to register land in an official manner approved by the official, executive and judicial authorities. For example, Bitland records would be at risk in case of a dispute or gain financial loans. The legality of the land registers is probable to be enquired, particularly in statutory courts, where some titles were not recorded via the official government channel and were not dealt by Lands Commission surveyors. Legitimacy in the registration process regarding the conformity of formal rules for land registration is debatable since it has not involved formal surveyors from the government. The translation stage proposed by the theory can provide some arrangements between the official authorities and Petland, which affects the gradual entry of the new system into the official records. If the community does not trust the system, it would continue using unwritten agreements with the chiefs. In that case, the registry has little durability in securing that their right of use is tracked, and records are secured from tampering [52].

7.7 Challenges and limitations of ANT

As with all other approaches to social-technical theory, in attempting to answer the question of how social orders are created and maintained, ANT faces epistemological, ontological and methodological challenges. Indeed, ANT's applicability has resulted that it does not seem to be a theory, because the approach is so descriptive and unable to offer any details of the way actors should be seen and how their actions should be interpreted. However, it is important to take into account the classic concept of theory to more scout this issue. A theory should answer the questions 'how and why' things take place by scouting their connections. ANT can offer a straightforward illustration of how things take place, but it is hard to use ANT to describe why things take place. More challenges meeting ANT such as hard to be used to examine empirical evidence because it is so wide and then hard to refute. It can, therefore, assist in illustration and offer an interpretation vocabulary [73]. To some extent, researchers are affecting the way that actors choose which is important to be considered, as it is expected that researchers who are using this approach are likely to use this sort of question. Therefore, researchers in the land registration field most probably will be part of the actors' network. They will affect this network and affected by the network according to the relations during research, particularly if the study includes a qualitative approach. Researchers should have enough knowledge of the way decisions are made [73].

Another weakness of ANT is describing it as a 'flat ontology'. It seems that no previous layers existed, instead only 'a single plane of endlessly entangled translations'. ANT's black boxes show a group of stable-for- now interconnections that may vary at any time – with no more theorisation [53].

One more defect is ANT's proposition of 'symmetry' between humans and things. ANT's reduction of humans as compared with technologies puts human impulse, wishes and morals out of the analytic scope and avoids ethical questions.

ANT application in developing countries is not necessarily a challenge, particularly the practical part, but it is important to work on the methodological constraints and the analytical challenge. ANT, therefore, presents an alternative view about the main blockchain operations that must be considered, and the way that networks can be composed. ANT enables the effective role of technology in digital government operations while at the same time aligning the interests and identities of different actors [67].

8. Conclusion

The study addressed the potential of using ANT as a tool to organise the introduction of BT in the context of digital government. This chapter highlighted the literature of both BT and ANT, and used the Sudan land registration system as an illustration.

The adoption of BT without taking into account, the many active actors in the network of relations related to it fails to explain the roles of human and non-human actors. ANT enables the roles to become clear and moves away from technology determinism to reveal important boundaries between BT's embedded social and technical characters. The flexibility in ANT to illustrate the flexible smart contracts based on local interests can help in reforming stable new and transformative networks that are persistent over longer periods.

By illustrating the necessity to negotiate actor interests in the process of enrolment, ANT could reduce the expected resistance against changing the existing power structures that favours a minority. ANT, therefore, allows for a balanced BT implementation and adoption approach that considers the important social, political and technical negotiations that accompany the introduction of transformative digital technologies.

The study was limited in using an illustration for one country. Further research is required to empirically test the use of ANT to introduce BT in digital government.

Author details

Reyan M. Zein[1*] and Hossana Twinomurinzi[2]

1 Sudan University of Science and Technology, Khartoum, Sudan

2 College of Business and Economics, University of Johannesburg, Johannesburg, South Africa

*Address all correspondence to: reyan.aziz@yahoo.com

References

[1] Zheng Z, Xie S, Dai H-N, Wang H. Blockchain challenges and opportunities: A survey. International Journal of Web and Grid Services. 2017;**2017**:1-24

[2] Tama BA, Kweka BJ, Park Y, Rhee RH. A critical review of blockchain and its current applications. ICECOS 2017—Proceeding 2017 Int. Conf. Electr. Eng. Comput. Sci. Sustain. Cult. Herit. Towar. Smart Environ. Better Futur. 2017. pp. 109-113

[3] Yli-Huumo J, Ko D, Choi S, Park S, Smolander K. Where is current research on blockchain technology?—A systematic review. PLoS One. 2016;**11**(10):1-27

[4] Ishmaev G. Blockchain technology as an institution of property. Metaphilosophy. 2017;**48**(5):666-686

[5] Postma D. A Critical Investigation of the Contribution of Actor-Network Theory towards a Critical Conception of Technology. Pretoria: University of Pretoria; 2006

[6] De Filippi P, Hassan S. Blockchain technology as a regulatory technology: From code is law to law is code. First Monday. 2016;**21**:12

[7] emadwiandr. Actor-network theory and the adoption of Mobile communications. Journal of Chemical Information and Modeling. 2013;**53**(9):1689-1699

[8] Fang Z. E-government in digital era: Concept, practice, and development. International Journal of Computer Internet. 2002;**10**(2):1-22

[9] Mingers J, Mutch A, Willcocks L. Critical realism in information systems research. Management Information System. 2013;**37**(3):795-802

[10] Ottens M. The Cadastral System as a Socio-Technical System The Cadastral System as a Socio-Technical System. In: Jt. FIG Comm. 7 COST Action G9 Work. Stand. Cadastr. Domain. 2004

[11] Walsham G. Actor-network theory and IS research: Current status and future prospects. Information System Quality Research. 1997;**1997**:466-480

[12] Furuholt B, Wahid F, Sæbø Ø. Land information systems for development (LIS4D): A neglected area within ICT4D research? In: Proceedings of the Annual Hawaii International Conference on System Sciences. 2015. pp. 2158-2167

[13] Arnaboldi M, Spiller N. Actor-network theory and stakeholder collaboration: The case of cultural districts. Tourism Management. 2011;**32**(3):641-654

[14] Zein RM, Twinomurinzi H. Towards blockchain technology to support digital government. Lecture Notes in Computer Science. 2019;**11709 LNCS**:207-220

[15] Azad B, Faraj S. E-government institutionalizing practices of a land registration mapping system. Government Information Quarterly. 2009;**26**(1):5-14

[16] Kompella L. Socio-technical transitions and organizational responses: Insights from E-governance case studies. Journal of Global Information Technology Management. 2020;**23**(2):89-111

[17] Sambuli N. New technologies and the global goals. UN Chronicle. 2019;**55**(4):32-34

[18] A. Gilding and Department. Actor-network theory and information systems research. International Journal Actor-Network Theory Technological Innovation. 1999;**1**(4):53-69

[19] Latour B. On actor-network theory. A few clarifications, plus more than a few complications. Logos (Russian Federaion). 2017;**27**(1):173-200

[20] Carroll N. Actor-Network Theory: A Bureaucratic View of Public Service Innovation. 2014. DOI: 10.4018/978-1-4666-6126-4.ch007

[21] Fioravanti C, Velho L. Let's follow the actors! Does actor-network theory have anything to contribute to science journalism? Journal of Science Communication. 2010;**9**(4):1-8

[22] Callon M. Techno-economic networks and irreversibility. The Sociological Review. 1990;**38**:132-161

[23] Callon M, Latour B. Unscrewing the big leviathan. In: Cetina KK, Cicourel A, editors. Advances in Social Theory and Methodology: Towards an Integration of Micro and Macro-Sociologies. London: Routeldge; 1981. pp. 277-303

[24] Islam N, Mäntymäki M, Turunen M. Understanding the role of actor heterogeneity in blockchain splits: An actor-network perspective of bitcoin forks. Proc. 52nd Hawaii Int. Conf. Syst. Sci. 2019;**6**:4595-4604

[25] Sarker S, Sarker S, Sidorova A. Understanding business process change failure: An actor-network perspective. 2006

[26] Thapa D, Omland HO. Four steps to identify mechanisms of ICT4D: A critical realism-based methodology. Electronic Journal of Information Systems in Developing Countries. 2018;**84**:6

[27] Cordella A, Shaikh M. Working paper series. Account and Finance. 1980;**20**(2):146-146

[28] Castells M. A network theory of power. International Journal of Communication. 2011;**5**(1):773-787

[29] Swan M, Dos Santos R. Smart network field theory: The technophysics of blockchain and deep learning. SSRN Electronic Journal. 2018. DOI: 10.2139/ssrn.3262945

[30] Swan M, de Filippi P. Toward a philosophy of blockchain: A symposium: Introduction. Metaphilosophy. 2017;**48**(5):603-619

[31] Ali S, Wang G, White B, Cottrell RL. A Blockchain-Based Decentralized Data Storage and Access Framework for PingER. In: Proc. of 17th IEEE Int. Conf. Trust. Secur. Priv. Comput. Commun. 12th IEEE Int. Conf. Big Data Sci. Eng. Trust. 2018. pp. 1303-1308

[32] Atzori M. Blockchain governance and the role of trust service providers: The TrustedChain® network. Journal of British Blockchain Association. 2018;**1**(1):1-17

[33] Dukkipati C, Zhang Y, Cheng LC. Decentralized, blockchain based access control framework for the heterogeneous internet of things. In: ABAC 2018—Proc. 3rd ACM Work. Attrib. Access Control. Co-located with CODASPY 2018. Janua; 2018. pp. 61-69

[34] Kamble S, Gunasekaran A, Arha H. Understanding the blockchain technology adoption in supply chains-Indian context. International Journal of Production Research. 2019;**57**(7):2009-2033

[35] Ølnes S, Jansen A. Blockchain technology as infrastructure in public

sector—An analytical framework. In: ACM International Conference Proceedings Series. 2018

[36] Di Ciccio C, Gabryelczyk R, García-Bañuelos L, Hernaus T, Hull R, Štemberger MI, et al, editors. Business Process Management: Blockchain and Central and Eastern Europe Forum: BPM 2019 Blockchain and CEE Forum Vienna, Austria, September 1-6, 2019 Proceedings. Cham; 2019

[37] Khan R, Ansari S, Jain S, Sachdeva S. Blockchain based land registry system using Ethereum Blockchain. In: Researchgate.Net. 2020

[38] Hughes L, Dwivedi YK, Misra SK, Rana NP, Raghavan V, Akella V. Blockchain research, practice and policy: Applications, benefits, limitations, emerging research themes and research agenda. International Journal of Information Management. 2019;**49**(February):114-129

[39] Helliar CV, Crawford L, Rocca L, Teodori C, Veneziani M. Permissionless and permissioned blockchain diffusion. International Journal of Information Management. 2020;**54**:102136

[40] Alketbi A, Nasir Q, Talib MA. Blockchain for government services-Use cases, security benefits and challenges. In: 2018 15th Learn. Technol. Conf. L T. 2018. pp. 112-119

[41] Miraz MH, Ali M. Applications of blockchain technology beyond cryptocurrency. Annals of Emerging Technologies in Computing. 2018;**2**(1):1-6

[42] Bennett RM, Pickering M, Sargent J. Transformations, transitions, or tall tales? A global review of the uptake and impact of NoSQL, blockchain, and big data analytics on the land administration sector. Land Use Policy. 2019;**83**:435-448

[43] Eder G. Digital transformation: Blockchain and land titles. In: OECD Global Anti-corruption Integr. Forum. 2019. p. 12

[44] Mintah K, Baako KT, Kavaarpuo G, Otchere GK. Skin lands in Ghana and application of blockchain technology for acquisition and title registration. Journal of Property, Planning and Environmental Law. 2020;**12**(2):147-169

[45] Anand A, McKibbin M, Pichel F. Colored Coins: Bitcoin, Blockchain, and Land Administration. In: Scaling up Responsible L. Gov. Annu. World Bank Conf. L. Poverty. 2016. pp. 1-16

[46] Kshetri N. Will blockchain emerge as a tool to break the poverty chain in the global south? Third World Quarterly. 2017;**38**(8):1710-1732

[47] Kshetri N, Voas J. Blockchain in developing countries. IT Professional. 2018;**20**(2):11-14

[48] Shuaib M, Daud SM, Alam S, Khan WZ. Blockchain-based framework for secure and reliable land registry system. Electronic Control. 2020;**18**(5):2560-2571

[49] Demuyakor J. Ghana go digital agenda: The impact of zipline drone technology on digital emergency health delivery in Ghana. Shanlax International Journal of Arts, Science and Humanities. 2020;**8**(1):242-253

[50] Chris Bates L. Bitland global White paper. C.S.O. Bitl. Glob. 2016

[51] Schmidt K, Sandner P. Solving challenges in developing countries with blockchain technology. FSBC Working Paper. 2017;**2017**:1-22

[52] Olsen BL. Beyond the hype: Exploring blockchain Technology in

Land Administration. International Business. 2018;**17**(9):9-2018

[53] Greenhalgh T, Stones R. Theorising big IT programmes in healthcare: Strong structuration theory meets actor-network theory. Social Science & Medicine. 2010;**70**(9):1285-1294

[54] Salavatian S, Hesampour M, Tohid S. Network theory : The case of IRIB. In: Anal. Netw. Media Organ. Audiences, ICTs Based Actor Netw. Theory Case IRIB Siavash. 2019. pp. 231-255

[55] Zevenbergen J. A systems approach to land registration and cadastre. International Congress on Surveyors. 2002;**1**:1-10

[56] Benbunan-Fich R, Castellanos A. Digitalization of land records: From paper to blockchain. In: International Conference on Information Systems 2018, ICIS 2018. 2018

[57] Thakur V, Doja MN, Dwivedi YK, Ahmad T, Khadanga G. Land records on blockchain for implementation of land titling in India. International Journal of Information Management. 2020;**52**(April):1

[58] Lemieux VL. Evaluating the use of blockchain in land transactions: An archival science perspective. European Property Law Journal. 2017;**6**(3):392-440

[59] Peiró NN, Martinez García EJ. Blockchain and land registration systems. European Property Law Journa. 2017;**6**(3):296-320

[60] Themistocleous M. Blockchain technology and land registry. Cyprus Review. 2018;**30**(2):195-202

[61] U. Nations. E-Government Survey 2018_FINAL.pdf. 2018

[62] Babiker M. Historical Overview of Land Policy in Sudan. In: Environment and Conflict in Africa: Reflections on Darfur. 2009. pp. 242-252

[63] Abdalla Y, Elhadary E. Challenges facing land tenure system in relation to pastoral livelihood security in Gedarif State, Eastern Sudan. Journal of Geography and Regional Planning. 2010;**3**(9):208-218

[64] ElHadary YAE, Obeng-Odoom F. Conventions, changes, and contradictions in land governance in Africa: The story of land grabbing in North Sudan and Ghana. Africa Today. 2012;**59**(2):59-78

[65] Besel RD. Opening the 'black box' of climate change science: Actor-network theory and rhetorical practice in scientific controversies. Southern Communication Journal. 2011;**76**(2):120-136

[66] Thorhildur J, Michel A, Niels BA. The generative mechanisms of open government data. In: ECIS 2013—Proceedings of the 21st European Conference on Information Systems. 2013

[67] Heeks R, Stanforth C. Technological change in developing countries: Opening the black box of process using actor-network theory. Devotional Study Research. 2015;**2**(1):33-50

[68] Callon M. Some elements of a sociology of translation: Domestication of the scallops and the fishermen of Saint-Brieuc Bay. Logos (Russian Federation). 2017;**27**(2):49-94

[69] Holmström J, Robey D. Inscribing organizational change with information technology: An actor network theory approach. Information Systems Research. 2002;**2002**:1-39

[70] Gao P. Using actor-network theory to analyse strategy formulation. Information Systems Journal. 2005;**15**(3):255-275

[71] Islam AKMN, Mäntymäki M, Turunen M. Why do blockchains split? An actor-network perspective on bitcoin splits. Technological Forecasting and Social Change. 2019;**148**:119743

[72] Christidis K, Devetsikiotis M. Blockchains and smart contracts for the internet of things. IEEE Access. 2016;**4**:2292-2303

[73] Cresswell KM, Worth A, Sheikh A. Actor-network theory and its role in understanding the implementation of information technology developments in healthcare. BMC Medical Informatics and Decision Making. 2010;**10**(1):67

Section 4

Blockchain in the Energy Sector

Chapter 9

Perspective Chapter: Blockchain Technology in the Field of Energetics – Organization of Effective Energy Market

Serge Chernyshenko, Valentin Afanasyev, Vardan Mkrttchian and Vsevolod Chernyshenko

Abstract

The article is devoted to the topic of blockchain applications in the field of energetics. A role of the technology in the digital economy development has been discussed; corresponding examples from Russian, Chinese, Estonian experience have been presented. A synergy between usage of blockchain and other information technology such as big data, intelligent avatars and the Internet of things has been demonstrated. Implementation of the blockchain technology into the functional scheme of the energy market has been proposed as a ways to ensure its decentralization and openness. Energy supply contracts can be negotiated directly between producers and consumers without intermediaries, so all the market actors have a high level of autonomy in purchasing and selling. The technology provides services for preparation and issuance of necessary invoices, as well as for making settlements for the entire transaction. Blockchain technology ensures economic and information security of transaction participants and includes convenient tools for realization of their market activities. Difficulties and risks, connected with the technology implementation, have been discussed, as well as main ways of its further development.

Keywords: blockchain technology, energy market, digital economy, decision making, distributed data, intelligent avatar

1. Introduction

In the mass consciousness, the term "blockchain" is associated with cryptocurrencies and financial mechanisms for avoiding centralized control over deposits. However, this view is too narrow. The blockchain is a technology for organizing distributed databases and transaction systems, associated with the databases, which should ensure reliable data storage and a possibility of their distributed processing [1]. The development of this technology supposes solution of many interdisciplinary problems in the field of cryptography, mathematics, Internet technologies and computer programming [2]. Blockchain, to put it simply, is a shared ledger, distributed over a special network, which fixes all transactions and guarantees correctness and safety of data.

A feature of the blockchain is that the confirmation of correctness of actions and accompanied data does not require involvement of a special third-party organization, as is the case of traditional transactions [3]. The validity of any actor, operating in the market, (trading organization or individual) is confirmed by validity of a node in the blockchain, which must be present in the network and visible throughout it.

There are several levels in the blockchain structure: data level; network level; consensus level; the level of incentives; contract level; and application level. The data level encapsulates kernel data and algorithms, associated with data encryption and timestamping. The data is an encrypted series of decimal digits that are decrypted using timestamps.

The network level includes methods of data distribution in the network and data validation. The consensus level includes methods of generating and distributing economic reasons, integrating by such way economic factors into the blockchain technology system. The level of contracts encapsulates various algorithms for concluding and implementing contracts (smart contracts). The application level includes other algorithms, that are not assigned to the listed levels, but are necessary in a particular blockchain models. The main innovations of the blockchain technology are the followings: the block structure; the use of timestamps; distributed node consensus mechanism; inclusion in the model of "economic reasons"; flexible and programmable smart contracts.

The following main areas of blockchain application can be distinguished now: financial system; insurance business; logistics; Internet of Things (IoT) [4]; public services [5]; social security; education [6]; digital copyright. We especially note that the implementation of the blockchain technology makes a powerful impact on the development of international financial and trade cooperation.

Evident nowadays leader in blockchain's applications is the financial sector. The Internet finance is moving to the forefront of the global financial system, being the driving force behind the introduction of new information technologies into the global economy. Along with the use of big data, mobile Internet, cloud computing, etc., blockchain is becoming the basis for the development of financial technologies.

Among the important areas, where the use of blockchain is becoming commonplace, one can also mention the energy industry [7] and, first of all, the energy market [8, 9]. Informatization of the energy market can give a significant synergistic effect for the development of national and global economies [10]. Blockchain creates prerequisites for the decentralization of the energy market [11], which is long overdue. The decentralization has especially advanced in the electricity market, where producers and consumers can be geographically close and do not need in a centralized intermediary [12]. This is especially true for alternative sources of energy (solar, wind, etc.), which do not have high power and stability and, correspondingly, product of which must be consumed locally [13].

The development of the blockchain will be effective only in conjunction with the development of other modern information technologies, which, on the one hand, facilitate the implementation of the blockchain, and, on the other hand, allow the effective use of the results of using the blockchain for effective decision-making and the overall development of the energy sector.

2. Synergy of blockchain and other IT

In modern conditions, blockchain technology, within the framework of general approaches to the development of the digital economy [14], should be combined with other information technologies [15]. These are, first of all, big data technology, which must be used when deploying ultra-long blockchains. Development of some blockchain services involves using artificial intelligence systems. And important to mention, that the blockchain is a technology, which is capable of integrating into IoT as its information kernel and means for communication between individual networks [16].

One of the most important functions of the blockchain is to support transactions and contracts. The objectivity and reliability of used artificial intelligence tools should minimize the possibility of error or misuse of these functions. When forming and implementing smart contracts, it must be guaranteed that the system saves all their parameters and leaves indelible timestamps.

Any transaction must be confirmed by both sides, and in many cases two intelligent bots (two "avatars") can take over this function. If one of the sides, after corresponding analysis, decides not to confirm the transaction, it is automatically canceled. In turn, the blockchain will provide sufficiently complete and reliable information for the avatars, which make the decision. The passed information can be filtered; the blockchain can have its own algorithms for filtering incoming information; it can be recognized as a spam, or as an informational recommendation, or as a direct command.

Avatar, as artificial intelligence (AI) software, is based on algorithms of self-learning neural networks and functions in close connection with an automatically generated knowledge base (KB) [17], in the development of which the blockchain is involved. KB includes both a block of basic knowledge and a constantly growing block of experimental knowledge, accumulated new information about observed specific processes and situations. In turn, the structure of KB can be based on blockchain technology [18, 19], which supports distributed and reliable storage of information, as well as information security of big data in the process of their collection, transmission and storage.

The use of AI systems are important also for authorizing and profiling blockchain users, as well as for monitoring their activities.

The development of IoT also needs the use of blockchain, as there is a problem of paying for centralized services, the support of which becomes unbearable for operating companies because of the geometric growth of IoT. Blockchain technology can provide collection and direct transmission of data without organizing a special centralized accounting system.

In addition, there is the problem of IoT privacy, since the centralized architecture involves centralized storage and transmission of all information, which can lead to large-scale data leaks. In this regard, distributed encryption in the blockchain makes client privacy more strong, since collecting and decrypting distributed data is technically much more difficult than hacking a central server.

Another issue stems from the fact that, under the current IoT architecture, clients can only perform network transactions on their own or trusted networks. This limitation, which greatly reduces the commercial value of the Internet of Things, can be circumvented by using the blockchain. For example, the "autonomous centralized remote control between centers", developed jointly by IBM and Samsung, uses a special blockchain registration to enable IoT devices to directly interact with each other and implement complex business logic.

3. Blockchain in energy

Data aggregation, their reliable distributed storage and possibility of imposing certain business functionality on them are in demand in the energy sector. First of all, the blockchain is used to ensure the decentralization of the energy system. Energy supply contracts can be negotiated directly between producers and consumers, so that both parties have a high degree of autonomy and do not need intermediaries. Drawing up contracts and making transactions is automated and provided with a sufficiently high information security system. All actions are recorded and can always be checked if the parties wish.

There are several areas of blockchain applications in the energy sector. First, as already mentioned, it is organization of the energy market. At present, it is quite complex organizationally and chaotic. At the global level, it is highly influenced by political factors. Some suppliers can appear or disappear on the market, depending on the political situation in their countries (for example, Libya, Syria, Nigeria), sanctions regimes arise and end (against Iran, Venezuela, etc.), new norms of centralized regulation are introduced (like EU energy packages), etc. Of course, blockchain technology cannot change the general rules, but it can mitigate the consequences of sudden changes and allow the market to adapt to new conditions. Integration on a single technological decentralized platform of numerous market actors makes it easier for the buyer and seller to find each other, while avoiding many risks. By using blockchain currencies (such as USDT), automatic checking the participants of the transaction and its compliance with corresponding customs regulations, it is possible, according to experts' opinion, to reduce the duration of the transaction to less than 5 minutes.

The application of blockchain in the oil and gas industry is focused on replacing traditional "paper" trading in commodities with more transparent and cheaper "electronic" trading. VAKT, the world's first blockchain platform for oil trading, was created in 2018 and is a good example to follow. At the end of 2017, BP and Shell proposed "to build a digital blockchain platform based on commodity trading." The following year, a blockchain platform called VAKT was successfully developed and put into production, and the founding companies immediately used it to trade North Sea oil.

The introduction of blockchain technology facilitates the recording of reliable data on carbon emissions, which is now a matter of great importance. There is currently limited transparency in the global market regarding the methods, used to measure and predict emissions, and there is a problem of trust in available information. Blockchain can provide greater confidence in data from global carbon inventories and registries.

Another area, promising for the use of blockchain, is the exploration and organization of oil and gas production. It is known that the corresponding business processes are multilevel and involve a wide variety of actors. Exploration is not limited to just searching for a deposit; in parallel, the cost of launching operation and the further profitability of production should be assessed. During the whole process, it is necessary to take into account numerous geological, geographical, physical, chemical, demographic, economic, technological and financial factors (starting from the depth and thickness of the field and finishing with the cost of credit funds). Data is collected by all participants of the process, (starting from geologists and finishing by financiers), and must reach the stakeholders in an adapted form. The blockchain system can provide significant assistance in this.

A well-organized distributed database facilitates the process and reduces costs. Already at the exploration stage, geologists receive from the blockchain the meteorological and seismic information, and intelligent avatars are able to suggest (basing on available, previously collected data) where, when and how to start exploration.

Thanks to the constant real-time putting new data into the blockchain, recommendations are constantly refined, which reduces the likelihood of errors in oil drilling modes and accelerates the progress of the project. Already in the course of oil and gas production, monitoring of critical parameters allows you to quickly conduct an economic analysis and calculate such parameters as the cost of energy resources, the life of the well, etc. Blockchain technology makes it possible to ensure targeted access to information, its use for various, in particular, scientific purposes, without the risk of leakage of its confidential part.

4. Distributed energy markets

One of the most promising ways to use the blockchain in the energy sector is the information support of the energy market in connection with the above-mentioned trend towards its decentralization. Flexible spatial distribution of generation and consumption of energy resources improves the quality of customer service, reduces operating costs and ensures reliable network operation. Given the growing demand for renewable energy and tightening control of carbon dioxide emissions, the production of electricity on microgrids with blockchain information support is becoming increasingly important in the development of the energy sector [20].

In the operation of a microgrid blockchain, it is especially important to provide an appropriate transaction mechanism and implement optimization of a microgrid power output. In order to achieve the lowest total cost of a microgrid blockchain, a two-way transaction mechanism and corresponding smart contracts are usually used.

There is no perfect centralized transaction information support model that would provide complete trust in terms of ensuring the interests of all parties. Blockchain technology largely combines the properties of reliability, transparency and security. Distributed decentralization helps organize low-cost and highly efficient power distribution work, solving problems of trust and energy efficiency [21, 22].

The microgrid blockchain can increase or decrease the contribution of a node according to real needs, which is carried out using a transaction mechanism and ensures energy optimization. Such a flexible mechanism is especially important for "renewable energy sources" such as photovoltaic, wind, tidal, etc. Renewable energy sources are unstable and cannot work effectively in isolation. For them, the stabilizing function of the blockchain is critical.

Let us consider the transactional architecture of the microgrid blockchain and propose one of its possible efficient architectures. The general structure of the energy blockchain consists of many microgrids and several large power grids connected on a point-to-point basis (P2P). Each node of the blockchain network can turn on and trade through the network. Each microgrid corresponds to one specific type of energy device: a photovoltaic array, a wind turbine, and so on. The microgrid also includes energy storage equipment (batteries). When energy generation is insufficient, batteries give out energy, and when generation is excessive, it is stored. The users of the microgrid blockchain are consumers of household energy and owners of electric vehicles.

In the blockchain, the distribution of energy among the numerous microgrid users is controlled by a special control block. The structure of energy transactions represents a "microgrid – user" relationship, i.e. both parties trade according to a one-to-many structure. However, since the generation of renewable energy is largely influenced by natural conditions, and consumer behavior is relatively stable, then there is often a mismatch between electricity generation and demand in the trading link of the microgrid. When electricity production exceeds demand, the excess capacity must be sold to the outside world; when there is not enough electricity to meet the demand in the microgrid, electricity must be purchased from the outside. If this kind of relations will be carried out only with large power networks, this will increase unnecessary costs due to fixed prices at "wholesalers" and creates need to perform electricity transactions arrhythmically.

Unlike the structure of blockchain transactions with a single microgrid, in a multicomponent blockchain, each microgrid is located in parallel, independently of each other. On the electricity trading market, formed on the basis of multi-grid blockchain, the parties to each transaction are equivalent, regardless of belonging to a microgrid. In this case, the mechanism of "bilateral auctions" is being implemented, which ensures a high degree of market adaptation to changing economic and natural conditions. Bilateral smart contracts do not require participation of third-party institutions; direct transactions are not only simplified, but they also provide a greater level of confidentiality and independence. The chain data structure makes the data immutable and easy to check, and makes it easier to monitor the market.

Designing a blockchain structure of the market, after defining the overall architecture, implies the ability to model and optimize the power of the microgrid blockchain. Two aspects are taken into account: power generation periods and energy storage schedules.

The main generating devices of the regional energy system are photovoltaic panels (pv) and wind turbines (wt). Conventional fuel generators are used as emergency power generation equipment. The electrical energy exchanged between microgrids comes from electrical reserves, so it is not within the scope of the general model. An energy storage equipment mainly consists of lead-acid batteries.

The microgrid calculates its own cost of electricity generation and the cost of transactions with large power systems. On the basis of the calculations, an objective function is built to optimize the sale or purchase of capacities through the macrogrid blockchain.

The process of trading energy between microgrids under the blockchain is divided into separate stages. The selling price of electricity in a large grid does not change with time, but the price of electricity in transactions between microgrids changes over time. Transaction participants naturally prefer to trade with microgrids with lower rates. To achieve more efficient electricity trading, it is necessary to develop a dynamic method of managing the blockchain as a whole, which this technology allows to do very efficiently.

The mechanism of bilateral auctions should provide flexible allocation of resources, ensure the coordinated and orderly conduct of multilateral transactions, and is widely used in multilateral trading platforms. The bilateral auction mechanism has basically two forms: direct bilateral auction and centralized bidding. Centralized trading should focus on the quotes of all trading participants for a certain period of time, and after several operations, reload the information. This transaction model requires the execution of transactions within a fixed time, while it requires a relatively large amount of calculations that require the involvement of significant computing power.

The direct two-way auction mechanism is more suitable for transactions between microgrids. The process of two-sided auctions is oriented towards the starting price and sale price models. The main control center of the blockchain balances supply and demand. When the total amount of electricity, offered on the market, is lower than the demand for it, the selling price will be increased accordingly, and vice versa.

When trading between blockchain microgrids takes place, no third-party institution is expected to be involved as a transaction coordinator. Instead, a smart contract technology is proposed, basing on the use of a special computer program that runs on the blockchain. The algorithmization of the bilateral auctions and the smart contracts allows to ensure following the business logic and to eliminate the risk of interference by third-party attackers. It provides trust between the parties of the transaction by logging the actions: all key information about the transaction is recorded into the blockchain for its further confirmation. Based on the price forming model, a proper form of the transaction is developed, and a smart contract is drawn up.

The contract design involves segmenting the lifecycle of the blockchain microgrid for 24 periods (hours) per day and calculating supply / demand in each period. All surplus will be distributed among the microgrids according to their needs, and corresponding smart contracts will be formed. The participants, whose requests were not satisfied, will receive energy from one of the major power grids at a fixed price.

The preparation and implementation of the smart contracts consists of several stages, such as registering requests, publishing offers, buying and selling, checking results, and settling transactions. Registration of requests and publication of offers are used for formation of quotations. The functional implementation of the purchase and sale is the mechanism of bilateral auctions, which has been described above. The function of checking results involves using digital signatures and recording information about the transaction in the blockchain. The transaction settlement function tracks the transmission of electricity and its payment, after which the smart contract is deactivated.

Development of the architecture of the decentralized and intelligent technology of blockchain transactions increases the efficiency of using the microgrid power and ensures the openness, transparency and security of transactions. Future development in this area should mainly focus on blockchain sharding technology and improvement of the speed and efficiency of blockchain algorithms, minimizing the imbalance between electricity supply and demand.

5. Some national blockchain usage models

The history of the use of blockchain technology in the energy sector can be traced back to April 2016, when residents of Brooklyn, New York, USA tried to conduct transactions to buy/sell solar energy using the blockchain platform. Since then, the approach has been used to support increasingly complex transactions in an ever-increasing demand for secure energy supplies in an increasingly decentralized energy environment.

There are different opinions regarding the prospects of the development of blockchain platforms, but an analysis of existing solutions in the power industry has shown that about 12 blockchain projects are currently under development and at the implementation stages. In Europe, the bulk of projects are focused on renewable energy sources. The extraction of own natural energy resources in the EU is steadily declining, as well as production of electricity from them, but the producing renewable energy sources is growing, the share of which in some countries already exceeds 50% [23].

Today, Germany leads the European countries in terms of the number of smart grid projects, followed by Denmark [24]. However, such small state as Estonia with population of 1.3 million also is very interesting because of its progress in digitalization. Despite the small number of citizens and a modest area of the territory, Estonia is 100% equipped with smart electricity meters, and the government of the country is actively using digital technologies [25].

The Estonian energy market is being actively transformed based on the blockchain; this is being done by the European company WePower in cooperation with the Estonian backbone network operator (TSO) Eltring. At the first stage of the project, data on the production and consumption of electricity in Estonia for the year (24 TWh) were transferred to Ethereum, which proved the technical feasibility of transferring such a volume of real data to the blockchain. This data was then converted into 38,973,240,000 Smart Energy Tokens, where each token means a smart contract for the purchase and sale of 1 kWh of electricity. Tokens can be traded and cashed out on the local wholesale electricity market, which is achieved by linking contracts to grid data via the blockchain [26].

Caspar Kaarlep (WePower) believes that the blockchain will help to increase the share of renewable energy sources in energy consumption: the system will clearly fix the source of origin of each kilowatt-hour in the general energy network and give buyers a guarantee that they are purchasing wind energy, and not generated by thermal power plants.

Let us note participation of China in financing the project on introduction blockchain technology to the Estonian power industry. In general, China belongs to one of the leading roles in practical implementation of this technology, which is facilitated by the quite centralized management of economy development, practiced in this country.

China set up the world's first energy blockchain lab in Beijing on May 15, 2016 [27]. Blockchain development has received the status of a self-regulating industry initiative, and blockchain development cooperatives have been opened. The State Electric Grid Corporation became the leader in the implementation of the technology. In November 2017, it submitted a patent application titled "Principles and Methods of Energy Management on Blockchain". The patent describes the structure of the blockchain, which consists of information blocks distributed among the nodes in order to avoid leakage of information, which is possible when it is stored centrally. The information in each block is stored in a binary tree structure, with each data change controlled at the root node of the tree. Through a simple check, the root node can detect that the data has been falsified. This patent indicated the direction of development of blockchain energy technologies in China.

The initial idea of the Chinese power system development was to improve the unified system of planning generation and supply of electricity. Indeed, centralized control reduces the cost of building and operating the power system, however, due to the uneven production of electricity, an imbalance in supply / demand occurs, and, in addition, transmission of electricity over long distances leads to large energy losses.

Since then, a lot of research has been carried out in the Chinese energy sector on the introduction of new network approaches. Distributed macro- and micronetworks, intelligent network agents for decision support, network monitoring systems, etc. were introduced. Not all the technologies have lived up to expectations, but there has been steady progress. Energy microgrids are already widely used, the concepts of the energy Internet and digital management of decentralized production, transmission, distribution and transactions of energy have been implemented. Most of the researches is devoted to the blockchain architecture of distributed energy consumption and optimization of power transmission.

One of the factors, stimulating the development of blockchain in the Chinese energy sector, is the intensive introduction of electric vehicles into operation. New energy vehicle sales in China totaled 1.367 million in 2020, up 13.3% year on year. New electric vehicle production in China was 1.366 million, an increase of 10% year on year, for a total of 4.92 million [28]. It is predicted that electric vehicles will account for more than two-thirds of global passenger car sales by 2040, rising from 3 million in 2020 to 66 million. With the growth of electric vehicles, the demand for charging equipment will inevitably increase. Many private charging stations can share chargers, set charging prices, and use blockchain for billing and payment.

The use of blockchain in energy is not limited to the electricity market in China. This technology is increasingly being used in other areas. Let us mention finance, IoS, logistics, public services, digital copyright, insurance business and social services. These areas cover, basically, all the most important aspects of the country life, so, many research institutions in China have taken up the development of blockchain technology in various aspects.

Currently, Russia is lagging behind Western countries and China in equipping smart electricity meters and deploying renewable energy sources. However, there is essential progress in the use of blockchain for settlements with consumers in the retail electricity market. Steps for the use of "smart" energy in Russia have been determined and a roadmap has been approved to eliminate administrative barriers to payments using blockchain technologies [29, 30].

The Russian electric power industry is a complex technological and economic system that includes more than 700 power plants operating at different levels (wholesale and retail markets), more than 1500 network organizations and millions of buyers (individuals and legal entities) consuming electricity according to different schedules. The functioning of the system requires prompt collection and analysis of a huge amount of information, timely adoption of managerial decisions, economic balancing the system, using collected experience and current data. Digitization of the industry is in progress, and software tools, needed for reliable storage and processing of large amounts of data (big data, blockchain, smart documents and others), are designing and introducing.

The new technologies requires an appropriate infrastructure, the development of stationary and mobile services [31]. In particular, for the use of blockchain technology in the electricity market, a necessary condition is 100% equipping of consumers with "smart" metering devices, integrated into the IoT/M2M (Internet of things and machine-to-machine communications) system, which collects information from the metering devices.

Results of the Russian IoT/M2M market research, conducted by J'son & Partners Consulting LLC (JPC), showed that as of 2018, the number of IoT/M2M devices, connected to the Internet, exceeded 23 million units. The long-term forecast from JPC assumes an increase in the number of connected devices by 2023 to 42 million units [32]. At the same time, it is obvious that the Russian IoT /M2M market does not take into account the need for intelligent electric energy metering devices ("smart" meters) as Internet devices [33].

By 2018, the total fleet of electricity meters in the Russian Federation amounted to about 70 million units. Almost 8–9 million metering devices are needed annually to replace existing and install new meters at industrial production facilities, housing

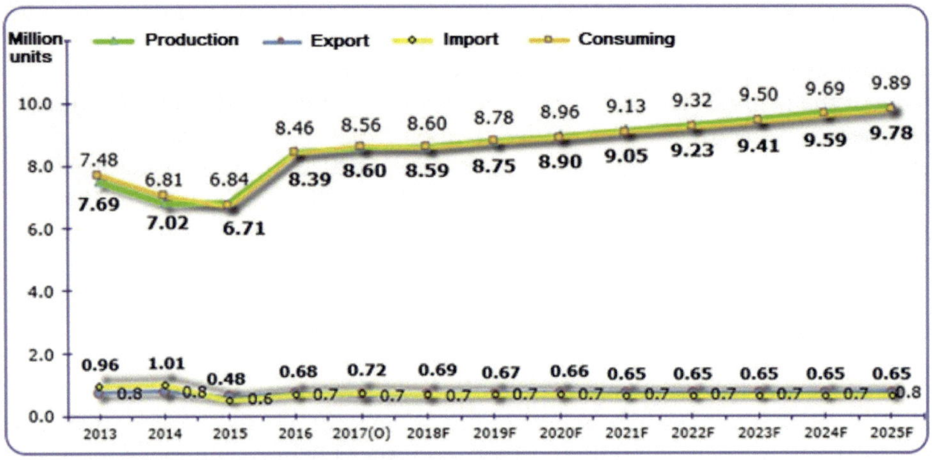

Figure 1.
Dynamics and structure of consumption of electricity meters in 2013–2017 and forecast until 2025 in million units (within the base development scenario).

and communal services, which is almost completely covered by domestic production [34] (**Figure 1**).

In 2018, the Federal Law No. 522-FZ "On Amendments to Certain Legislative Acts of the Russian Federation in Connection with the Development of Electricity (Power) Metering Systems in the Russian Federation" emphasizes that "the subjects of the electric power industry, consumers of electric energy (capacity) and other owners of electrical energy metering devices are obliged to carry out information exchange of data obtained in the course of providing commercial accounting of electrical energy (capacity) in retail markets and for the provision of utility services for electricity supply, necessary for mutual settlements for the supply of electrical energy and capacity, as well as for those associated with these supplies services, free of charge in the manner prescribed by the rules for the provision of utility services to owners and users of premises in apartment buildings and residential buildings, established in accordance with housing legislation, the rules for organizing electricity metering in retail markets" [35]. The norms of the Federal Law are aimed at unifying the various requirements for presenting a huge amount of information, a standard minimum set of functions for the intelligent electricity metering system, as well as the metering devices themselves.

The retail electricity market in the Russian Federation is represented by various participants: generating, grid, sales and management companies, settlement centers and ordinary consumers of electricity. Disagreements often arise between market participants: for example, between grid and sales companies on the volume of productive supply and the amount of electricity losses. Similar disagreements may arise between management companies and consumers due to the rather complex and non-transparent process of accruing volumes for general house needs (lighting entrances, operation of elevators, etc.). In addition, disagreements may arise due to different periods and methods of taking readings from metering devices. All this leads to disputes, often they are not resolved for a long time. The uncertainty of local data has a negative impact on financial calculations and, in general, on the energy market.

Large-scale work has begun in the Russian electric power industry to ensure transparency of payments and reduce non-payments, using the possibilities of digitalization. Thus, a group of companies PJSC "Rosseti" (Rosseti) is implementing a comprehensive digital transformation program, including the design and implementation of pilot projects to test the latest blockchain technologies [30].

A resident of Skolkovo, a company from Yekaterinburg "B41 Blockchain Development", developed by order of Rosseti an automated system for accounting and payment of electricity based on a completely domestic blockchain platform—Nodes Plus Blockchain. The project partner was Sberbank PJSC. The financial institution, acting as a payment bank, provided the server capacities of the SberCloud platform.

In 2019, the developer successfully tested a system for collecting information based on the blockchain platform and presented the results to Rosseti. The project included 12.5 thousand metering devices for electricity, cold and hot water on the basis of two residential complexes in Yekaterinburg [36].

The implementation of systems, based on blockchain technology, in the Russian power industry is focused on obtaining a number of benefits. Decentralization is assumed, when all network participants are directly involved in maintaining the system's performance, and there is not a dedicated center. Automatic distribution of data among its participants guarantees safety and immutability of the information, entered into the blockchain. Transaction transparency is ensured; all participants have access to the entire history of their transactions, up to the very first one. The speed of transactions increases, since they occur directly between users, regardless of their location and without involvement of intermediaries. Additionally, the absence of intermediaries reduces transaction costs.

6. Problems and risks

Naturally, there are some problems and limitations, associated with the use of the blockchain. One of them is the limited memory of the blockchain; for example, the memory size of the Bitcoin blockchain in mid-June 2018 was 171 GB. Another matter is a decrease or even lack of confidentiality, because each network user has a wallet address, assigned personally to him or her, and all network members can trace, what transactions were made from it.

In some cases, the use of blockchain is associated with significant energy consumption. Thus, according to a number of data, by the end of 2018, the Bitcoin mining took about 0.5% of the total energy consumption of the entire planet. In addition, one more weakness is the possibility of significant delays in confirming transactions due to problems with mempools (queues of transactions waiting for confirmation by miners in the network) [37].

At the technical level, breakthroughs are needed. The problem of network bandwidth is acute, since the blockchain copies all the information generated earlier into new blocks, so the amount of information in the next block is greater than in the previous one. Potentially, the amount of information in the blockchain tends to infinity, and eventually this problem will have to be solved once by revising the blockchain model.

One of the advantages of the blockchain—its consistent decentralization, creates certain difficulties. Decentralization hinders government oversight and, in many countries, it hinders the development of technology. Regulatory (in particular, fiscal) authorities should recognize the technology. To win their trust, it is necessary to make some efforts, both technological and in PR.

Blockchain transactions, like all commercial transactions, are associated with certain risks, and they have a number of significant features in this special case. These features of risks should be taken into account, when a business, related to the use of blockchain, is insured. In order to pay the insurance indemnity, the insurance company requires data, confirming the insured event, and checks the accuracy of this information. The big data and blockchain technologies should provide such possibility; corresponding data should be recorded, protected from falsification, stored indefinitely, and available to the insurer.

These properties, if it they reliably ensured, create new opportunities for the insurance business. Insurance information can be recorded on the blockchain, providing technical opportunity for data exchange between insurance companies that have joined the blockchain. Such data as customer's insurance history, claims records, etc. allows to assess better the level of risk for a particular customer. The exchange of information can be organized more widely, for example, between insurance companies and hospitals to prevent health insurance fraud. The use of network technologies creates opportunities for more accurate insurance analytics, which facilitates the creation and maintenance of insurance contracts.

7. Conclusion

The article discusses the main possibilities of using blockchain technology in the energy sector and proposes a general scheme of blockchain networks for information support of the energy market (primarily, the electricity market). One of the main functions of the blockchain is to ensure the trust of market participants in each other, due to the clear recording of all actions, the security of information and the transparency for participants of transactions [38]. Market actors have a high degree of autonomy in purchases and sales. The technology provides services for the preparation and issuance of invoices, as well as the organization of settlements for the entire transaction. It ensures the economic and information security of the participants in the transaction, as well as the efficiency of its implementation. Weaknesses and risks, associated with the use of blockchain in the economy, are analyzed and ways to overcome them are discussed.

Like any technology, blockchain is constantly evolving. Further research can be carried out in several directions [12].

Among the algorithms included in the blockchain, the consensus protocol, which is used to coordinate operations between individual nodes in distributed systems, is in particular need of further development. This is a key technology in the blockchain, which should be in line with the requirements of reliability (tolerance to failures and malicious attacks) and efficiency (high speed of convergence, that is, the speed at which the system reaches consistency or "steady state" [39]). The biggest challenge in implementing consensus agreements is to ensure a balance between safety and efficiency; currently, the speed of consensus algorithms (when organizing complex transactions) needs to be improved.

The underperformance of blockchain networks, in course of growing, is another serious problem. One solution is to use increasingly modern high-performance computers. However, an even more significant factor is the use of special intelligent software tools. According to the OpenAl survey, due to introducing and sequential training of artificial intelligence systems, the real computing power of computer networks has been growing exponentially since 2012, and the data processing speed has doubled every 3.5 months (much faster compared to Moore's law for processors, according to where the speed of computing doubles every 18 months). A convenient tool that goes well with blockchain technology is the intelligent the blockchain technology is an intelligent self-learning network agent—an avatar [7, 17], which automatically, during periods of low network congestion, processes data from the blockchain and generates proposals for concluding transactions.

Also, to improve performance, work is underway to improve network protocols, specially "sharpened" for the blockchain architecture.

The third problem, the optimal solution of which should be obtained in the course of further development of the technology, is to ensure the dynamic structural flexibility of blockchain networks. Three main models are currently in use: public, corporate, and private networks. The public chain is open to all users, the corporate chain serves a specific alliance of users, and the private chain is used by only one legal entity. Different blockchain products are being developed for the different models. At the same time, there is a great need to provide support both inter-network interaction and changing the model of a particular network: its decentralization in the transition to a public model, or centralization in the opposite case.

Business interactions between different organizations (i.e. interactions between different chains) can be a big problem. Accordingly, "cross-chain technologies" are currently being developed to solve this problem, and it is one of the main directions of blockchain development nowadays. The collected experience shows that, apparently, it is impossible to solve the problem without elements of centralization in the issue of harmonizing protocols between blockchains. Either generally recognized (on state or corporate level) standards of blockchain input/output interfaces are needed, or universal state-supported networks should be designed to provide a friendly information environment for interaction between separate blockchains.

An example is the HKTFP network, maintained by the Hong Kong Monetary Authority. The advantage of the model is that the public platform allows to objectively resolve information conflicts between blockchains, and also supports both a simple user registration and a system of basic business scenarios. It is important that such an environment helps to solve the above-mentioned problem of interaction blockchains with fiscal and other central authorities.

Finally, the blockchain management mechanism should be improved; there are all prerequisites for this. The blockchain contains the necessary logic to automate the voting process (concerning updating network rules, changing protocols, distributing quotas, determining the operator of the blockchain supernode, etc.). The standard model is the "chain voting", but its details vary considerably across networks, and this, again, makes difficult the inter-networking coordination. Significant work remains to be done to develop a detailed and most effective standard.

In the development of technologies for efficient and reliable data storage, the quantum model can become the main competitor of the blockchain technology. However, its future is largely connected with the success of the widespread introduction of quantum computing tools. With the existing models of computer devices, the blockchain is still out of competition, so we can expect new successes in the development of this technology.

Acknowledgements

In remembrance of Prof. Serge Chernyshenko who tragically passed away before the book's completion, he brought tremendous knowledge, passion, and dedication to this project, and his absence is profoundly felt by all of the authors here.

Author details

Serge Chernyshenko[1*†], Valentin Afanasyev[1], Vardan Mkrttchian[2]
and Vsevolod Chernyshenko[3]

1 State University of Management, Moscow, Russian Federation

2 HHH University, Sidney, Australia

3 Financial University under Government of Russia, Moscow, Russian Federation

*Address all correspondence to: serge.v.chernyshenko@gmail.com

†Deceased.

References

[1] Mougayar W. The Business Blockchain: Promise, Practice, and Application of the Next Internet Technology. London: Wiley; 2016. p. 214

[2] Namasudra S, Deka GC, Johri P, Hosseinpour M, Gandomi AH. The revolution of blockchain: State-of-the-art and research challenges. Archives of Computational Methods in Engineering. 2020;**28**:1497-1515

[3] Lu Z. Expectation: Innovative application of financial technologies in the banking industry. Chinese Financial Computer. 2018;**1**:10-13

[4] Mkrttchian V, Gamidullaeva L, Finogeev A, Chernyshenko S, Chernyshenko V, Amirov D, et al. Big Data and Internet of Things (IoT) Technologies' Influence on Higher Education: Current State and Future Prospects. International Journal of Web-Based Learning and Teaching Technologies (IJWLTT). 2021;**16**:137-157. DOI: 10.4018/IJWLTT.20210901.oa8

[5] Kizabekova A, Chernyshenko V. E-government avatar-based modeling and development. In: Mkrttchian V, Aleshina E, Gamidullaeva L, editors. Avatar-Based Control, Estimation, Communications, and Development of Neuron Multi-Functional Technology Platforms. Hershey: IGI Global; 2020. pp. 19-34

[6] Mkrttchian V, Kharicheva D, Aleshina E, Chernyshenko V, Gamidullaeva L, Panasenko S, et al. Avatar-based learning and teaching as a concept of new perspectives in online education in Post-Soviet Union Countries. International Journal of Virtual and Personal Learning Environments. 2020;**10**:66-82

[7] Di Silvestre M. Blockchain for power system: Current trends and future applications. Renewable and Sustainable Energy Reviews. 2019;**19**:109-131

[8] Ya AV, Lyubimova NG, Ukolov VF, Shayakhmetov SR. Impact of blockchain technology for modification of the supply chain management in energy markets. International Journal of Supply Chain Management. 2020;**9**(3):757-762

[9] Andoni M. Blockchain technology in the energy sector: A systematic review of challenges and opportunities. Renewable and Sustainable Energy Reviews. 2018;**10**:143-174

[10] Chernyshenko SV, Ruzich RV. Energy and information aspects of self-organisation of ecological systems: Mathematical models and interpretations. In: Proceedings of 29th European Conference on Modelling and Simulation. ECMS-2015. Albena (Varna), Bulgaria; 2015. pp. 94-99

[11] Hao Y, Li Y, Guo Y, Chai J, Yang C, Wu H. Digitalization and electricity consumption: Does internet development contribute to the reduction in electricity intensity in China? Energy Policy. 2022;**164**:112912

[12] Afanasyev V, Chernyshenko V, Kuzmin V, Voronin V, Mkrttchian V. Advanced information technology for development of electric power market. International Journal of Advanced Manufacturing Technology. 2021;**115**:2-4. DOI: 10.1007/s00170-021-07324-8

[13] Wu H, Hao Y, Ren S. How do environmental regulation and environmental decentralization affect green total factor energy efficiency:

Evidence from China. Energy Economics. 2020;**91**:104880

[14] Brynjolfsson E, Kahin B. Understanding the digital economy: Data, tools, and research. Journal of Documentation. 2003;**59**(4):487-490

[15] Mkrttchian V, Chernyshenko SV. Organizational knowledge of digital economy in transformation, in Big Data, and in Internet of Things. In: Encyclopedia of Organizational Knowledge, Administration, and Technology. Hershey, USA: IGI Global; 2020. pp. 463-476. DOI: 10.4018/978-1-7998-3473-1.ch035

[16] Wenbin U. Principles, models and proposals of banking trading blockchain. Hebei University Journal (Edition on Philosophy and Social Sciences). 2015;**6**:159-160

[17] Mkrttchian V, Chernyshenko S, Aleshina E. Avatar-based control and development of neuron multi-functional platforms for transformation processes in the digital economy. In: Mkrttchian V, Aleshina E, Gamidullaeva L, editors. Avatar-Based Control, Estimation, Communications, and Development of Neuron Multi-Functional Technology Platforms. Hershey: IGI Global; 2020. pp. 231-247

[18] Asharaf S, Adarsh S. Introduction to blockchain technology. In: Decentralized Computing Using Blockchain Technologies and Smart Contracts: Emerging Research and Opportunities. Hershey: IGI Global; 2017. pp. 10-27

[19] Mkrttchian V. Avatars-based decision support system using blockchain and knowledge sharing for processes simulation a natural intelligence: Implementation of the multi chain open source platform. International Journal of Knowledge Management. 2020;**17**:72-92

[20] Kshetri N. Blockchain's roles in meeting key supply chain management objectives. In: International. Journal of Information Management. 2018;**39**:80-89

[21] Kamble SS, Gunasekaran A, Subramanian N, et al. Blockchain technology's impact on supply chain integration and sustainable supply chain performance: Evidence from the automotive industry. Annals of Operational Research. 2021. DOI: 10.1007/s10479-021-04129-6

[22] Saberi S, Kouhizadeh M, Sarkis J, Shen L. Blockchain technology and its relationships to sustainable supply chain management. International Journal of Production Research. 2019;**57**(7):2117-2135

[23] Kupriyanovsky VP, Konev AV, Grinko OV, Pokusaev ON, Namiot DE. On the way to the energy internet: New regulations, business models, economic and technical conditions. International Journal of Open Information Technologies. 2019;**7**:3

[24] Reiger A. Building a blockchain application that compiles with the EU general data protect regulations. MIS Quarterly Executive. 2019;**18**:263-279

[25] Transition of the Estonian energy market to blockchain and other global initiatives to introduce technology. [Internet]. Available from: https://forklog.com/blockchain-initiatives-review-181027 [Accessed: January 27, 2020]

[26] Estonia's energy market tokens using blockchain. Available from: http://digitalsubstation.com/blog/2018/10/30/energorynok-estonii-tokeniziruyut-spomoshhyu-blokchejna/ [Accessed: January 27, 2020]

[27] The world's first energy blockchain lab was established [Internet]. Available from: the world's first energy blockchain

laboratory was established-polaris wind power grid (bjx.com.cn) [Accessed: January 21, 2022]

[28] 2020 China's car ownership data: 4.92 million new energy vehicles [Internet]. Available from: https://baijiahao.baidu.com/s?id=1688291565032294902&wfr=spider &for=pc [Accessed: September 7, 2021]

[29] Federal Law No. 522 of December 27, 2018 "On Amending Certain Legislative Acts of the Russian Federation in Connection with the Development of Electricity (Power) Metering Systems in the Russian Federation" [Internet]. Available from: http://www.consultant.ru/document/cons_doclAW_314661/ [Accessed: January 27, 2020]

[30] The development strategy for the group of companies of PJSC Rosseti until 2030 [Internet]. December 27, 2019. Available from: https://www.rosseti.ru/press/news/index.php?ELEMENT [Accessed: January 27, 2020]

[31] Chernyshenko S, Simonov A. The Russian market of mobile location based service. Industrie Management. 2012;**2**:51-54

[32] Chausov I, Kholkin D. In anticipation of the dreadnought: p2p-energy is looking for its main caliber [Internet]. Available from: https://medium.com/internet-of-energy/e97c252aef8a [Accessed: January 27, 2020]

[33] The Russian market of M2M / IoT (inter-machine communications and the Internet of Things). Results of 2018, forecast until 2022 [Internet]. 15 May 2019. Available from: http://json.tv/ict_telecom_analytics_view/rossiyskiy-rynok-m2miot-mejmashinnyh-kommunikatsiy-i-interneta-veschey-itogi-2018-g-prognoz-do-2022-g-20190515023757 [Accessed: January 27, 2020]

[34] The market of electricity meters in Russia continues to grow. Electrical, electronics and optics [Internet]. June 6, 2018. Available from: http://www.indexbox.ru/news/rynok-schetchikov-ehlektroehnergii-prodolzhaet-rasti/ [Accessed: January 27, 2020]

[35] Federal Law No. 522 of December 27, 2018 "On Amending Certain Legislative Acts of the Russian Federation in Connection with the Development of Electricity (Power) Metering Systems in the Russian Federation" [Internet]. Available from: http://www.consultant.ru/document/cons_doc_LAW_314661/ [Accessed: January 27, 2020]

[36] RUSNET launches a pilot blockchain project for billing in the Urals, Yekaterinburg. [Internet]. Dec 16, 2019. Available from: https://ekb.rbc.ru/ekb/freenews/5df772cc9a794781fbf5e65a [Accessed: January 27, 2020]

[37] Pereira J, Tavalaei MM, Ozalp H. Blockchain-based platforms: Decentralized infrastructures and its boundary conditions. Technological Forecasting and Social Change. 2019;**146**:94-102

[38] Vyalkov D. When everyone trusts everyone. Blockchain in the electric power industry. Expert Ural. 2019;**1-3**:816

[39] Belozyorov VY, Chernyshenko S. Quadratic model of inter-population interaction: Investigation of stability areas. Applied Mathematics and Computation. 2014;**230**:43-56

Section 5

Blockchain in Health Science

Chapter 10

Blockchain from a Modern Perspective: An Evolution to Health Science

Aryan Chaudhary, Keshav Kaushik and Sunil Kumar

Abstract

This paper gives a thorough evaluation of the literature on blockchain applications in healthcare. The evaluation included 42 papers that presented current information on the existing implications and gaps in the usage of blockchain technology for enhancing healthcare systems. According to the SLR results, blockchain is being utilized to produce unique and sophisticated solutions to enhance the prevailing standards of medical data handling, sharing, and processing. In the healthcare business, blockchain technology is experiencing conceptual evolution, adding considerable value through enhanced efficiency, access control, technical innovation, privacy protection, and data management process security. The findings also indicate that the current limits are mostly related to model performance, as well as the constraints and costs involved with implementation. An integrated approach is offered to cover prospective areas where future researchers might bring considerable value, such as regulatory compliance, system architecture, and data protection. Finally, the SLR believes that further research can help to enable the wider implementation of blockchain applications to handle crucial challenges like as medical diagnostics, legal compliance, preventing fraud, and enhancing patient care in remote monitoring or medical emergencies.

Keywords: healthcare, biomedical, data sharing, medicine, distributed ledger technology, distributed systems, health information exchange, security, interoperability, transparency, privacy

1. Introduction

Blockchain was at first proposed as a technique to fuel Bitcoin however has since extended to where it is currently alluded to as a basic innovation for different decentralized applications. Blockchain is being advanced as a suitable device for taking care of delicate information, especially in the medical services, clinical exploration, and protection ventures. Medical services might be seen as a framework made out of three significant parts: (a) focal providers of medical care, like doctors, medical caretakers, emergency clinic organizations, and specialists; (b) basic administrations engaged with clinical benefits, like clinical exploration and health care coverage; and (c) essential recipients of clinical and wellbeing administrations, like patients or the

IntechOpen

overall population. In this review, we characterize the medical services framework as contact-based and innovation based remote checking administrations offered by constituent support suppliers with an end goal to advance, keep up with, or recuperate recipients' wellbeing Privacy and security breaks in medical services are supposedly rising many years, with more than 300 breaks kept in 2017 and 37 million clinical data affected somewhere in the range of 2010 and 2017 [1]. The developing digitalization of medical services has likewise raised worries about the safe stockpiling, possession, and sharing of people's very own wellbeing records and associated clinical information. Blockchain has been proposed as an answer for significant medical care concerns, for example, secure clinical record trade and consistence with information protection regulation. Blockchain has been proposed as an answer for significant medical services concerns, for example, secure clinical record trade and consistence with information protection regulation.

For instance, utilized bibliometric philosophies to offer an outline of blockchain perspectives and exploration patterns connected with the utilization of blockchain in medical care. Given an exhaustive portrayal of the numerous frameworks that have been made to use blockchain in medical care. Investigated many cases of blockchain innovation use in medical services, the issues experienced, and expected arrangements. Analyzed compromises and plan choices made by analysts in a few circumstances including blockchain innovation.

Security, protection, and adaptability of wellbeing records capacity and sharing are basic in the medical care business and clinical offices since right information is expected for conclusion and appropriate wellbeing decisions concerning the patient's condition. The information provided by clinical specialists for sufficient patient follow-up should be moved with security and should be cutting-edge as per the patients' wellbeing conditions. Telemedicine and e-wellbeing administrations are likewise the main spaces of medical services in which information is imparted to mind suppliers in distant regions to analyze and treat patients using media transmission innovation. Telemedicine is a remarkable progression that has turned into a fundamental part of medical services framework. Because of the touchy information of patients, security, responsiveness, and protection in these web-based patient checking frameworks may be an enormous concern. Along these lines, solid and secure information exchanges are firmly connected with sound and exact correspondence among patients and clinicians.

Moreover, interoperability between the medical care business and various exploration bunches is the most troublesome hindrance to defeat for effective and safe information exchange. The cutoff points to sound collaboration of gatherings taking part in information trade incorporate a variety of medical care records, different information sharing arrangements, information responsiveness, and certain moral guidelines, which must all be tended to when useful information is exchanged among various associations. Many kinds of examination have been directed lately, and current answers for the medical services industry are proposed utilizing state of the art advancements like Artificial Intelligence, Internet of Things (IoT), AI, profound learning, and PC vision for more effective and powerful medical services framework to help people. Among these technologies, blockchain technology has had a significant influence on providing more safe and secure healthcare infrastructure, such as healthcare record-keeping services, biomedical fields, medical supply chain management, telemedicine, genetic data, and e-health data sharing services. Blockchain is the buzzword of the year, and this technology is gradually making its way from finance to supply chain logistics. Blockchain in healthcare infrastructure gives significant opportunity to drive digital transformations of medical records, pharmaceutical supply chains, smart contracts for payment distribution, and a plethora of other techniques to exploit this technology in the healthcare business (**Figure 1**).

The consensus mechanism approves all transactions in the blockchain network. The network's consensus serves as an agreement among all nodes to validate new blocks in the chain. Blockchain, on the other hand, is a system in which numerous linked computers maintain a secure and up-to-date distributed ledger with no central authority. **Figure 2** depicts the peer-to-peer blockchain transaction network with a flow diagram:

Blockchain offers the capacity to defend medical care and clinical information by utilizing the dispersed organization's security and protection abilities. Patients benefit from clinical and clinical information saved money on a safe blockchain for regular treatment and subsequent meet-ups with their doctors, regardless of whether they are in distant districts. Patients' delicate information becomes alright for their security and reliable for doctors to further develop medical care applications and diagnostics.

The reception of blockchain innovation as a stage for trading clinical information among clinical subject matter experts and medical care suppliers with better protection and security is supported. Vora et al. [2] introduced a blockchain component as a progressive design for safeguarding and keeping patients' electronic wellbeing records to save delicate information and handle significant information security worries by utilizing a blockchain programming foundation all through an emergency clinic framework. Blockchain innovation has additionally exhibited its worth in biomedical examination, clinical information fields, and clinical production network the board. Clinical arrangements, plans, and conventions might be saved money on a blockchain network prior to starting a clinical assessment, and delicate information associated with clinical preliminaries becomes state-of-the-art, secure, and openly accessible. To keep up with network straightforwardness, brilliant agreements on the blockchain might be introduced, duplicated, and executed at different phases of clinical review. A blockchain-based telemonitoring framework for the identification and therapy of malignant growth cancers for patients in remote spots was introduced in [3]. Savvy contracts and blockchain were utilized in the proposed system to guarantee the realness and security of patient information at specific clinical foundations as well

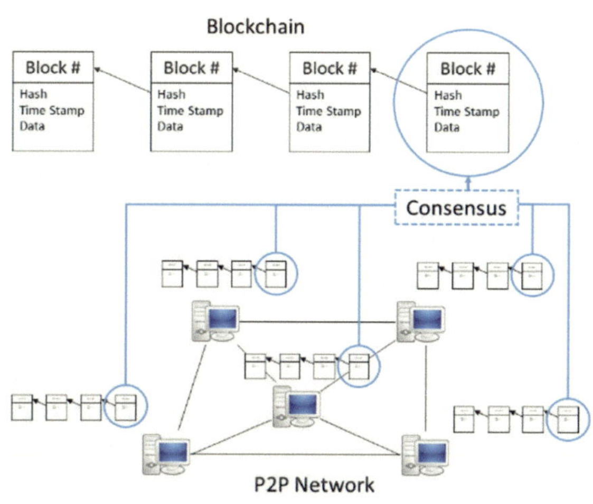

Figure 1.
Key elements of blockchain.

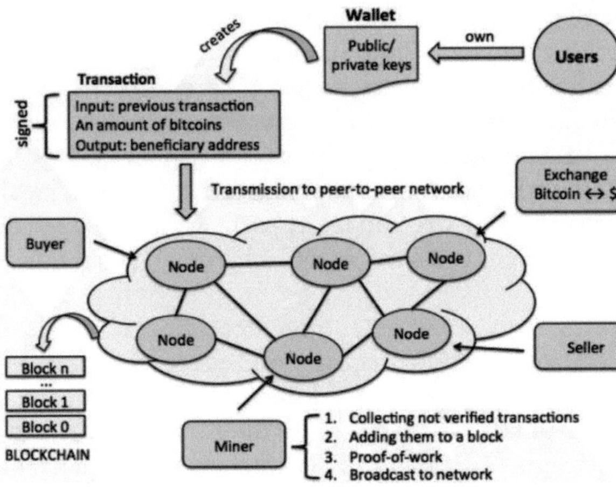

Figure 2.
P2P transaction flow network.

as at their homes. In [4], a blockchain-based framework named DermoNet is proposed for helping dermatology patients with online dermatological meeting systems by means of tele dermatology observing. Proactive Elderly is a blockchain-based network introduced in [5] to help maturing individuals' dynamic living. Blockchain innovation might be an amazing and appropriate choice for complex clinical therapy cycles like persistent diseases, surgeries, and maturing. Besides, drug organizations, medication creators, and biomedical scientists are utilizing blockchain to carry out complex hereditary analysis in the medical care market.

2. Applications of blockchain in healthcare

Blockchain innovation was at first utilized in banking and monetary applications as cryptographic forms of money, yet its true capacity has now extended to incorporate medical services and natural applications [6]. Electric Health Records (EHR), clinical production network the board, clinical innovative work, genomic market, drug medication, neuroscience studies, biomedical turn of events, and Telemedicine and E-Healthcare are instances of how blockchain innovation is being utilized in the medical care space. Blockchain gives a solid and secure framework for putting away and sharing information in all subdomains of the medical care business, permitting doctors and medical care suppliers to involve recorded information for a wide range of exchanges and tests. The utilization of blockchain in the medical care business is depicted beneath considering a few late examinations in the field and outlined in **Figure 3**.

2.1 Electronic health records (EHR)

The interest for medical services record digitization has developed over the course of the past 10 years as clinical experts and medical services suppliers need basic admittance to patient information for expedient direction. Electronic wellbeing records (EHR) are advanced renditions of information that approved medical care experts might access out of the blue with more prominent security. The electronic wellbeing record (EHR) contains a patient's clinical history, conclusion reports,

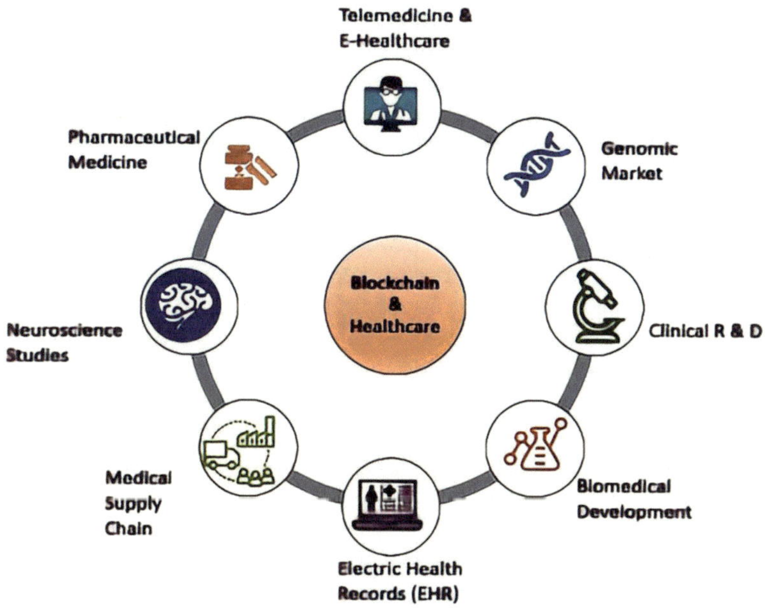

Figure 3.
Applications of Blockchain in the medical industry.

remedy information, treatment plans, research facility and experimental outcomes, etc. The most predominant utilization of blockchain in medical care frameworks is to work on the security and unwavering quality of electronic wellbeing records.

Indeed, even before blockchain innovation, the imperative of electronic wellbeing records was that patient information was scattered across different medical care suppliers in view of the patient's conditions, and earlier information was not accessible even in EHR frameworks. Numerous scholastics have proposed blockchain as a clever methodology for putting away patients' EHRs to keep their set of experiences and present data secure for a lifetime and retrievable whenever.

A model "MedRec" was created in [7] for conveying a modern and permanent record history of patients, with particular blockchain benefits for taking care of validation, trustworthiness, security, privacy, and sharing of information records. This model works on a decentralized administration framework and ensures that patients have a permanent and effectively open history of their clinical information, which they can basically trade with their different medical care suppliers for therapy. With a few limitations while taking on records with EHR, for example, loss of command over records, information provenance concerns, and information following, clinical information sharing could become pivotal. To address these issues, [8] launched MeDShare, a blockchain-based platform for the safe and trustless sharing of data and electronic health records among various cloud service providers, healthcare professionals, and research institutions with better data provenance and privacy.

2.2 Healthcare medical supply chain

In the medical care area, there is an earnest requirement for more successful production network the board. Coronavirus showed significant defects in the United States' medical care supply chains, with emergency clinics compelled to manage deficiencies of individual defensive hardware (PPE), ventilators, and urgent drugs. Besides, medical organizations have been feeling the squeeze lately to expand productivity and address rising store network costs. Truth be told, it has been expressed that reinforcing emergency clinic store network the executives might save emergency clinics generally $12 million every year. Medical clinics, medical services suppliers, drug organizations, and different partners in the medical services area can further develop medical services production network the executives to not just dispose of store network shortcomings and superfluous production network costs, yet additionally to incorporate dexterity and strength into the medical care esteem chain. With a more adaptable and strong medical care inventory network, upstream and downstream accomplices will actually want to answer all the more successfully to unexpected occasions like regular fiascoes or a worldwide pandemic, working on persistent results and saving lives.

2.3 The importance of healthcare supply chain management

The essential objectives of medical services store network the executives are to further develop perceivability and proficiency all through the inventory network, and as of late, this has extended to incorporate the essential objective of further developing store network dexterity and strength — fundamental in these seasons of expanded vulnerability and unpredictability in both organic market conditions. At the point when wellbeing inventory network the board is done accurately, supply accomplices are better prepared to perceive and deal with bottlenecks, potential interferences, and different worries that might emerge anyplace along the store network. It can possibly work on tolerant consideration and security while additionally diminishing waste and inefficient cost. Checking and supporting the progression of prescriptions, clinical supplies and hardware, and clinical benefits from maker to patient are the significant errands associated with medical services inventory network the board.

Inventory network quality administration and arranging, production network mechanization and advancement, and provider relationship and chance administration are all important for it. At the point when organizations utilize advanced instruments and innovation to oversee medical services supply chains, they utilize significant bits of knowledge got from multi-source information to persistently adjust and improve production network frameworks and cycles.

One part of the drug climate where blockchain may profit from its one of a kind characteristics and center ideas is the prescription production network. While embracing innovation driven arrangements and use cases, the drug production network is given extraordinary thought. The overall fake and phony market is worth up to $200 billion every year, and fake prescriptions are a significant wellspring of concern on the grounds that the worldwide underground market offers such items to people without moral endorsement. The danger to human existence presented by fake drugs and medicines is developing and ought not be disregarded. Duplicating is a major issue in the drug area, and a substitute cure should be created. Blockchain innovation can assist the medical care industry with further developing its store network by decreasing extortion involving against duplicating advancements and working on quality control in the assembling and dissemination of drug things and prescriptions.

Blockchain-based methods can follow the entire life expectancy of meds and medicines, down to a solitary portion, from assembling to dispersion. Blockchain innovation for safe computerized stamping of drug things is now being utilized by firms like as Block pharma, TIERION, CHRONICLED, and the Centers for Disease

Control and Prevention to offer a solid and dependable track of drugs all through their life expectancy (CDC).

2.4 Improving healthcare supply chain management using digital technology

A developing number of firms are carrying out inventive innovation to help and change medical services production network the board. Coronavirus has started computerized change projects inside the drug area and then some. There are currently imaginative and easy to-utilize arrangements accessible to help undertakings in separating information storehouses, a well established issue in medical care, and creating applications that permit them to interface production network information with clinical information.

The advanced stock organization, or cloud-based production network organizing innovation, makes it simpler for organizations to associate their production network frameworks to electronic wellbeing records (EHR) and other medical services frameworks. Firms can normalize, smooth out, and robotize tedious business processes, bringing down costs while working on tolerant results, and laying out a clinically incorporated inventory network that shows restraint focused and proactive by interfacing inventory network information with electronic clinical records and other related informational collections.

Multienterprise work the board programming is one more important innovation for medical services inventory network the executives. These applications empower patient-driven organization all through the medical services esteem chain. They help groups in teaming up across interior and outer areas to better knowledge into store network tasks and further develop store network flexibility. Through the trading of expressed needs and designated commitment, they help supply accomplices in creating trusted and commonly fruitful business organizations. Exchanging accomplices might share information and knowledge by means of normalized, secure, and consent passages and team up to handle the issues influencing medical care supply chains today.

3. The benefits of healthcare supply chain management digitization

The connected store network, or inventory network 4.0, is solid and versatile, able to do quickly answering changing circumstances and recuperating from surprising interferences. Medical services organizations are carefully modernizing medical care production network the board by embracing the most current medical services store network innovation, simplifying it for themselves as well as their exchange accomplices to streamline and improve the production network. They are using this innovation to acquire continuous, definite access into their stock chains from starting to complete the process of, permitting them to recognize and address potential hardships and obstacles early and turn quickly on a case by case basis.

They might use medical services store network investigation to all the more accurately gauge interest, advance stock preparation and the executives, and answer all the more successfully to changing economic situations by joining inventory network and clinical information. Computerized inventory network the executives innovations can help raise efficiency and lessen time to advertise by advancing expanded correspondence and joint effort all through the medical services production network. Besides, by upgrading trust and straightforwardness all through the stock organization, they further develop store network flexibility, aiding the production of a medical services esteem chain that can more readily answer and recuperate from future pandemics and other general wellbeing disasters.

Work the executives for provider networks across a few ventures. Resolve provider episodes up to 80% faster.

3.1 Healthcare supply chain management in the future

A computerized production network organization, intended to deal with the intricacy of medical services production network the board, gives associations in this quickly developing area with the straightforwardness and start to finish perceivability they expect to deal with a convoluted and extending organization of exchanging accomplices by means of constant data streams, specialized devices, and cooperative work processes. It can help them in distinguishing and settling organic market concerns right off the bat, before patients and their primary concern endures. Since it is cloud-based, network individuals benefit from the versatility and adaptability expected to incorporate and utilize mechanical cycle computerization, man-made consciousness and AI, prescient investigation, and other state of the art store network 4.0 advancements.

Work the board programming likewise helps firms in producing cooperative advancement and creating patient-driven inventory network organizations. Supply accomplices might team up more successfully to organize patient results in original ways and make new advanced working models for the fate of medical care. Medical organizations might utilize work the board instruments and a computerized supply organization to ensure that all patients get the treatments they require, when and when they require them. Individuals from the organization are engaged to team up for everyone's best interests, and worldwide populaces benefit from expanded admittance to life-saving drugs and better treatment made conceivable by computerized medical services store network the board.

3.2 Blockchain technology could enhance the caliber of clinical research

Since it empowers information recording, sharing, and care, blockchain can possibly impact clinical examination. Without a doubt, it involves a decentralized secure global positioning framework for any information connections that might happen with regards to clinical preliminaries, as well as a distributed comprehensive organization that empowers information sharing on the examination side while likewise guaranteeing all vital straightforwardness and protection worries on the patient local area side.

Therefore, this technique might encourage more confidence in clinical exploration, whose standing has been seriously hurt by ongoing scandals [9, 10]. Blockchain innovation might be seen as an establishment for improved clinical examination strategies as well as a push toward more transparency to construct certainty inside research networks and among exploration and patient networks.

3.3 Building reliable clinical studies: At each step, keep track and timestamp

Information sacredness and trustworthiness are two crucial qualities of information at the utilitarian level, "the information level." as far as information sacredness and accuracy, Blockchain guarantees that occasions are kept in their exact sequential control, significantly forestalling deduced reproduction study.

To begin with, the cryptographic approval of every exchange guarantees information honesty [11]. This is basic for guaranteeing information truthfulness — forestalling information creation, information "beautification," and, here and there, information innovation. Second, one of the principal abilities of Blockchain is information discernibility and accuracy: every exchange is timestamped [12]. This information is available to people in general; any client can get a duplicate of the time-stepped information. **Figure 4** portrays the mind boggling progressions of heterogeneous information and metadata that circle in a clinical report, proposing countless medical care partners and all records whose presence might be confirmed utilizing Blockchain. Accordingly, the presence of information becomes evident while the information stays private.

Rundown of non-thorough instances of key data that can favorably "sit on the top" of the Blockchain:

- Before the clinical preliminary starts, the information sharing technique, including the schedule, dataset documentation, and information sharing arrangement, if any, should be given with the goal that this data might be timestamped in sequential request in the unfalsifiable Blockchain.

- Assents and clinical preliminary convention, including sort of study, essential and optional results, and consideration and avoidance models, can be bundled into information structures saved money on the Blockchain before the clinical preliminary starts. The information structures are then in coordinated correspondence with assents, the convention, and its changes, giving hearty affirmation of their reality. This element can support the counteraction of normal issues related with non-discernible clinical preliminary conventions, for example, particular announcing of results connected with specific detailing of damage, under-revealing of non-critical results, and confuses between arranged results in the convention and last distribution. These are irrefutable wellsprings of bias. We may moreover record data, for example, the method of information assortment, attribution strategy, dates of withdrawals to separate among ahead of schedule and late ones, and dates of repeating events in the Blockchain metadata set.

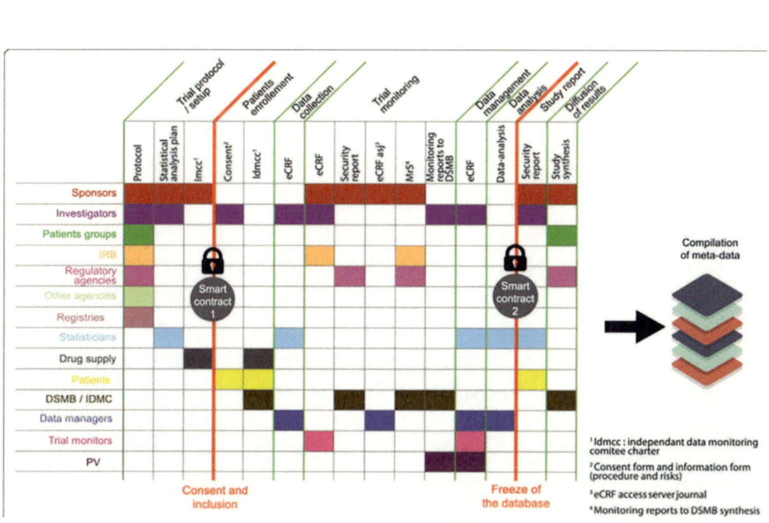

Figure 4.
Clinical trial complex data workflow encoded in Blockchain.

- The measurable examination procedure is fundamental and is timestamped before the investigation is finished and, on account of a dazed preliminary, before the information is unblinded. This technique covers measurable procedure, the meaning of unsafe events, and any various variable alterations. For instance, test size is a significant variable to consider while deciding the force of a review. Consider timestamping an assortment of metadata on the Blockchain, for example, test size, type I and type II blunders, anticipated occasion rate, and treatment effect of revenue. Timestamping will be an achievement in the Blockchain, validating the recently determined example size.

- To stay away from insightful errors [13, 14], the logical code [15] ought to be distributed and disclosed. Considering that contents create and that a decent condition of the code is used to handle the information, this particular condition of the code should be "frozen" and timestamped to guarantee that the circumstances under which the information was checked on and broke down are reproducible. Various arrangements consider the cooperative sharing of numerous forms of code, with "git" being by a long shot the most well known.

4. Data sharing and privacy by design in community-driven medicine

At the "local area level," Blockchain is much of the time characterized as "trustless," which could give the appropriate conditions to information trade. As a general rule, trust is innate into the actual convention. Blockchain is a shared framework that is "protection by-design." With how much trust it can impart, it ought to be viewed as an introduction to the time of local area driven approaches.

Surveys regularly uncover that more than 80% of customers are glad to share their clinical data [16], as long as protection and security are protected. Users do not need an outsider to trust the framework because of the receptiveness of the Blockchain data set — claimed by nobody, straightforwardly editable by everybody, and with solid crypto-situated consistency of the data set exchange. In this manner, the data set gives a wide way to information client control or differential security, information sharing, and local area driven clinical examinations [15]: in a believed climate, clinical exploration groups can "swarm select" individuals to be signed up for conventions utilizing local area the board strategies, and individuals can likewise elect to take part in such examinations.

4.1 Smart contracts for clinical trial phase control

We might chain together unmistakable clinical preliminary stages with the end goal that each step is subject to its ancestor, as well as protecting clinical preliminary stage compilable metadata on the Blockchain. Blockchain innovation, known as Smart Contracts, give devices to these "cutting" and "tying" processes, and can guarantee a degree of straightforwardness, recognizability, and command over clinical preliminary successions.

"Savvy Contracts are PC conventions that straightforwardness, confirm, or implement the exchange of an agreement," as indicated by Wikipedia [17], and their execution might be performed utilizing cryptographic hash chains. Shrewd Contracts, by and by, permit the approval of a stage with the main condition that each first step has been totally checked. For instance, the chain of progressive blocks could check that the planned system was followed, and the material introduced to distributers would incorporate the actual distribution as well as the arrangement of blocks that involve the Smart Contract, the right execution of which demonstrates verification that the review was all around led.

Figure 4 portrays a Smart Contract as a line of code which holds an automatic official understanding between however many gatherings depending on the situation, without the contribution of a confided in delegate, and that executes algorithmically as per the terms determined by the contracting parties. Brilliant Contracts empower patient incorporation with the sole condition that they have assented, or information examination with the checked and kept up with that the data set is frozen. Every one of the clinical preliminary stages portrayed in the image can be connected together in the succession portrayed, bringing about a more straightforward review and forbidding deduced remaking or information enhancing.

5. A trial of theory for consent gathering

We used a Blockchain system to acquire participant permission for a clinical trial in a proof-of-concept experimental investigation ([18] (under review), [19]). Indeed, the US Food and Drug Administration reports that nearly 10% of the trials they monitor have issues with consent collection, including failure to obtain written informed consent, unapproved forms, an invalid consent document, failure to re-consent to a revised protocol, and a lack of institutional review board approval for protocol changes [20, 21].

To be more specific, in a fictitious experimental trial, we captured and stored each patient consent on the Blockchain and sought for consent renewal with each protocol amendment. We acquired a unique master document that contains all of the consent gathering data in a single data structure or software package called Chainscript [22], each tied to a version of amended protocol versions. In actuality, these data have been "hashed," or structured into a cryptographic version of the original permission and procedure document data. Because of the rigorous one-to-one correlation between hashed data and effective consent data, this master document represents a secure, strong evidence of existence of the whole consent-collection process. This proof of existence can also be verified on any dedicated public website.

5.1 Blockchain in genomics

Genomic medication opens the universe of hereditary information to give exact conclusion, guess, and treatment of various genetic diseases. A person's hereditary data is assessed utilizing genomic strategies to evaluate ailment defenselessness and applicable treatment decisions for redid medication.

However, the increment of hereditary information raises various issues, including information access, security, and protection. This is where Blockchain enters the image.

5.2 Genetic information

The human genome contains around 3 billion base matches, equivalent to roughly 1.5 GB of material. A few inherited illnesses are unprecedented, while others, like hypertension, diabetes, and Alzheimer's infection, have been connected to a hereditary weakness. Researchers can gain a superior handle of the cycles hidden a large number of these uncommon infections and normal clinical problems by utilizing genomic planning.

The planning information can then be utilized to plan pertinent arrangements and drives. Researchers should gather however much data on sicknesses as could reasonably be expected from people who are impacted by them. Unfortunate information access and interoperability has forever been and keeps on being a worry in medical care. Over 10 years after the human genome project was done, specialized upgrades have made having one's genome sequenced and hereditary data examined extensively more reasonable.

Ten years prior, the expense of genome sequencing would have been $10,000,000. Today, the expense has been altogether diminished to around $1000. This implies how much hereditary information will keep on developing when genome sequencing is economical.

Frequently, hereditary information might not have an obviously perceived proprietor, making it powerless against unpredictable dispersal and raising protection worries for the genuine proprietor. In the absence of viable strategies for security and validation, the gigantic number of accessible hereditary information raises serious difficulties that can be settled by the progressive blockchain innovation.

Blockchain is an openly available report of digitalized exchanges and information put away in blocks. Blockchain lays out a decentralized organization of distributable information that might be traded among interconnected data set frameworks. Blockchain innovation utilizes a timestamped permanent arrangement of information obstructs that are accessible to anyone with a connected framework, anyplace on the planet, for however long they are appropriately approved.

This innovation permits buyers decision over how their information is conveyed, protecting their security. Clients can encode their information utilizing unbalanced encryption (Public Key Cryptography) for expanded security. However, when selling or donating data, the receiver is given a private key to decrypt the data, guaranteeing that no unauthorized third party gets access.

Blockchain enables direct connection between data sources (users) and purchasers (pharma companies, research institutes). Furthermore, cryptographic keys protect users' anonymity throughout data sharing.

These purchasers may then use the data sets to research genetic trends in a certain population, allowing for the creation of medications and other therapies depending on the genetic profiles assessed. Blockchain decreases the danger of data change and manipulation by using cryptographic blocks, offering a genuine database for research and the development of novel diagnoses and treatment options for uncommon conditions.

When DNA sequencing data is added to a blockchain database, it generates a block with a timestamp. To avoid data manipulation, each block is cryptographically connected to the other blocks in the chain. Furthermore, with smart contracts, blockchain technology offers data owners control over what information is shared and with whom it is shared.

6. Blockchains in medical fraud detection

Drug medicine store network the executives is a huge utilization of blockchains in the clinical business. Supply the board is a significant subject to safeguard in all businesses, however it is particularly significant in medical services attributable to its rising intricacy. This is on the grounds that each disturbance in the medical services production network affects a patient's prosperity.

Since there are such countless moving components and people engaged with supply chains, they are vulnerable and have holes for deceitful assaults.

Blockchains, by bringing better information straightforwardness and improved item discernibility, give a completely safe stage to take out this issue and, in specific circumstances, forestall misrepresentation occurring.

Since a blockchain record must be affirmed and changed by means of a shrewd agreement, modifying the blockchain is troublesome.

6.1 Blockchains in neuroscience

The amount of information and investigation committed to blockchain applications is expanding, and neuroscience is without a doubt included [23]. Current cerebrum innovations are endeavoring to construct another worldview that disposes of mechanical association with the encompassing framework and permits clients to oversee contraptions and information with mental guidelines. Such brain gadgets might peruse examples of mind action and make an interpretation of them into orders for working outer hardware, as well as survey an individual's present mental state in light of cerebrum movement information.

Brain interface gadgets incorporated with various delicate sensors, calculation processors, and remote association handle the one of a kind obligation of perusing and deciphering cerebrum driving forces. They read the electrical movement of the cerebrum, which is then deciphered and moved to the controlling device. Every one of this happens in a solitary device that the client wears on their head. To store such mind motivations on the brain interface, compex calculations and huge information will utilize blockchain reasoning. Neurogress is one of the organizations that has affirmed that it would utilize blockchain innovation. The organization was established in 2017 in Geneva and is centered around creating brain control frameworks that permit clients to work automated arms, drones, brilliant machines, and AR/VR (expanded reality/augmented reality) items with their viewpoints.

The control component of Neurogress is predicated on utilizing AI to expand its cerebrum understanding exactness, which requires keeping 90% of mind information to prepare the AI used by the framework. At the end of the day, "tremendous information of client cerebral action" is expected, with the Human Brain Project requiring "exabytes (1 exabyte = 1 billion gigabytes) of memory" to act as an illustration of the sort of capacity limit required. It's not really shock, consequently, that Neurogress means to take on blockchain, which it says "actually handles the issue of information stockpiling security and protection."

At the point when client information is recorded on a decentralized blockchain, it becomes "invulnerable to programmer attacks" and consequently more private. All the while, the reception of blockchain innovation makes the Neurogress framework "open and straightforward to future Neurogress stage administrations shoppers." The innovation will "ensure security and classification of individual information" since any atypical conduct will be obviously detectable.

Thus, obviously blockchains are a kind of data innovation with different key future applications, including cerebrum expansion, mind recreation, and mind thinking. Yet again digitizing a full human mind obviously requires the utilization of certain media to store it, and here is where blockchain innovation becomes possibly the most important factor. One idea is to store mindfiles, which would work as information building blocks in private figured chains and be shareable in a distributed organization document framework with verifiable forming.

This perspective about blockchain is presented as an information handling yield computational framework with different characteristics that take into consideration man-made reasoning, human expansion, and conceivable coordination. To affirm the source and reality of a record, blockchain permits a connected organization of PCs to shake hands at timestamp spans. If we somehow happened to develop a cerebrum without any preparation, this kind of trust component might permit organizations of neurons to store and recover data with accuracy and trust of what is emotional versus objective of a specific occasion.

As a blockchain application, multifaceted verification connected to an individual idea chain can empower the solid development of an evaluated self information lodge for people. Such an information center brings down human information storehouses while as yet permitting every person to hold command over their security or sharing of their experience, possibly for money related gain without the need of an outsider or unified power. Eventually, using an improved adaptation of this innovation, when at least two individuals experience similar second from emotional perspectives, we might reassemble their encounters to be more level headed about what's going on.

In a perfect world, this would consider the creation of virtual copies of earlier recollections, as well as the capacity to see somebody's viewpoint subjectively. Someone else's perspective Once we have a more adaptable information on novel profound mappings, Considering tangible encounters as adding to a specific memory, this would take in tactile information onto the blockchain representing things to come (i.e., sight, smell, etc.). Besides, the innovation to produce. This is a reality that is being built. We will actually want to catch our tangible information in the not-too-far off future. Wearable innovation encounters, the current status of mind and nerve inserts, and biofeedback imaging, and whatever other sensors that give a multifaceted finger impression novel to a specific human's transient experience is recorded. Involving these innovations as an establishment, examination might be directed to improve navigation, learning, review, and rehabilitative medicines.

7. Future perspectives

The clinical business can benefit significantly from blockchain innovation. Comparatively to how the web changed medical services and presented telemedicine, blockchain innovation is probably going to take clinical science to the powerful later on by bringing down the expenses of checking, design, and having a focal server for information, as well with respect to the organization dealing with the clinical information. Utilizing blockchains in medical care settings would significantly cut handling time since, when a patient signs up for a review, the entire assortment of information will be accessible without a moment's delay inferable from the conveyed record's accessibility.

Moreover, clinicians will not need to stress over the patient giving a genuine clinical history since they will actually want to get to the first, bona fide, and quality source-reported information progressively, decreasing any potential clinical history mistakes. Likewise, on the grounds that the information is straightforward, patients will not have to look for a second assessment from another specialist. Having patient records on a blockchain organization will permit individuals to find out about and interface with individuals all around the world who have comparable ailments as they do, which will not just help their wellbeing yet will likewise leave patients feeling acknowledged, upheld, and with a more grounded will to battle the sickness. Patients will have complete command over their information and will actually want to pick who gets it. As Richie Etwaru contended at a book send off in 2016, the coming period can be portrayed as Freedom-as-a-Service.

8. Conclusions

Blockchain innovation addresses a huge chance for clinical exploration: it can help with the improvement of additional straightforward and auditable procedures and, gave a bunch of center metadata is characterized, can support the straightforward and to some degree algorithmic check of clinical preliminary uprightness. At last, Blockchain can prompt the development of a local area driven Internet of wellbeing information, uniting specialists and patient networks, interpersonal organizations, and Internet of Things information streams on a worldwide scale, with individual granularity, decentralization, and security, as well as straightforward connections to empower simpler and more straightforward examination.

The viable use of blockchain innovation in the medical services space will help an enormous number of people, clinical experts, medical services suppliers, R&D trained professionals, medical care substances, and biomedical scientists by permitting them to really scatter gigantic measures of information, share clinical information, and impart suggestions while keeping up with more prominent security and security. The compelling arrangement of blockchain innovation in medical care clinical settings will without a doubt propose new examination ways for biomedical exploration improvement. Conversely, in accuracy medication applications, the protected, secure, and adaptable social occasion, stockpiling, and trade of clinical information would help with the improvement of practical methodologies for sickness conclusion and therapy. Neurotechnology is still in its beginning phases, and a couple of associations have ventured to such an extreme as to approve blockchain innovation's contribution.

In any case, the topic of how secure individual cerebrum information would be on a blockchain stays unanswered. Albeit the decentralized and straightforward construction of blockchains will keep information from being controlled or taken, a considerable lot of the typical stresses over huge scope information gathering remain:

That delicate information might be offered to outsiders for questionable showcasing objectives, and people might in any case be by implication conspicuous (as with bitcoin) by means of pseudonymous characters or information designs.

Thus, this blockchain-based medical care structure will connect with individuals more in their medical services, ultimately working on their personal satisfaction in a more proper manner.

Author details

Aryan Chaudhary[1*], Keshav Kaushik[2] and Sunil Kumar[3]

1 Nuvem Resources Pvt Ltd, Kolkata, India

2 School of Computer Science, University of Petroleum and Energy Studies, Dehradun, India

3 School of IT, AURO University, Surat, India

*Address all correspondence to: researchhead@nuvemresources.com

References

[1] Anushree T, Amandeep D, NajmulIslam AKM, Mäntymäki M. Blockchain in healthcare: A systematic literature review, synthesizing framework and future research agenda

[2] Vora J et al. BHEEM: A blockchain-based framework for securing electronic health records. In: 2018 IEEE Globecom Workshops (GC Wkshps). Dec 2018. pp. 1-6. DOI: 10.1109/GLOCOMW.2018.8644088

[3] Shubbar S. Ultrasound Medical Imaging Systems Using Telemedicine and Blockchain for Remote Monitoring of Responses to Neoadjuvant Chemotherapy in Women's Breast Cancer: Concept and Implementation. Ohio, United State: Kent State University; 2017

[4] Mannaro K, Baralla G, Pinna A, Ibba S. A blockchain approach applied to a teledermatology platform in the Sardinian region (Italy). Information. 2018;**9**(2):44. DOI: 10.3390/info9020044

[5] Ianculescu M, Stanciu A, Bica O, Neagu G. Innovative, adapted online services that can support the active, healthy and independent living of ageing people. A case study. International Journal of Economics and Management Systems. 2017;**2**:321-329. Accessed: Mar. 29, 2020. [Online]. Available from: https://www.iaras.org/iaras/home/caijems/innovative-adapted-online-services-that-can-support-the-active-healthy-and-independent-living-of-ageing-people-a-case-study

[6] Kuo T-T, Kim H-E, Ohno-Machado L. Blockchain distributed ledger technologies for biomedical and health care applications. Journal of the American Medical Informatics Association. 2017;**24**(6):1211-1220. DOI: 10.1093/jamia/ocx068

[7] Azaria A, Ekblaw A, Vieira T, Lippman A. MedRec: Using blockchain for medical data access and permission management. In: 2016 2nd International Conference on Open and Big Data (OBD), Vienna, Austria. IEEE; Aug 2016. pp. 25-30. DOI: 10.1109/OBD.2016.11

[8] Xia Q, Sifah EB, Asamoah KO, Gao J, Du X, Guizani M. MeDShare: trust-less medical data sharing among cloud service providers via blockchain. IEEE Access. 2017;**5**:14757-14767. DOI: 10.1109/ACCESS.2017.2730843

[9] Available from: http://retractionwatch.com/category/by-author/don-poldermans/

[10] Available from: https://fr.wikipedia.org/wiki/BIA_10-2474

[11] Pérez-Marco R. Bitcoin and Decentralized Trust Protocols. Paris, France: CNRS, Univ. Paris 13; 2016

[12] Gipp B, Meuschke N, Gernandt A. Decentralized Trusted Timestamping using the Crypto Currency Bitcoin. Japan: National Institute of Informatics Tokyo;

[13] Giles C. Financial Times. London, England, UK: Financial Times, Bracken House; 2014 [October 17, 2014] (Thomas Piketty's exhaustive inequality data turn out to be flawed)

[14] IOM (Institute of Medicine). Evolution of Translational Omics: Lessons Learned and the Path Forward. Washington, DC: The National Academies Press; 2012

[15] Sandve GK, Nekrutenko A, Taylor J, Hovig E. Ten simple rules for reproducible computational research. PLoS Computational Biology. 2013;**9**(10):e1003285

[16] Chu S. Apple watch release news: Survey finds 80 percent of US employees would give health data from wearables to employers. iDigitalTimes, 2 Feb 2015. [Accessed: 7 July 2015]

[17] Available from: https://en.wikipedia.org/wiki/Smart_contract

[18] Benchoufi M, Porcher R, Ravaud P. Blockchain protocols in clinical trials: Transparency and traceability of consent. F1000Research. 2017;**6**:66

[19] Nugent T, Upton D, Cimpoesu M. Improving data transparency in clinical trials using blockchain smart contracts [version 1; referees: 3 approved]. F1000Research. 2016;**5**:2541. DOI: 10.12688/f1000research.9756.1

[20] Office of Scientific Investigations, Metrics. US Food and Drug Administration. 2014

[21] Barney JR, Antisdel M. Common problems in informed consent. Human Research Protection Program (HRPP). Jul-Sep 2013;**4**(3):134-140

[22] Chainscript documentation. Chainscript is developed by a Blockchain solutions provider, Stratumn SAS. 2017. Available from: http://chainscript.io. [Accessed: 25 Jan 2017]

[23] Swan M. Blockchain thinking: The brain as a decentralized autonomous corporation. IEEE Technology and Society Magazine. 2015;**34**:41-52

Chapter 11

A Simulation Model of a Blockchain-Based Decentralized Patient Information Exchange System for Parkinson's Disease Patients

Armando de Jesús Plasencia Salgueiro and Arlety García García

Abstract

Parkinson's disease (PD) is a progressive disorder of slow progress of the nervous system produced by the absence of levels of dopamine, which can incite unrestrained instinctive movements of the body and psychological affections. For the development of a practical, low-cost, and general diagnosis system of the symptoms to support PD patients, the implementation of an IoT health monitoring system that uses smartphones for data collection is necessary. However, data can be processed in Cloud Computing (CC) for analysis and comparison, but to reduce the latency of retrieving data from sensitive applications, Fog Computing (FC) plays a vital role. Nevertheless, these technologies IoT, CC, and FC have several limitations and are vulnerable to security threats. Blockchain technology enhances IoT challenges in a network in terms of security and availability. This chapter implemented a Decentralized IoT Fog-based Solutions and Blockchain using Ethereum Smart Contract for the authentication system. The smart contract is programmed using Solidity to allow Things to communicate with each other automatically without intermediaries and to store data in a public/private blockchain. The validation of the system was simulated them using the simulations tools Cisco Packet Tracer, iFogSim, and Remix Ethereum. The obtained results proved the feasibility of the proposed system.

Keywords: Parkinson's disease patients, blockchain, fog computing, decentralized IoT, simulation

1. Introduction

Neurological and mental ailments regularly present with side effects that are mind boggling, abnormal, fluctuant in illness evolution, and show high fluctuation between patients. Current diagnostic and efficacy evaluation methods frequently rely on in-clinic visits and patients', caregivers', or clinicians' subjective evaluations. Most of the time, in-clinic evaluation methods are expensive, take a long time, and only allow for a limited number and quality of observations. They also frequently exhibit high inter- and intra-rater variability. In the early stages of the disease, when there is a lag between the onset of the pathological process and the onset of symptoms, the diagnostic process may be affected by the aforementioned issues with traditional methods of diagnosis [1].

Neurological and psychiatric illnesses typically last a long time and cause significant changes in symptoms over time. As a result, the primary challenges in evaluating distinct diseases during periodic in-clinic visits are recall and reporting biases. Sensor-based smart technologies are rapidly developing remote monitoring of patients in their daily lives, which may assist clinicians in facilitating early diagnosis and evaluating and adjusting interventions. Use of recently developed smart sensor technologies for patient monitoring has become increasingly popular [1].

Parkinson's disease is a degenerative disorder of slow progress of the nervous system caused by the lack of levels of dopamine, which can provoke uncontrolled involuntary movements of the body and psychological affections [2].

An incremental number of sensors, such as motion (acceleration and gyroscope), location (the Global Positioning System, or GPS), environment (barometer, temperature, and light), and health (heart rate) sensors, are included in the modern smartphones and wearables. Smartphones have the potential to replace in-clinic evaluations for a variety of valuations due to their extensive array of sensors, ability to collect ecological momentary assessments (EMA), and information about social interaction (such as social media, messaging, and phone calls). Digital biomarkers (DBs) are terms used to describe the health-related data gathered during clinical trials. In order to gain a deeper comprehension of particular diseases, DBs can provide information that is useful, objective, and ecologically valid. Additionally, DBs make it possible to conduct frequent assessments of larger target populations over longer time periods, which may provide an in-depth understanding of the variation in daily life between and within individuals due to disease [1].

A few commitments empowering the utilization of cell phones as a valuation device have been as of late presented. Commercial devices make up the first set. By displaying notifications about a user's heart rate, number of steps taken, and type of activity, these apps primarily aim to provide feedback on the user's daily activities. However, the majority of these devices do not support high-frequency data collection and only provide limited access to the raw data. Applications and platforms created by investigators are the second category. The primary goals of these mostly open-source platforms are to make it possible to share and reproduce data as well as collect data for investigation applications. However, these software packages are much of the time restricted by a frequently constricted concentration to a few explicit clinical signs or concerning protection perspectives. Additionally, these periodically updated platforms render some unstable for the rapidly expanding smartphone ecosystem [1].

An example is a System developed by [3]. It builds on the fact that the medication treats Parkinson's disease gait abnormalities. It works by putting a smartphone in the pocket without requiring any special skills—a common occurrence in our daily lives. The information about a person's gait can be continuously detected by a smartphone without the user's active participation. The system makes passive sensing possible in this way. The system has two ends: a smartphone end and a cloud-server end. The smartphone sends the raw gait data to the cloud server from the smartphone end. After that, it is the job of the cloud server to look at the data and send the results to the smartphone. The smartphone notifies the user of the next medication time or reminds them to take their medication according to a drug schedule set by the healthcare provider. The system assists patients in avoiding missed, anticipatory, or additional doses through this approach. Although several platforms can collect context-driven data, the trade-off between privacy, optimization, stability, and research-grade data quality is not finding an optimal solution [1].

This paper objects to the solution of the abovementioned problems, to propose a decentralized authentication system utilizing blockchain technology and fog computing. With the characteristics and features of blockchain technology such as smart contracts, it addresses authentication using a decentralized database and communication between fog devices (nodes). The proposed system achieves authentication and communication without a central authority typical for this technology.

The main contributions of this paper can be summarized as follows:

1. Suggesting a secure decentralized user authentication that utilizes blockchain technology, smart contract, and secure ledger.

2. The system proposes to handle authentication requests using the username, password, Ethereum address, user email, and data from a biometric sensor.

3. The system is scalable and developed under the conception of scale to multiple IoT devices.

4. Proposing a methodology for the simulation of secure decentralized fog/edge architecture for healthcare systems under an IoT conception.

The rest of this paper is organized as follows: Section 2 introduces authentication systems, fog computing, and blockchain technology in Smart Healthcare Systems. Section 3 discusses the related works.

Section 4 presents the proposed authentication system. Section 5 provides the implementation details of the proposed system, followed by details of the experimental setup using simulation for validation of the proposed system.

Section 6 concludes the paper along with future directions.

2. Methods

2.1 Authentication systems

Authentication has become increasingly important in the world, with individuals, corporations, and businesses making use of it to control access to data and information [4]. In Smart Healthcare Systems, Authentication has great significance for your implication in patient safety and the growing introduction of digital monitoring patients systems in General Health Systems.

Other authentication mechanisms, such as biometric authentication (inherent factors) and token-based authentication (possession factors), are increasingly becoming popular and used transversely in services. Basic authentication methods, such as username and password, are widely used across platforms and services [4].

Alphanumeric and special character usernames and passwords are used in knowledge-based authentication systems. These are widely used across platforms and services and are regarded as conventional. Because they are simple to remember and can be processed quickly, they are popular. Due to their widespread use, password-based authentication systems are vulnerable to a variety of attacks, including dictionary, shoulder surfing, brute force, and dictionary attacks. Additionally, password length, character count, and password strength are limitations. Some weaknesses in text-based passwords have been fixed by introducing graphical passwords. This technique is resistant to attacks, for example, word reference attacks on account of its huge secret key space [4].

It has long been known that biometric authentication is superior to password-based authentication, thereby reducing some of these vulnerabilities. Typically, it is divided into physiological (fingerprints, DNA, face) and behavioral (voice, signature, etc.) characteristics. On smartphones, laptops, and other smart devices, face recognition and fingerprints have largely been used to control user access and authenticate on social media, banking apps, and services. Additionally, biometric authentication comes with drawbacks and dangers. This is large because biometric data cannot be changed and are managed in a centralized database or module. The reuse of stolen biometric data, among other security risks, may result from the discovery of these biometric data [4]. Although these methods are popular, they are not widely used as the traditional username and password.

2.2 Cloud, edge, and fog computing in smart healthcare systems

2.2.1 Cloud-based solutions

A network, cloud servers, and a mobile device make up cloud-only medical architectures. The issue of high latency is exacerbated by these components' potentially large distances between them (**Figure 1**). A comparison of distributed, or fog, cloud architectures, and traditional cloud architectures has recently been included in a number of medical monitoring solutions. For real-time emergency situations such as fall detection and stroke mitigation, which both require immediate medical response times, cloud-only solutions have retrieval times that are too high. Frequently sending

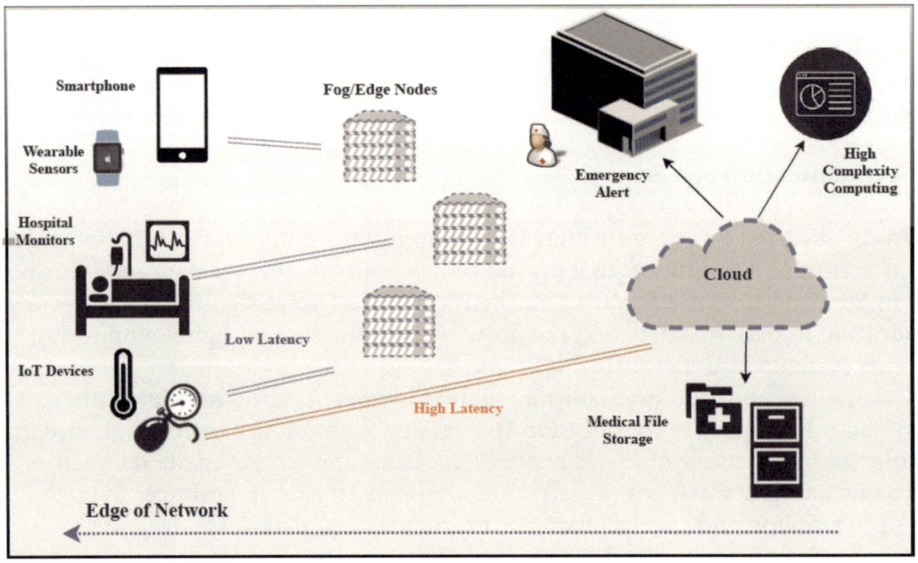

Figure 1.
General fog/edge architecture for healthcare systems [5].

data to the cloud for computation results in increased power consumption and costs, which is especially true in today's world when sensors generate a lot of data. When compared with a distributed computing architecture with multiple computing nodes in various locations, the typical cloud service demonstrated high latency and low sustained performance. Additionally, cloud-based solutions lack a low-cost mobile environment, which is essential for many patient monitoring scenarios [5].

2.2.2 Edge and fog-based solutions

Data processing is moved closer to the network edge with edge and fog-based solutions, resulting in faster response times and improved energy efficiency. Rather than continually moving information to the cloud for figuring activities, which represents the energy costs, information can be mined and handled by on edge devices and servers nearer to the client. In situations involving health monitoring, low latency of edge and fog solutions enables prompt arrival of emergency medical assistance. Privacy and security remain major issues due to a large amount of data that is typically sent to cloud services, particularly in situations where a patient's medical data could be hacked. Improved privacy can be achieved by disseminating data across a fog rather than concentrating it in a single area of the network. Device usability is also important because accurate data transmission depends on these sensors being easy enough for untrained personnel to use. The particular needs that are addressed are [5]:

- Cost
- Low Latency
- High-Level Security/Privacy
- Location Awareness
- Energy Efficiency
- Usability

The fog is typically referred to as a decentralized distributed computing system in which various entities own multiple fog devices and organizations can participate from a variety of locations, including hospitals, schools, airports, and smart hubs. According to the investigators, fog computing is a virtualized environment that is tasked with the delivery, storage, and computing of resources in cloud computing centers. It is not entirely outside the network space. These are a variety of fog nodes with limited computing and storage capabilities. Fog computing is widely regarded as an extension of the cloud that is close to devices that collect data for resource-constrained IoT. These devices are referred to as fog nodes and have storage, a network connection, and computational power. They are situated in various geographical locations that have network connectivity. These fog nodes are located close to devices that collect data [4].

The main characteristics of fog computing are given below [4]:

- Adaptability: These are made up of a lot of network sensors and other fog devices that handle computing tasks and provide storage resources.

- Real-time communications: Real-time communication between fog nodes and cloud-based data is provided by fog computing.
- Physical distribution: Fog computing offers services and applications hosted in multiple locations that are decentralized.
- Less latency: The proximity of fog computing to the edge devices helps position responsiveness to host fog devices in multiple locations and reduces the amount of time spent computing information with the edge devices.
- Compatibility: Fog modules are made to work with a lot of different platforms and service providers.
- Cloud integration and web analytics: The fog's location between the cloud and the edge devices is crucial for data processing and computing near them.
- Heterogeneity: Edge devices and fog nodes are made by different companies and have different features; therefore, they need to be hosted in accordance with their features.

2.3 Blockchain technology

Blockchain is a distributed ledger (register) technology that achieves immutability, traceability, anonymity, security, transparency, and decentralization through the use of consensus algorithms and cryptographic methods. All nodes on the blockchain verify and record confirmed transactions with a timestamp. The distribution of the ledger among the blockchain network's members provides transparency. Security of the information kept in the record is ensured and cannot be obstructed. One of the most important aspects of blockchain is the smart contract. It is a short piece of code that is part of the blockchain and can be executed automatically if the conditions are met. Blockchain is known to be of three types: These include consortium, public, and private blockchains. A private blockchain is more trusted by participants because it is managed by a single organization and governs its activities, such as who can participate in the blockchain. Any entity can participate in a public blockchain, which does not require permission. Privacy, security, and performance are among its drawbacks due to its highly decentralized nature. A permissioned blockchain created by a group of different businesses is known as a consortium blockchain. Nodes are the entities in this kind of blockchain. Which organizations or entities participate in the blockchain is controlled by the consortium [4].

The Internet of Things (IoT) has been the primary beneficiary of blockchain technology's features, such as the decentralized characteristics that enable devices to connect and share sensitive data securely in an IoT environment. These applications include auctions, mutual authentication, and blockchain technology. Blockchain is able to provide authentication systems with a dependable computational platform and secure storage thanks to these characteristics. A consensus algorithm helps the devices in a blockchain network establish trust. This makes it possible for devices to keep cryptographic keys (both private and public keys) and a digital signature. This makes it easier for devices in the network to communicate and transact. These transactions are tamper-proof on the blockchain network because they cannot be changed [4].

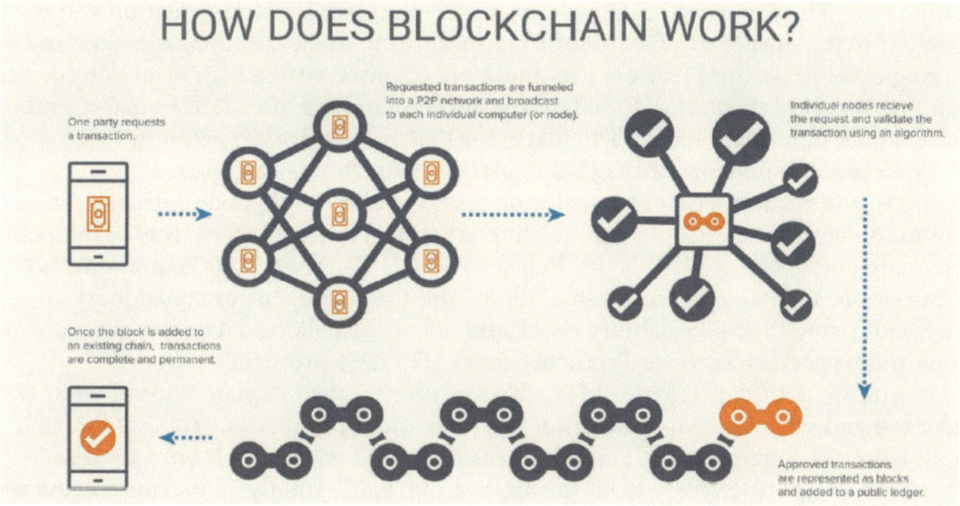

Figure 2.
Functional diagram of Blockchain [6].

A blockchain is designed to resist data modification. It is "an open, distributed ledger that can efficiently and permanently record transactions between two parties." A blockchain is typically managed by a peer-to-peer network that collectively adheres to a protocol for inter-node communication and block validation in order to function as a distributed ledger. The data in any given block cannot be changed after it has been recorded without changing all subsequent blocks, which requires the majority of the network. Blockchains may be considered as secure by design and represent a distributed computing system with high Byzantine fault tolerance (any fault presenting different symptoms to different observers), despite the fact that their records can be changed. As a result, a blockchain has been advertised as providing decentralized consensus [6]. **Figure 2** represents the functional diagram of Blockchain.

3. Results

3.1 Related works

Fog computing and blockchain technology have been used in IoT, sales, cloud computing, and other systems, primarily in decentralized ways, in a number of works proposed for authentication. In order to comprehend the proposed scheme, a review of schemes including cloud computing, fog computing, blockchain technology, IoT, and authentication has been provided in this section.

3.2 JTrack

JTrack [1] was developed as an online server-side dashboard and an Android-based smartphone application. The main components of the JTrack application fall into the following categories: Human Activity Recognition (HAR), location data, sensor data, smartphone, and application usage monitoring, and both active (with user interaction) and passive (without user interaction) monitoring options are available for each component. The dashboard side is a web-based platform for study creation and management that incorporates DataLad infrastructures to make data management and sharing easier. Because JTrack is a modular open source with a high level of optimization, it is a practical solution for clinicians and researchers to collect, manage, and share digital biomarker data, particularly for Parkinson's disease patients.

The main components of the JTrack platform are shown in **Figure 3**.

In [1], the authors proposed a solution with the aim of QR-Code Authentication to provide a secure way of activation. Additionally, the JTrack platform was developed in accordance with Google Developer Policies and GDPR. At no stage is any sensitive information, such as a person's name, phone number, contacts, or actual location, recorded. Using the MD5 stability checksum, all of the collected data were transferred using the Hypertext Transfer Protocol Secure (HTTPS) protocol.

Regarding patient privacy, all JTrack users receive clear explanations of what was recorded and why. During installation and activation, permission requests for each module must be approved. All members may likewise pause and leave a review whenever straightforwardly from the application. Additionally, clinicians can maximize control over the collected data with remote configuration and one-step recording without having to collect any identifying information [1].

Along with Firebase integration for performance and crash reports, automatic restarting is implemented to reduce data loss caused by crashes or reboots. The optional recorded data, which are not active by default, include information about the phone's manufacturer, model, and operating system version. This information can be used to analyze and deal with cross-sensor variability [1].

3.3 Continuous patient monitoring with a patient-centric agent

Internet of Things (IoT) applications in the modern healthcare system include devices, services, and wireless sensors that detect physiological signs with wearable or ingestible sensors that stream data to remote and often Cloud-based servers. Secure

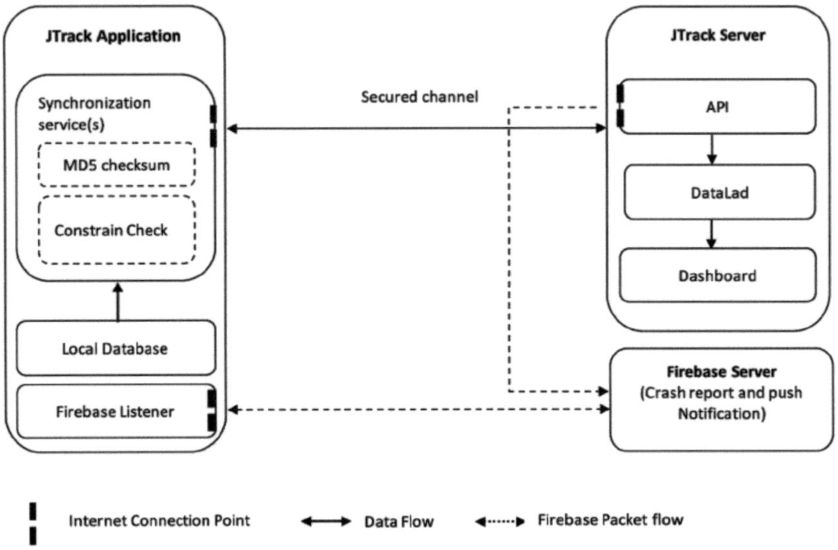

Figure 3.
JTrack platform overview [1].

continuous monitoring of patients' physiological signs has the potential to augment traditional medical practice [7].

The process of integrating demographic, health record, and geographic location data with physiological data obtained from wearable or implantable medical devices (IMDs) is known as remote patient monitoring (RPM) [7].

The aggregation and indexing of huge streams of continuous data while maintaining patient privacy are one of the challenges of designing remote patient monitoring systems that are effective, efficient, and secure. One's personal space, which also includes the ability to control data and set others' access levels, is referred to as privacy [7].

In addition to operational environments, insiders and outsiders pose threats to healthcare information's confidentiality, integrity, and accessibility. By gaining unauthorized access to confidential data, insiders such as healthcare professionals and support staff, service providers and outsiders such as hackers pose a threat to the security of health information [7].

Unauthorized actions have the potential to alter patient information and even result in death. Patient and medical professional trust in the system can be damaged by privacy breach [7].

Health information systems can also be threatened by resource misuses such as personal use of systems and software disruption caused by viruses, worms, and malware. Threats to the confidentiality and integrity of patient data include communication infiltration, interception, embedded malicious code, and repudiation. The security of health information can also be compromised by accident, technical infrastructure failure, and operational errors [7].

These threats have not yet been addressed by existing RPM architectures, as described in the following section. As a result, RPM software and devices require architectures that provide increased attack protection [7].

Efforts to ensure privacy in RPM have been made in recent years; however, most approaches focus on a single link in the architecture that chains data from patient sensors to healthcare professionals through intermediary devices and servers [7]

An effective and efficient RPM needs to address issues of rapid storage at appropriate security levels, user authentication, access control, mobility management, and sustainability of patient health data [7].

In order to ensure that appropriate levels of the trade-off between effectiveness and privacy can be established for rapid, secure data storage and access, user authentication, role-based access, and sustainability, an advanced End-to-End eHealthcare architecture that addresses RPM healthcare data management issues has been developed. A Patient-Centric Agent for End-to-End data stream coordination and a Blockchain component for distributed data storage are key architectural features [7].

In [7], it was suggested that the inclusion of a Patient-Centric Agent (PCA) can decrease RPM challenges. The End-to-End data flow is inaccurate for the Patient-Centric Agent (PCA). The level of storage, security, and access required at any given time is determined by the PCA. The patient sensors and devices, Blockchain nodes, and healthcare service provider devices are all coordinated by the PCA. If a stream of data should be stored in a Blockchain, the PCA manages the process and determines whether it should. The PCA executes on a machine with mass memory limit and high handling power.

There are two levels to the proposed architecture. The data streaming and storage solution is provided by the lower tier, while the Healthcare Control Unit (HCU) manages the primary healthcare provider. **Figure 4** depicts six systems that make up

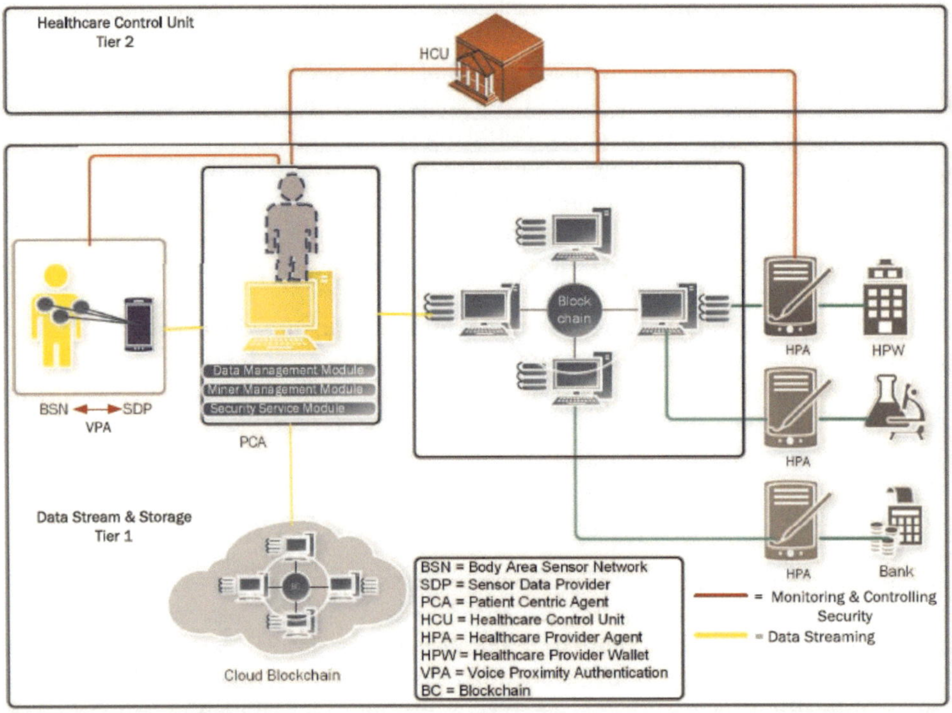

Figure 4.
The tier-based remote patient monitoring architecture [7].

the lower tier. Patient Centric Agent (PCA), Blockchain, Healthcare Provider Agent (HPA), and Healthcare Provider's Wallet (HPW) are all examples of Body Area Sensor Network (BSN). In **Figure 4**, a Sensor Data Provider, such as a smartphone, connects BSN to the Patient-Centric Agent (PCA) [7].

The Healthcare Control Unit, the Cloud, and the Blockchain network are all connected to PCA. Medical services Supplier Specialist interfaces Blockchain, Medical care Control Unit, and Medical services Wallet at the medical services supplier end. The functional view of the architecture is shown in **Figure 5**, and the architecture is explained by the communication links that connect the various segments below. The architecture is built to handle a lot of patients at once [7].

3.4 Fog-enabled blockchain-based authentication system

3.4.1 System architecture

In [4], the blockchain-based components of the decentralized authentication system are described. **Figure 6** illustrates the system's architecture. The components of the proposed system are outlined below:

- Ethereum Smart Contract:

  This authentication system's contract is used to handle user registration and authentication. The agreement would require data such as the email, password,

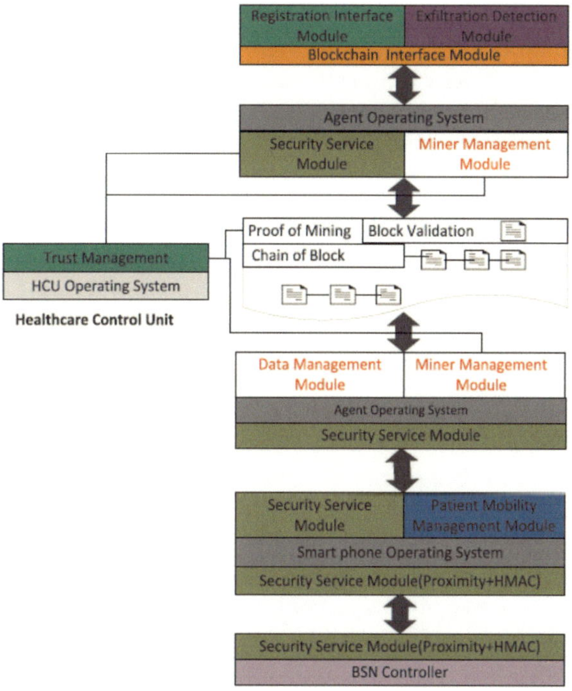

Figure 5.
Conceptual view of the tier-based health monitoring architecture [7].

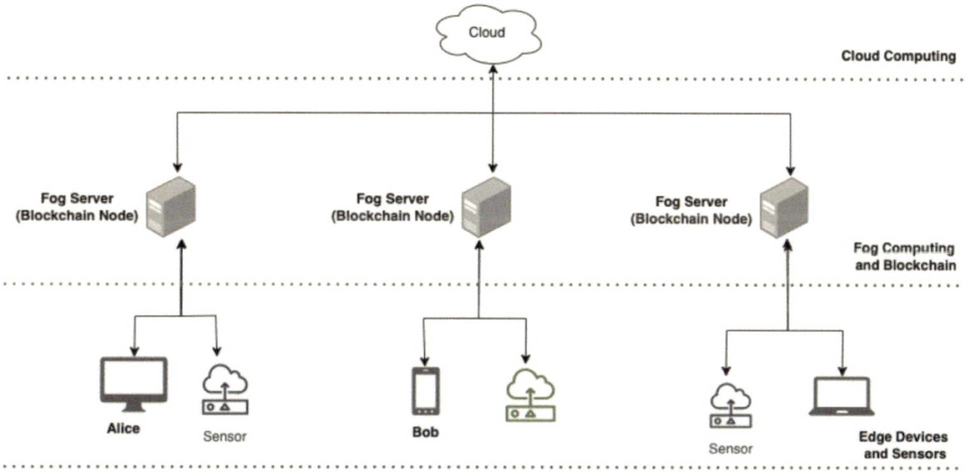

Figure 6.
The fog-enabled blockchain-based authentication system's system architecture [4].

and the UserEthAdr to enlist clients upon enrollment and to validate clients in the ensuing collaboration with the framework [4].

- Fog Node:

  Devices that serve as blockchain nodes and servers are known as fog nodes. Every node has a duplicate of the BlockC, LDG, and SmContract. When a User registration or authentication transaction takes place on a node, the BlockC information there is updated. To host or be a part of the BlockCN, the fog device or fog server must meet sufficient requirements [4].

- Edge Devices:

  During registration and authentication, the user's end devices are mapped to nodes. The BlockC cannot be hosted on these devices due to a lack of resources [4].

- Cloud:

  IoT data are stored, hosted, and computed in the cloud, which is a large storage unit. Data generated by IoT or edge devices must be processed and analyzed by this cloud server [4].

3.5 Automated decentralized IoT-based blockchain using ethereum smart contract for healthcare

In Ethereum, smart contracts are used to automate participant interactions and the execution of data from Things or any other type of data. Nodes use it to test, debug, verify, and test transactions. Consensus algorithms are at the heart of the blockchain, ensuring its security and integrity [8].

The proposed blockchain of Things architecture in the work [8] consists of five layers and is based on Ethereum smart contract including Things, gateway, Fog, Cloud, and Application layer (see **Figure** 7), demonstrated as follows:

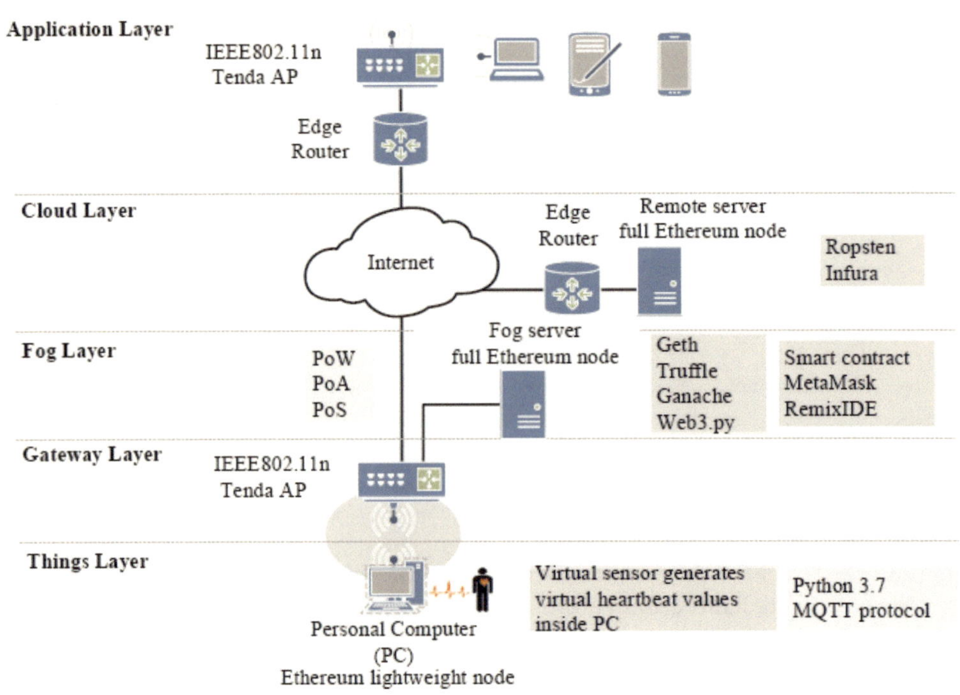

Figure 7.
Proposed IoT-based blockchain architecture for healthcare network [8].

Things layer: it consists of a virtual sensor programmed in python version 3.7 programming language [8].

Gateway layer: To transfer data to higher layers, such as the cloud and fog layers, the virtual sensor is connected to an Access Point (AP) as a gateway device with WiFi connectivity (IEEE802.11n Tenda AP) [8].

Fog layer: It stores IoT data on a private blockchain by collecting it from the Things layer's virtual sensor. This layer is a 24-hour power-on with enough storage space to store a local copy of the entire blockchain [8].

Cloud layer: This layer is built on top of a public testnet blockchain that was made for testing and mining. It uses Ether, which has no value, which can be bought from faucets that look like the mainnet network [8].

Application layer: Things' data can be monitored by doctors, patients, and their families. However, because blockchain data are immutable and unchangeable, they are unable to alter or delete them. Additionally, physicians can quickly respond to an emergency by monitoring data in real time even if it is not processed in blockchain [8].

4. Proposed system

The proposed decentralized authentication system, which is required for user authentication, is described in detail in this section. Some of the assumptions that are taken into consideration when developing the proposed methodology are outlined before the proposed system is presented in detail. The blockchain-based authentication systems described above frequently employ these conventions, such as [4]:

- There are mobile and immobile devices connected via multiple networks in the fog computing environment.
- The blockchain technology is accessible to registered user devices.
- To be able to serve as a node or a server and host the blockchain, the fog device must meet certain requirements.
- The registration and authentication of users ought to be carried out by the smart contracts.
- Nodes should not have to rely on other nodes to do their jobs.

Conceived of a smart contract, ledger, and Ethereum blockchain-based decentralized authentication system. Fog nodes will function as blockchain nodes in this system and host a decentralized digital ledger in which each fog node possesses a copy of the smart contract.

5. Experimental setup

An evaluation based on several experiments is presented to validate the proposed system. The proposed authentication system is implemented through some simulators such as Cisco Packet Tracer [9], iFogSim [10], and Ethereum smart contracts utilizing

Solidity [11]. This smart contract is tested, and the simulation is run through Remix Ethereum. This IDE offers various features such as the creation, deployment, testing, and debugging of smart contracts. The network layout of the system and simulations are executed in Ethereum Remix IDE and Cisco Packet Tracer.

Data from the metrics are collected such as the time needed to send packets from user devices to the fog nodes using Cisco Packet Tracer, the energy consumed, the cost of execution in the cloud, the total time required for module migration using iFogSim, and the transaction cost, execution cost, and miner fees are recorded for both registration and authentication requests through the blockchain network.

Simulations are run through Cisco Packet Tracer to replicate the fog network with nodes and determine the time needed to send packets from user devices to the fog nodes.

5.1 Cisco packet tracer simulation

The network is replicated in Cisco Packet Tracer [9]. Fog servers (nodes) and user devices are used to run simulations and tests on a variety of packets and protocols (HTTPS, SSH, SMTP, ICMP), and multiple requests are made on both wired and wireless networks to find out how long it takes to send authentication packets on the network. For the purpose of this experiment, these packets were chosen because they are frequently used for secure communications over computer networks [4].

Figure 8 represents the developed configuration of the proposed network scheme in Cisco Packet Tracer. The development was taken as a reference to the configuration given by [12], also the conceptions developed in [13]. **Table 1** shows the relation of components of the proposed network.

The results for simulation showed that all the packets are delivered in 40.049 s in a wired and wireless network, considering delivering SMTP packets, SSH packets, and HTTPS and ICMP. The time required to authenticate various packets using the proposed method is shown in the obtained results. This demonstration can quickly and effectively handle authentication requests, scale to multiple devices in a fog

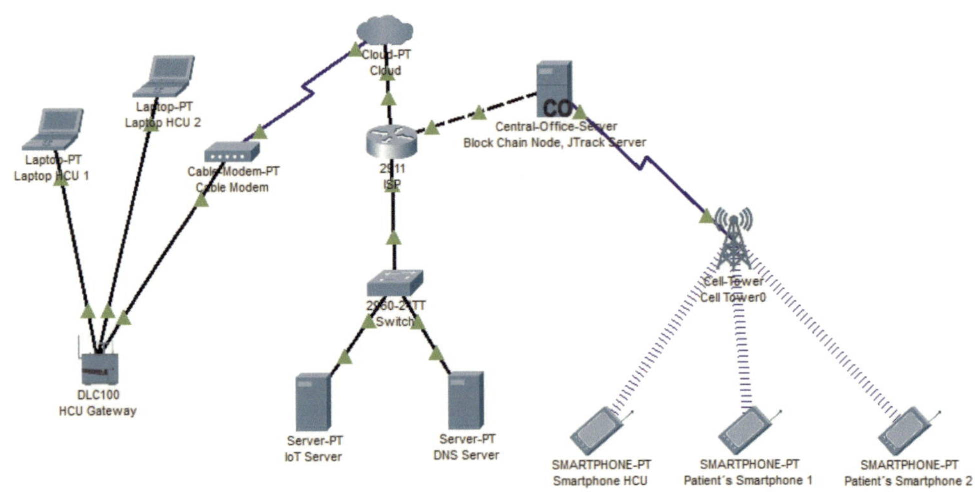

Figure 8.
Configuration of proposed network scheme in packet tracer.

No	Devices	Function
1	Router (2911)/ISP	Used to connect a cellular network to HCU
2	Cable modem	Used to HCU gateway to cloud
3	HCU gateway	Used for smart devices registration
4	IoT server	Used to control smart devices registered on it
5	DNS server	Used to access smart devices by the domain name
6	Central office server (Fog Server)	Used to connect a cell tower to a router and vice versa (JTrack Server). Block Chain Node
7	Cell tower	Used to connect the smartphone to the internet
8	Smartphone	Used to monitor Parkinson's Disease Patients
9	Laptop	Connect to the HCU gateway to access the Patient's smartphones

Table 1.
Devices used for the simulation.

computing environment, and is suitable for use on edge devices. The used version of Cisco Packet Tracer was 8.1.1.

5.2 Simulation results in iFogsim

iFogSim simulation toolkit is developed upon the fundamental framework of CloudSim. CloudSim is one of the wildly adopted simulators to model Cloud computing environments. A customized Fog computing environment with a large number of Fog nodes and Internet of Things (IoT) devices (such as sensors and actuators) can be simulated with the help of iFogSim. However, the classes in iFogSim are annotated in such a way that users without prior knowledge of CloudSim can quickly and easily define the policies for Fog computing's infrastructure, service placement, and resource allocation. When simulating any application scenario in the Fog computing environment, iFogSim employs the Sense-Process-Actuate and distributed dataflow models. It makes it easier to evaluate end-to-end latency, network congestion, power consumption, operational costs, and customer satisfaction with QoS [14].

iFogSim2 is a simulator [15], which is an extension of the iFogsim simulator and addresses distributed cluster formation among Edge/Fog nodes of various hierarchical levels, microservice orchestration, and service migration for various mobility models of IoT devices.

The new iFogSim2 simulator components are loosely coupled to support various simulation scenarios. As a result, the components (Mobility, Clustering, and Microservices) can be used solely for simulation or integrated for more complex scenarios [15].

In the Healthcare solution of Parkinson's Disease, body-connected IoT devices perceive the health context of the users through a Client application module. Generally, the IoT devices are usually connected to smartphones, but in this particular case, the sensors are inside the same smartphone.

For the corresponding application, the smartphones serve as the Application gateway node. These nodes prepare the sensed data from IoT devices. The application's data analysis and event management operations are carried out in upper-level Fog computational nodes unless the Application gateway node's resource availability meets the requirements. Application gateway nodes select appropriate computational nodes in the second scenario to deploy additional application modules and initiate actuators based on the results of those modules [14].

In the IoT-enabled Parkinson's Disease Patient monitoring healthcare solution using iFogSim was simulated the necessary Fog environment. The application model for the IoT-enabled healthcare solutions is represented in **Figure 9**.

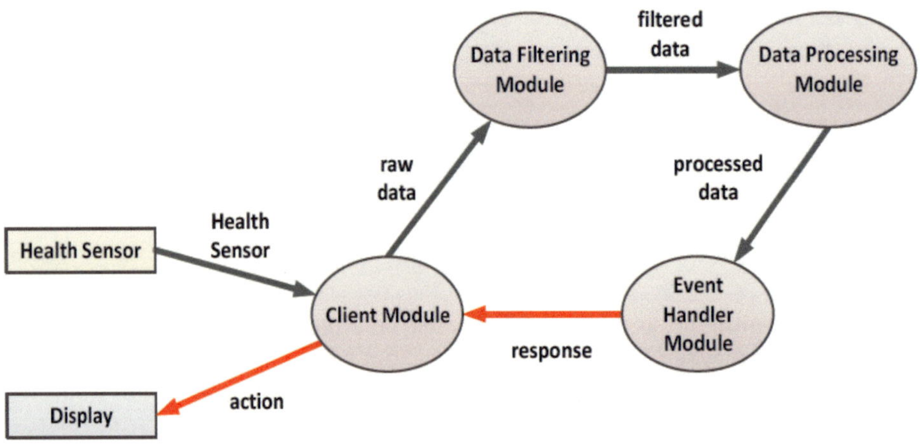

Figure 9.
Application model for IoT-enabled healthcare case study [14].

The system and the application along with the required guidelines to model them in iFogSim are an n-tire hierarchical Fog environment. As the rank of Fog levels goes higher, the number of Fog devices residing at that level gets lower. Fog devices form Clusters among themselves and can be mobile. IoT devices (smartphones) are connected to lower-level Fog devices [14]. The sensing frequencies of IoT devices are different. The application model consists of four modules with a sequential unidirectional data flow. The requirements of the application modules are different, and each application module can request additional resources from the host Fog devices to process data within the QoS-defined deadline. The results of the simulation are shown in **Figure 10**.

5.3 Simulation smart contract through remix Ethereum

Ethereum is a decentralized, open-source blockchain with smart contract functionality. The Ethereum network is a well-established blockchain network that allows people to actively develop new and innovative applications and products that directly tie to Ethereum and use its native cryptocurrency, ether. This allows people to use a blockchain network that already has nodes and avoid the need to set up self-hosted nodes for testing. However, some limitations include long transaction times and high gas prices [16]. For the simulation of blockchain in IoT communications, Ethereum was the single simulator. This meant that Ethereum was used in hashing and encoding. In smart devices, Ethereum was used as the base network for storing data via smart contracts and generating hashes [16]. For all these properties, Ethereum was selected for the simulation.

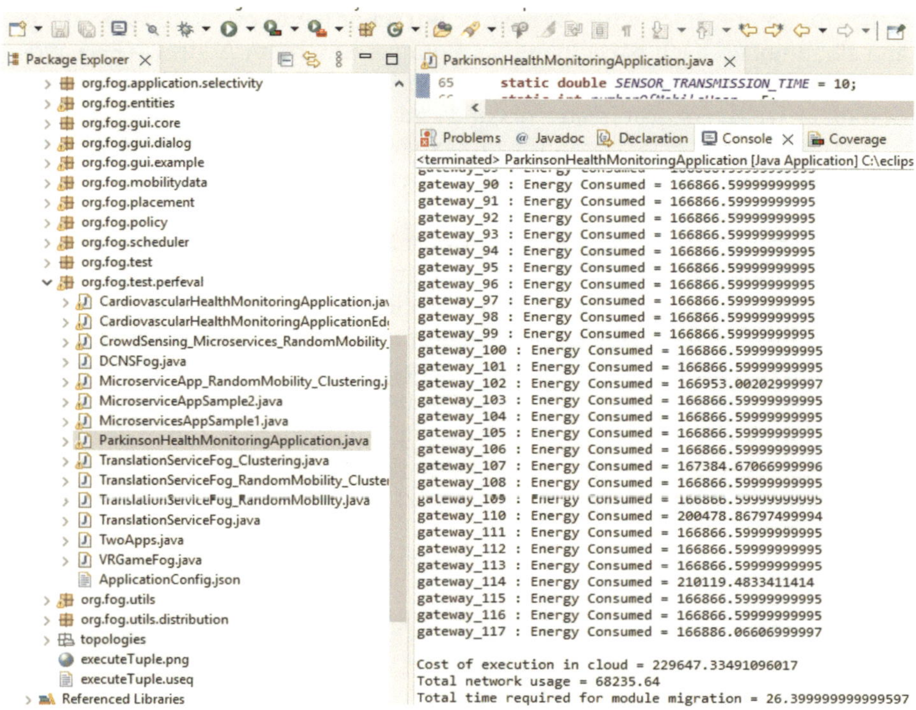

Figure 10.
Results of simulation IoT-enabled Parkinson's Disease Patients monitoring healthcare solution using iFogSim.

Figure 11 depicts the Ethereum-enabled fog nodes of the proposed system architecture. Five main participants with Internet access to Ethereum smart contracts make up the architecture: fog nodes, users, administrators, and cloud servers that store IoT data. Although IoT devices have a unique Ethereum address and public and private

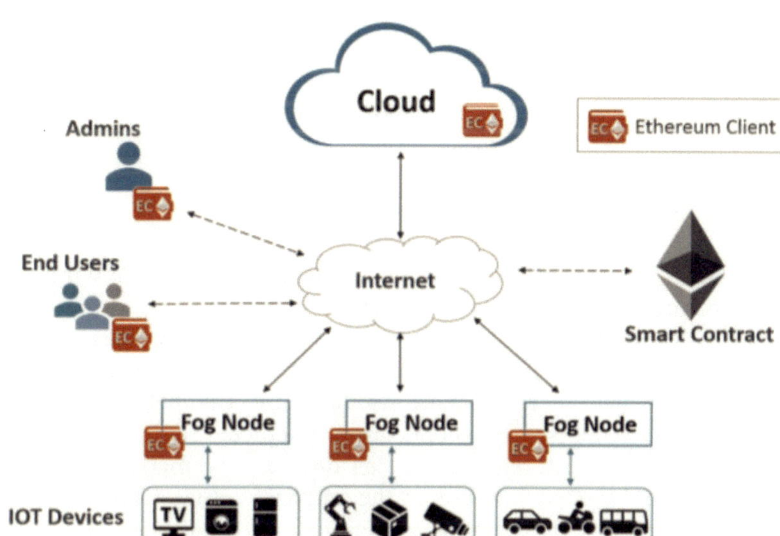

Figure 11.
System architecture with fog nodes that can use Ethereum [17].

keys, they do not interface with the smart contract or have connectivity. In the case of fog and cloud nodes, each participant has a unique Ethereum Address (EA) and interacts directly with the smart contract via an Ethereum client or a front-end application or wallet for administrators and end users [17].

Ethereum Remix IDE is chosen for this experiment for its features, which allows for the development, administering, and deployment of smart contracts in a virtual blockchain environment.

The simulation model is displayed in **Figure 12**. In step 1, User A presents data and sends a registration request to the blockchain-enabled fog node via the edge device. User A is registered as a new user in step 3, while user data are stored on a distributed ledger by the blockchain-enabled fog node in step 2. User B presents data and sends an authentication request to the blockchain-enabled fog node through the edge device in step (4). At the following stage (5), the blockchain-empowered fog node affirms that Client B's information is legitimate and exists in a distributed ledger. The user is then verified or denied. The exchange cost, execution cost, and miner fees are recorded for both registration and authentication demands through the blockchain network [4].

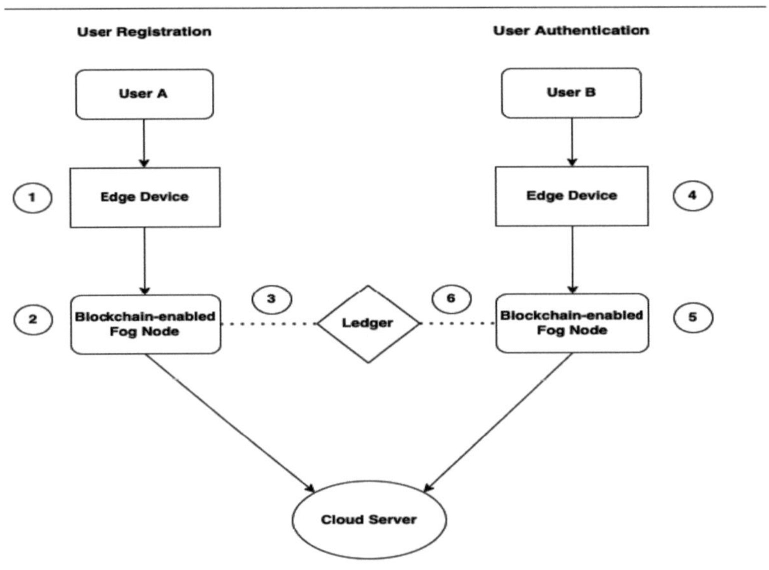

Figure 12.
The simulation model [4].

Figure 13 shows the sequence diagram of exchange messages showing a successful authentication of the end user to the IoT Device.

In order to carry out the authentication scheme, the system entities interact in two primary ways; namely: interactions between the on- and off-chains, as shown in **Figure 13**. A sequence diagram for a secure session connection between an end user and an IoT device following successful authentication is depicted in **Figure 13**. The admin first creates the smart contract, registers the IoT devices, and maps them to a fog node using the *addDeviceFogMapping* function in the on-chain activity. The Ethereum Address (EA) is unique to each fog node and IoT device. Using the *addUserDeviceMapping* function, the administrator can also grant access permissions to specific IoT devices to end users. On Github, the entire smart contract logic, lists, rules, and code for the authentication registry are also made available to the public [18].

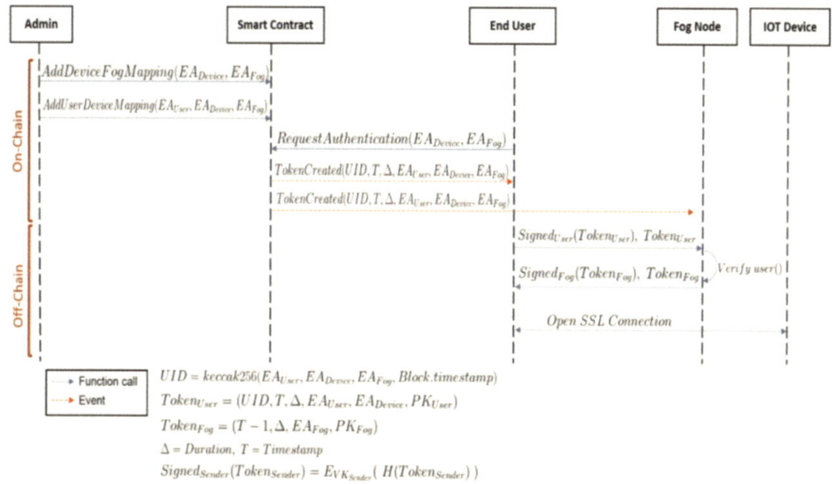

Figure 13.
Sequence diagram of exchange messages showing a successful authentication of the end user to IoT Device [17].

The end user first sends an authentication request to the smart contract using the request Authentication function, specifying the EA of the IoT device, before attempting to access that device. The user's saved list of authorized IoT devices will be reviewed by the smart contract. A rejected request event will be generated in the event that the user is not authorized to access that device. Otherwise, the smart contract will issue an acceptance event and an access token if the user has permission to do so, $TokenCreated = (UID, T, \Delta, EA_{User}, EA_{Device}, EA_{Fog})$. By definition, the event is broadcasted to all users and fog nodes [17].

The rest of the exchange messages showing a successful authentication of the end-user to the IoT Device are detailed and described in [17]. A view of the implementation of the given solidity software code given in the available Github is shown in **Figure 14**.

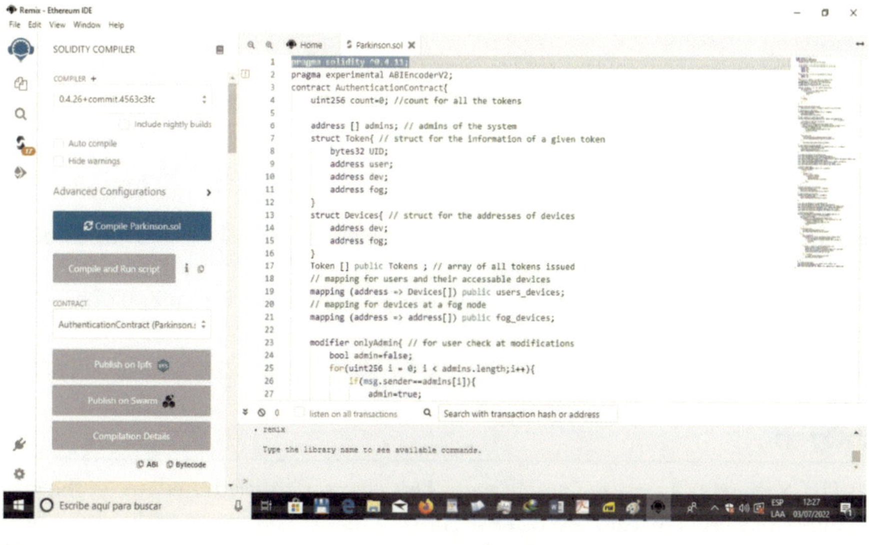

Figure 14.
Implementation of the simulation smart contract through the IDE Remix Ethereum.

6. Conclusion

In this chapter, a decentralized, Fog Computing, user authentication system was proposed for the development of a practical, low-cost, and general diagnosis system of the symptoms of PD patients using an Ethereum smart contract. It aimed to address user authentication in a decentralized environment and address fog computing security problems inherited from IoT and Cloud/Fog computing. It provided a solution for immutability and scale-ability problems in fog computing.

The system was simulated for validation and design using the simulation tools Cisco Packet Tracer, iFogSim, and Remix Ethereum. The obtained results proved the feasibility of the proposed system.

In future work, it is necessary to execute the process of building a fully functional system prototype involving real-based presented IoT devices connected to fog nodes equipped with Ethereum client to be connected with the real public Ethereum network, which is hosting the smart contract code. This makes it possible to complete a real diagnosis system of symptoms to support PD patients, necessary for the implementation of a real IoT health monitoring system that uses smartphones for data collection and machine learning algorithms for data processing.

Conflict of interest

"The authors declare no conflict of interest."

Author details

Armando de Jesús Plasencia Salgueiro[1*] and Arlety García García[2]

1 Nacional Center of Animals for Laboratory (CENPALAB), La Habana, Cuba

2 Youth Island University "Jesús Montané Oropesa", Nueva Gerona, Cuba

*Address all correspondence to: aplasencia278@gmail.com

References

[1] Sahandi Far M et al. JTrack: A digital biomarker platform for remote monitoring of daily-life behaviour in health and disease. Frontiers in Public Health. 2021;**9**(763621):11

[2] de Jesús A, Salgueiro P, Shichkina Y, García AG, Rodríguez LG. Parkinson's disease classification and medication adherence monitoring using smartphone-based gait assessment and deep reinforcement learning algorithm. Procedia Computer Science. 2021;**186**: 546-554. DOI: 10.1016/j.procs.2021.04.175

[3] Zhang H, Xu C, Li H, Rathore AS, Song C, Yan Z, et al. PDMove: Towards passive medication adherence monitoring of Parkinson's disease using smartphone-based gait assessment. Proceedings of the ACM on Interactive Mobile Wearable and Ubiquitous Technologies. 2019:23. DOI: 10.1145/3351281

[4] Umoren O et al. Securing fog computing with a decentralised user authentication approach based on blockchain. Sensors. 2022; **22**(3956):21

[5] Hartmann M et al. Edge computing in smart health care systems: Review, challenges, and research directions. Transactions on Emerging Telecommunications Technologies. 2022;**33**(3710):28

[6] E. LLC. Lab Design Guide For Artificial Intelligence (AI), Internet of Things (IoT), Autonomous Vehicles, AR/VR, Blockchain and Industry 4.0 Labs. Dubai, United Arab Emirates: EdNex; 2022

[7] Uddin A et al. Continuous patient monitoring with a patient centric agent: A block architecture. IEEEAccess. 2018; **6**(32700):27

[8] Al-Joboury IM et al. Automated Decentralized IoT Based Blockchain Using Ethereum Smart Contract for Healthcare. Baghdad: Springer Nature Switzerland; 2021

[9] Jesin A. Packet Tracer Network Simulator. Birmingham: Packt Publishing; 2014

[10] Awaisi KS, Abbas A, Khan SU, Mahmud R, Buyya R. Simulating Fog Computing Applications using iFogSim Toolkit. In: Mobile Edge Computing. Cham: Springer; 2021. pp. 565-590

[11] Mukhopadhyay M. Ethereum Smart Contract Development. Birmingham - Mumbai: Packt Publishing; 2018

[12] Thera D. davidthera/iot-simulation-with-cisco-packet-tracer, 28 Jun 2020. [Online]. Available from: https://github.com/davidthera/iot-simulation-with-cisco-packet-tracer. [Last access: 30 May 2022]

[13] Thera D. Internet of things simulation using CISCO packet tracer. [Master of Science in Computer Engineering thesis], İzmir Institute of Technology. 2020

[14] Mahmud R, Buyya R. Modeling and simulation of fog and edge computing environments using iFogSim Toolkit 433 Wiley STM. In: Srirama B, editor. Fog and Edge Computing: Principles and Paradigms. Chapter 17 Introduction to Fog and Edge Computing. 2019. DOI: 10.1002/9781119525080.ch17

[15] Mahmud R et al. iFogSim2: An extended iFogSim simulator for

mobility, clustering, and microservice management in edge and fog computing environments. 2021. p. 17. arXiv: 2109.05636v2

[16] Zheng J et al. An in-depth review on blockchain simulators for IoT environments. Future Internet. 2022; **14**(182):22

[17] Almadhoun R, Kadadha M, Alhemeiri M, Alshehhi M, Salah K. A User Authentication Scheme of IoT Devices using Blockchain-Enabled Fog Nodes. In: 2018 IEEE/ACS 15th International Conference on Computer Systems and Applications (AICCSA), Aqaba, Jordan. 2018. pp. 1-8. DOI: 10.1109/AICCSA.2018.8612856

[18] Kadadha M. Authentication at a scale. 28 Apr 2018. [Online]. Available from: https://github.com/mkadadha/AuthenticationAtAScale1.git. [Accessed: 30 Apr 2022]

Chapter 12

COVID-19 Data Analytics Using Extended Convolutional Technique

Anand Kumar Gupta, Asadi Srinivasulu, Kamal Kant Hiran, Tarkeswar Barua, Goddindla Sreenivasulu, Sivaram Rajeyyagari and Madhusudhana Subramanyam

Abstract

Health care system, lifestyle, Industrial growth, economy and livelihood of human-beings worldwide effected due to triggered global pandemic by COVID-19 virus originated and first reported from Wuhan city, Republic Country of China. COVID cases are difficult to predict and detect on its early stages due to that its spread and mortality is uncontrollable. RT-PCR (Reverse Transcription Polymerase Chain Reaction) is still first and foremost diagnostic methodology accepted worldwide, hence it creates a scope of new diagnostic tools and techniques of detection approach which can produce effective and faster results compared to its predecessor. Innovational through current studies that complements to the existence of COVID-19 to findings in Chest X-ray snap shots, the proposed research's method makes use of present deep getting to know models (U-Net and ResNet) to method those snap shots and classify them as the positive patient or the negative patient of COVID-19. The proposed technique entails the pre-treatment phase through dissection of lung, getting rid of the environment which does now no longer provide applicable facts and can provide influenced consequences; then after this, preliminary degree comes up with the category version educated below the switch mastering system; and in conclusion, consequences are evaluated and interpreted through warmth maps visualization. The proposed research method completed a detection accuracy of COVID-19 round 99%.

Keywords: COVID-19, classification algorithms, CNN, feature selection, ECNN, data pre-processing

1. Introduction

The disease referred to as "the extreme acute breathing syndrome coronavirus 2 (SARS-CoV-2)" was determined in year end of 2019. As per reports, this disease was originated in China, have become the reason of disorder referred to as "Corona Virus Disease 2019" or "COVID-19". The WHO (World Health Organization) has declared this disorder as a "deadly disease" in March 2020 [1, 2]. As per the reviews delivered and up to date with the aid of using worldwide health organizations, authorities/ entities and governments, pandemic has affected tens of thousands and thousands of human beings globally [3]. The maximum severe contamination due to COVID19 is associated with the lungs which include pneumonia. The signs and indications of the disorder may range & consist of excessive body temperature (high fever), dyspnea, coryza and cough. These instances can be normally recognized through the usage of lung x-ray evaluation of the irregularities [4].

Throughout this hasty period, several scholars have tried towards expand numerous transmission gear and diagnosing systems. Such as, the RT-PCR (Reverse Transcriptase-Polymerase Chain Reaction), which is still the vital testing technique to discover extreme severe breathing disease SARS-COV2 [1, 2, 4] and in addition to COVID-19 [3]. Though RT-PCR is considered to be the best method of screening so far, it still has limitations. The working system of Reverse Transcriptase-Polymerase Chain Reaction (RT-PCR) is complex and time-consuming [2, 4–11]. Thus, attempts were attempted to detect COVID-19 thru lung x-ray images which include CT (Computed Tomography) or lung x-beam photographs. It is said the investigative significance and accurateness of CT lung photographs over RT-PCR in COVID-19 [2] are highly accepted. The discoveries display that a lung CTs are an excessive sensitivity for the analysis of COVID [2].

2. Literature survey

The call for quicker analysis of COVID-19, more than one research carried out to focus on layout answers and clinical records concerning this exceedingly transmittable disease. Some picture identification, examination, clarification, and conclusion strategies were indexed on this segment. The DL (Deep Learning) method [12] has been projected and has efficaciously received satisfying outcomes in phrases of accurateness in diverse arenas [3, 13–17]. The instance research of COVID-19 examination of CT-scans had been offered with the aid of using authors together with Xu et al. [18], Srinivasulu [19], Qing et al. [20], Srinivasulu and Gangadhar [21]. Authors Xu et al. [18] mentioned that the COVID-19 well-known shows its' traits can change from different varieties of virus-related pneumonia, like viral influenza-A pneumonia. The study's goal has become to the broaden a preliminary testing outline for COVID-19 with the aid of using automatic respiratory CT-scans (CT photographs) of COVID-19, pneumonia, and ordinary instances. They hired 628 CT-scans test pattern photographs earlier than expansion, and their version acquired the accurateness of 90%. The writers' approach consists of the image pre-processing, dissection of more than one region (patches) accepting V-Net (Volumetric Network) [22] based separation version V-Net-IR-RPN [23], that has skilled for pulmonic tuberculosis resolution.

Our method includes 3 essential experiments to assess the overall performance of the predication and determine of an effect on of the distinctive levels of the procedure. Respective test follows the workflow. The distinction among trials are the dataset used from various repositories. In all occurrences, identical photographs of COVID-19 effective instances had been used. Meanwhile, 3 distinctive datasets for poor instances had been utilized. In that direction, Experiment 1 and 2 included comparing effective vs. poor instances datasets, and Experiment three entails Pre-COVID generation photographs (photographs from 2015 to 2017).

3. System methodology

3.1 Existing system

There are approachable in superficial learning strategies, for example, The Convolution Neural Organization and Intermittent Neural Organization. CNN computation Drawbacks: The disservices are:

- Little precision
- In flood time complexity
- In flood executing time
- In flood fault prone
- Insignificant data size

Computation downside:

- Little precision
- In flood time complexity
- In flood execution time
- In flood fault prone
- Little data size

3.2 Proposed system

There are available in deep learning method like Extended Convolutional Neural Networks i.e., in Deep Learning Technique.

ECNN algorithm advantages:

- In flood accurateness
- Fewer time consumption
- Little performance time
- Little mistake degree
- Big data scope

4. Results

Basic idea is to execution, is to assure that the Omicron disease severer affected role collected statistics functioned in the way that can compel preparation, subdivision from their first outlook.

4.1 ECNN algorithm

Two trials of one or the other CC or MLO seen should be adjusted utilizing the picture enlistment method. At that argument, a dissimilar picture was received by removing the former trial out of the existing trial and subsequently scaled to the full-range force. The territorial pictures from the refined district proposition are trimmed from the three pictures and scaled to 224 × 224 × 3 for each picture, which are utilized for ECNN is floodlight extraction. The three channels are rehashed from one-channel grayscale pictures (e.g., the current sweep of 224 × 224 × 1) since the pertained ECNN and ECNN models expect 3-channel pictures. Multi-measurements of three-state in floodlights (from earlier sweep, current output, and contrast pictures) are made to prepare a CNN model. For instance, The ECNN is floodlights utilizing ResNet-60 V3 of 2048 × 3 measurements for each view (CC or MLO) of a subject's side (left or right bosom). Remember that earlier sweep consistently relates to the ordinary (sound) status in any event, for a destructive subject. Assume we code sound and carcinogenic as 0 and 1 individually, at that point the ground realities (yields) compared to the three states (earlier, current, distinction) of the destructive view are [0 1 1]. This coding instrument can be handily stretched out to at least two earlier sweeps.

4.2 Algorithm

The following is the ECNN algorithm steps:

The Omicron disease infection data index, i.e., the absolute 522 pictures, our experiment involved the related following steps:

1. Introduces mandatory collection.

2. Introduces training dataset.

3. Executes in the floodlight ordering of change data.

4. Alignment with 70-time segments and 2 yield.

5. Introduces Keras (Keras is a Deep Learning library).

6. Resets ECNN.

7. Enhances ERNN part & about regulation of loss calculation function.

8. Improvement of yield part.

9. Adds the ECNN.

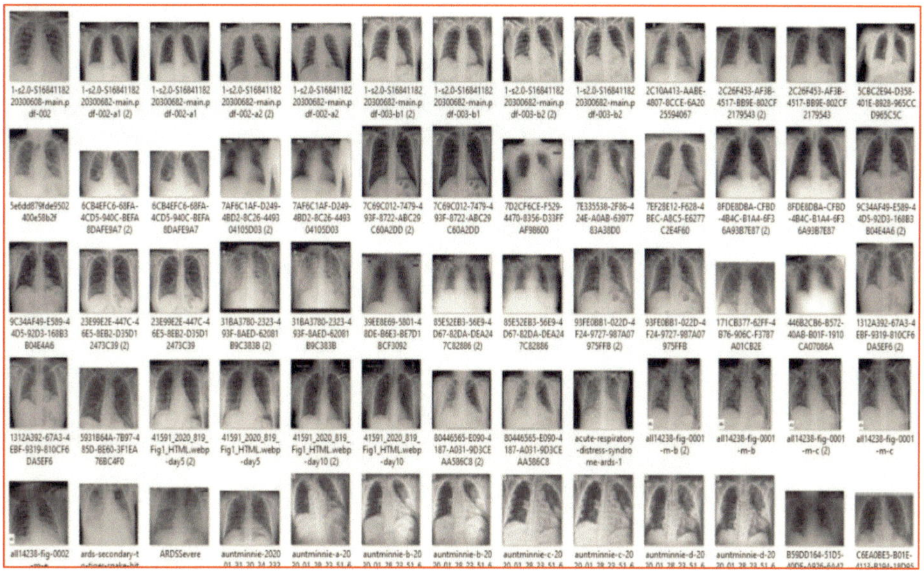

Figure 1.
Input dataset of the projected prototype for COVID-19 disease detection.

10. Adjusts ECNN in the assessment dataset.

11. Loads the Omicron disease infection test image data of the year 2020.

12. Become a predicted Omicron disease infection in Dec 2019.

13. Imagine aftereffects with anticipated or genuine Omicron disease infection.

INPUT DATSET: Here the input dataset is having 16 columns with target class, i.e., severity level of the COVID-19 disease consisting the database of 282 sample x-ray images (**Figure 1**).

5. Results: here are the result of in finding COVID-19 disease detection by integrating ECNN

Figure 2 shows the execution flow of the ECNN code on COVID-19 database analyzing time taken, accuracy, loss, Val_Loss, Val_Accuracy with respect to epochs.

The proposed model achieve the accuracy of 99% on the database collected and used from Kaggle, and UCI repositories (**Figure 3**).

Figure 4 displays the CPU and related resources occupancy of computer during ECNN code execution on COVID-19 database.

5.1 Evaluation methods

The following are measurements of evaluation methods or metrics.

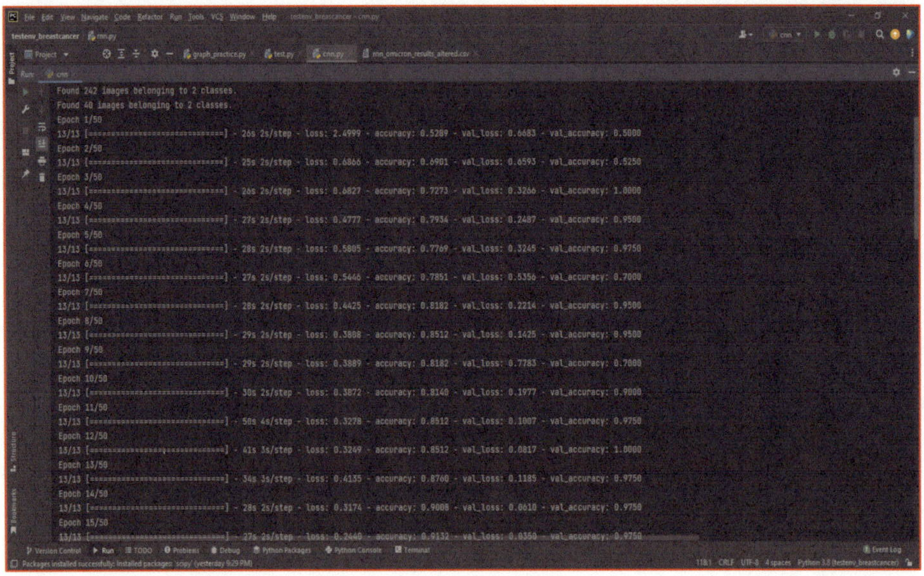

Figure 2.
ECNN code execution flow.

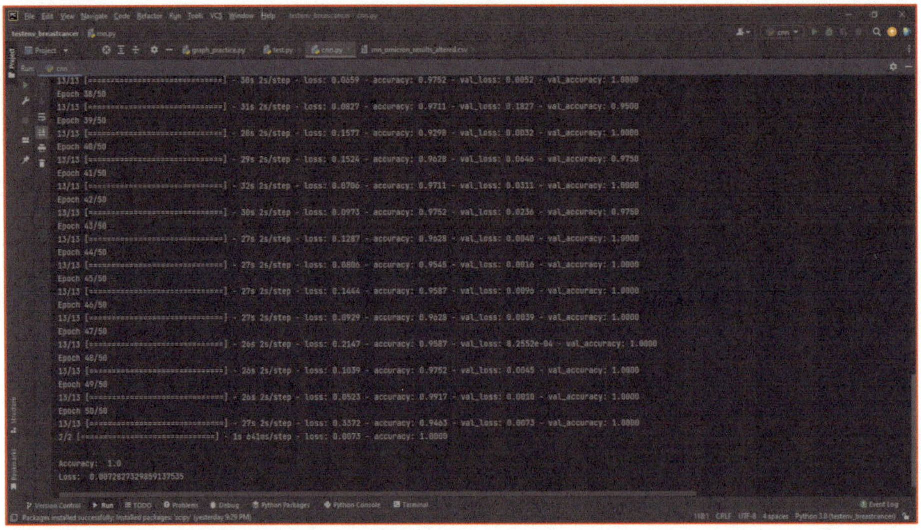

Figure 3.
Final results of COVID-19 using ECNN approach.

$$Quality = \frac{BP + VM}{BP + VP + BM + VM} \quad (1)$$

$$Preciseness = \frac{BP}{BP + VP} \quad (2)$$

$$Callback = \frac{BP}{BP + VM} \quad (3)$$

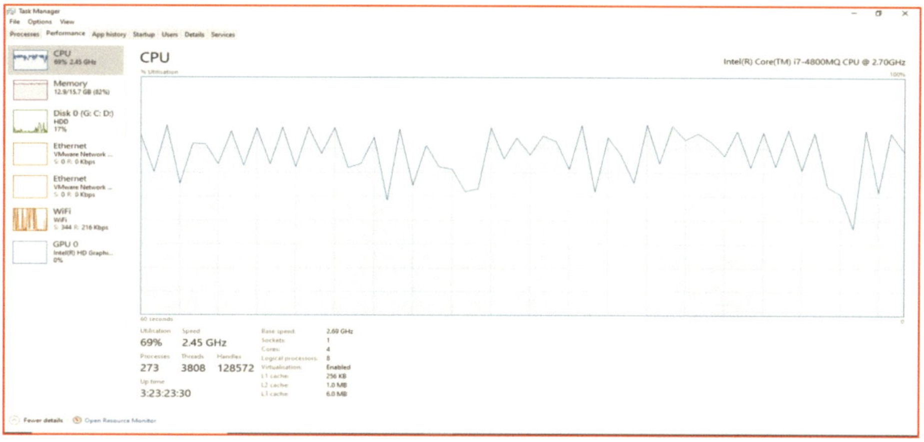

Figure 4.
Processor and related resources occupancy of computing device.

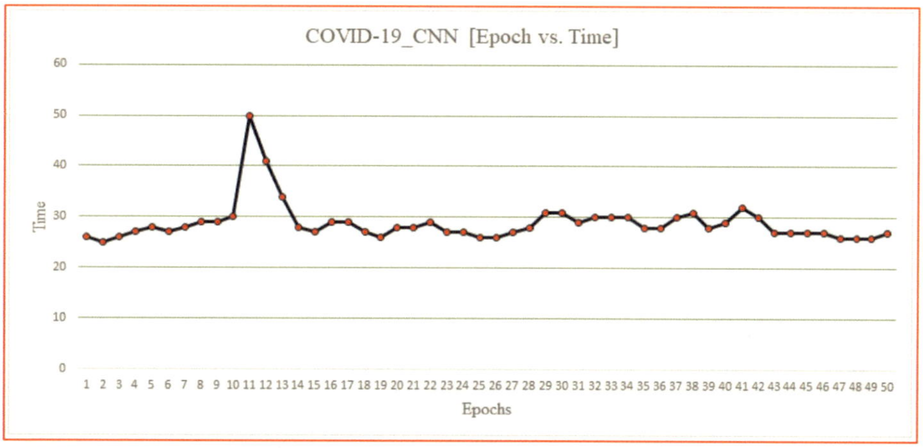

Figure 5.
COVID-19 ECNN graph comparing epochs vs. time.

$$F - measure = \frac{2xPrecisenessxCallback}{Preciseness + Callback} \quad (4)$$

Data Input: Our experiment was carried over on a database of 282 x-ray images.

Figure 5 demonstrates the time taken to complete iteration of epochs.

Explains the loss ratio with respect to each epochs during execution (**Figure 6**).

Demonstrates the accuracy achieved against each epochs during execution (**Figure 7**).

Demonstrates the loss reduction and accuracy gain of the training model of ECNN with respect to each iteration (**Figure 8**).

Value loss and value accuracy gained of the ECNN model during training (**Figure 9**).

At a glance representation of the comparison among epochs, loss, accuracy of the ECNN model (**Figure 10**).

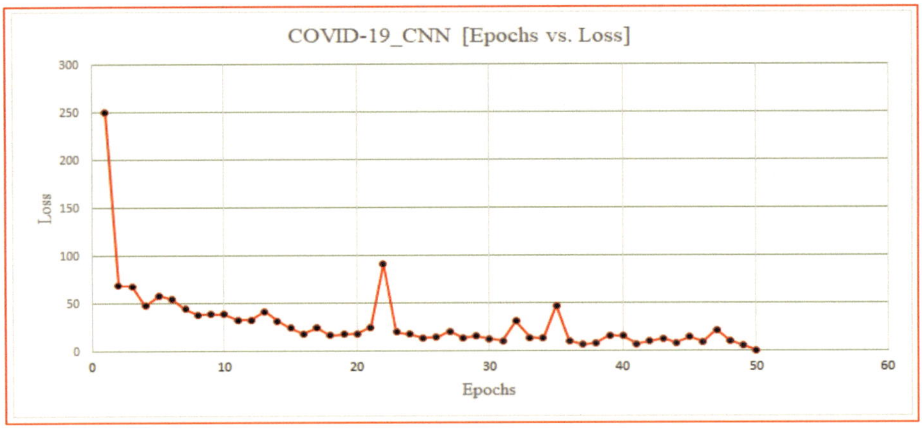

Figure 6.
COVID-19 ECNN graph comparing epochs vs. loss.

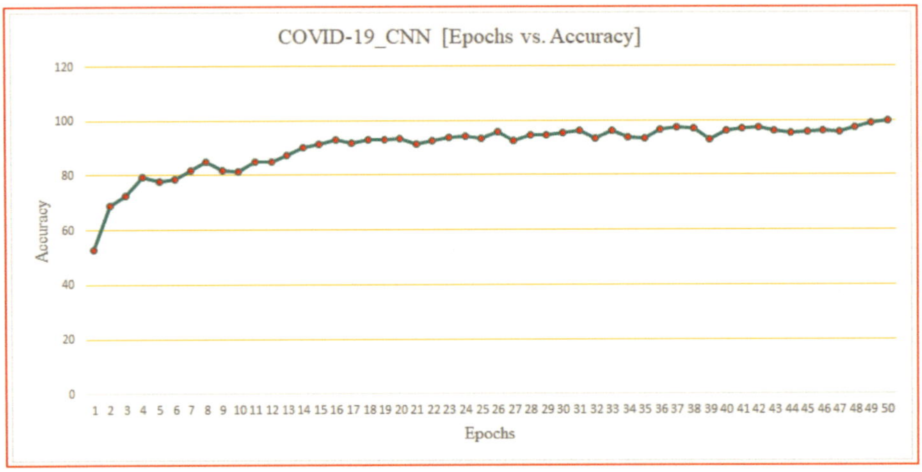

Figure 7.
COVID-19 ECNN graph comparing accuracy vs. epoch.

6. Conclusions

This method suggests how the present prototype may be beneficial for more than one tasks, specifically if it's far taken into consideration that the modified U-Net prototypes do now no longer have higher overall performance. Also, is proven how x-ray images' noise may produce predisposition withinside the prototypes. Most metrics display photographs without dissection as higher for categorizing COVID infections. Additional evaluation suggests that even though benchmarks are higher, those prototypes are primarily created on totally seen diagonosis throughout lung's x-ray as clean proof of COVID, so actual correct prototypes ought to be centered on lungs elements for classification. In this situation, dissection is desired for dependable outcomes through lowering this bias. Transfer getting to know changed into crucial of the outcomes offered. As proven categorized models, the use of this approach wants among 40 and 50 epochs to converge, even as segmentation prototypes without

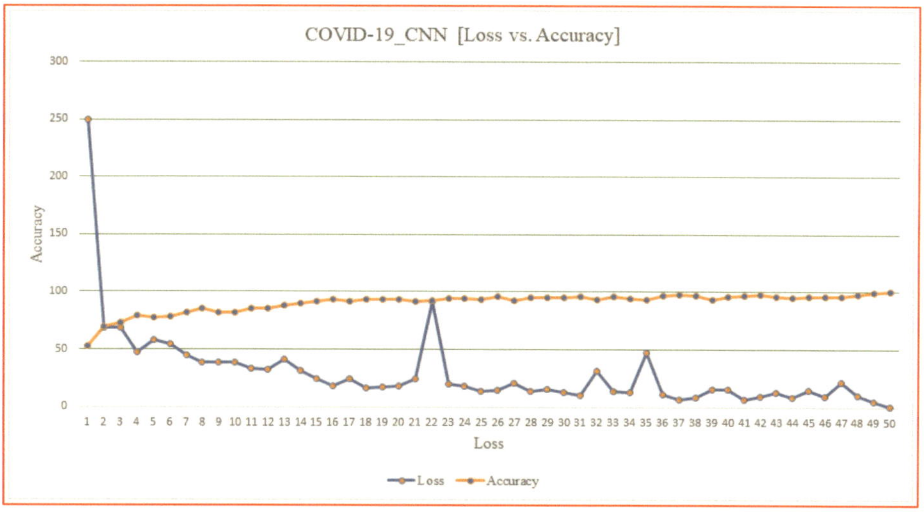

Figure 8.
COVID-19 ECNN graph comparing loss vs. accuracy.

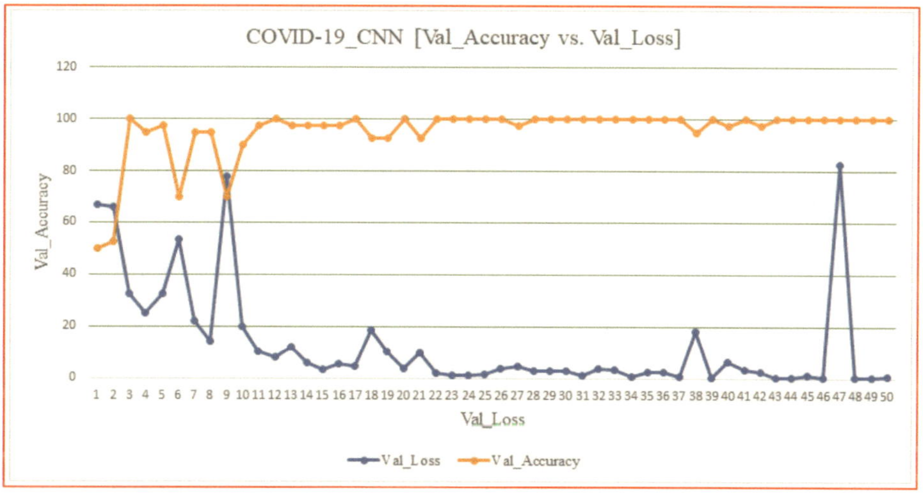

Figure 9.
COVID-19 ECNN graph comparing Val_Accuracy vs. Val_Loss.

modification was approximately 282. The sequence of prototypes was obtainable to decide COVID-19 Disease in Chest X-ray photographs with a general accuracy of 99% through categorizing COVID-positive and COVID-negative images. In the meantime, solitary for the COVID label, the method achieved an average of 98.58% accuracy withinside the take a look at the database for a threshold of 0.4. Changing the edge suggests a growth withinside the accuracy of prototypes as much as 99%.

The segmentation work suggests an excessive opportunity of imparting more statistics to element in the all experimentations, concluding the unconventional outcomes through dissecting lungs and including statistics mixed with surrounding noise. The noise is related to wires used in medical equipment's, patient's gender and/or age, making photographs without lungs have extra information for classification in those

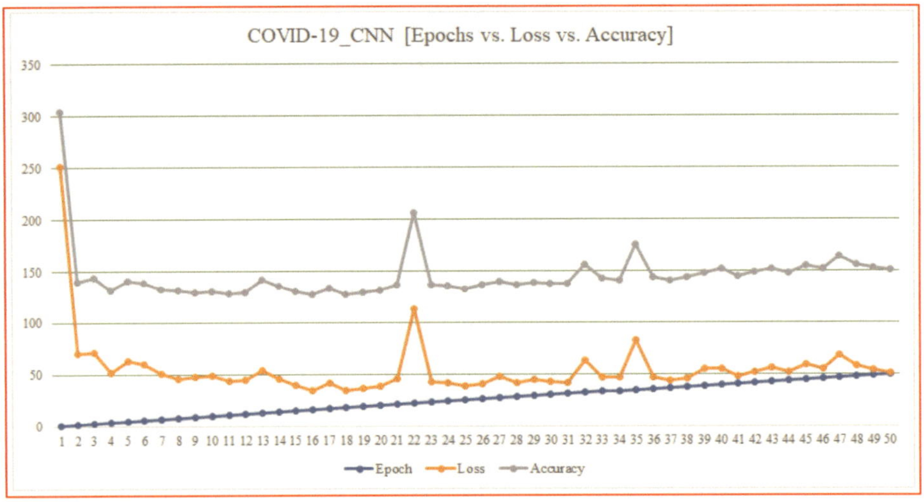

Figure 10.
COVID-19 ECNN graph comparing epoch vs. loss vs. accuracy.

situations. Either destiny efficacy or the use of prototypes without lungs should have to be the very best possibilities of mislabeling photographs due to errors. Further researches are required of section diagnosis recognized by the expert radiotherapist to make sure that any noise is a causing object for biased results. It is likewise critical to spot that the outcomes offered do now no longer always suggest the identical overall performance in each database. For example, the used database was collected of Asian victims; different international sufferers might also additionally display minor facts seize modifications or diagnosis, assuming a higher type is wanted the use of international databases. In addition, setting apart the databases through gender will offer extra statistics at the prototype's scope, because the tiny tissues of the chest might also additionally cover elements of the lungs, & it is far unidentified in case or not that it is taken into consideration a partiality with inside the forecast of the prototypical.

Author details

Anand Kumar Gupta[1,2]*, Asadi Srinivasulu[2], Kamal Kant Hiran[3], Tarkeswar Barua[4], Goddindla Sreenivasulu[5], Sivaram Rajeyyagari[6] and Madhusudhana Subramanyam[7]

1 Azteca University, Mexico

2 BlueCrest University, Monrovia, Liberia

3 Symbiosis University of Applied Sciences, Indore, India

4 Capgemini Ltd, New Delhi, India

5 Andhrapradesh, Tiruapti District, India

6 Shaqra University, Shaqra, Saudi Arabia

7 K.L University, Guntur, Andhra Pradesh, India

*Address all correspondence to: ganand40@yahoo.co.in

References

[1] Arias-Garzón D, Alzate-Grisales JA, Orozco-Arias S, et al. COVID-19 detection in X-ray images using convolutional neural networks. Machine Learning with Applications. 2021;**6**

[2] Reshi AA et al. An efficient CNN model for COVID-19 disease detection based on X-ray image classification. Hindawi Complexity. 2021;**2021**:12

[3] Sitaula C, Aryal S. New bag of deep visual words based features to classify chest x-ray images for COVID-19 diagnosis. Health Information Science and Systems. 2021;**9**

[4] Sarki R, Ahmed K, Wang H, Zhang Y, Wang K. Automated detection of COVID-19 through convolutional neural network using chest x-ray images. PLoS One. 2022;**17**(1):e0262052

[5] Aggarwal CC. Neural Networks and Deep Learning. Springer International Publishing AG, Part of Springer Nature 2018; 2020. pp. 351-352

[6] Ai T, Yang Z, Hou H, Zhan C, Chen C, Lv W, et al. Correlation of chest CT and RT-PCR testing for coronavirus disease 2019(COVID-19) in China: A report of 1014 cases. Radiology. 2020; **296**(2):E32-E40

[7] Apostolopoulos ID, Mpesiana TA. Covid-19: Automatic detection from Xray images utilizing transfer learning with convolutional neural networks. Physical and Engineering Sciences in Medicine. 2020;**43**(2):635-640

[8] Medical Imaging Databank of the Valencia region BIMCV .2020. BIMCV-Covid19–BIMCV

[9] Bravo Ortíz MA, Arteaga Arteaga HB, Tabares Soto R, Padilla Buriticá JI, Orozco-Arias S. Cervical cancer classification using convolutional neural networks, transfer learning and data augmentation. Revista EIA. 2021;**18**(35):1-12

[10] Cucinotta D, Vanelli M. WHO declares COVID-19 a pandemic. Acta Biomedica: Atenei Parmensis. 2020;**91**: 157-160

[11] Rustam F, Reshi AA, Mehmood A, et al. COVID-19 future forecasting using supervised machine learning models. IEEE Access. 2020;**8**:101489-101499

[12] X-ray (Radiography)-Chest. 2020. Available from: https://www.radiologyinfo.org/en/info.cfm?pg?chestrad#overview

[13] Cennimo DJ. Coronavirus disease 2019 (COVID-19) clinical presentation. Medscape. 2020;**8**:101489-101499. Available from: https://emedicine.medscape.com/article/2500114-overview

[14] Cohen JP. Github Covid19 X-ray dataset. 2020. Available from: https://github.com/ieee8023/covid-chestxray-dataset, 2020

[15] Wang W, Xu Y, Gao R, Lu R, Han K, Wu G, et al. Detection of SARS-CoV-2 in different types of clinical specimens. Journal of the American Medical Association. 2020;**18**:1843-1844

[16] Ai T, Yang Z, Hou H, Zhan C, Chen C, Lv W, et al. Correlation of chest CT and RT-PCR testing in coronavirus disease 2019 (COVID-19) in China: A report of 1014 cases. Radiology; **2020**:200642

[17] Liu Y, Whitfield C, Zhang T, Hauser A, Reynolds T, Anwar M. Monitoring COVID-19 pandemic through the Lens of social media using natural language processing and machine learning.

Health Information Science and Systems. 2021;**9**

[18] Guan W et al. Clinical characteristics of coronavirus disease 2019 in China. New England Journal of Medicine. 2020; **382**(18):1708-1720

[19] Srinivasulu A. Early prediction of Covid-19 using modified recurrent neural networks. Journal of Infectious Diseases and Treatment. 2021;**07**(08). Available from: https://infectious-diseases-and-treatment.imedpub.com/ and https://www.primescholars.com/articles/

[20] Qing L, Cai W, Wang X, Zhou Y, Feng DD, Chen M. Medical image classification with convolutional neural network. In: 2014 13th International Conference on Control Automation Robotics & Vision (ICARCV). 2014. pp. 844-848. DOI: 10.1109/ICARCV.2014.7064414

[21] Srinivasulu A, Gangadhar Ch, et al. Association of vaccine medication for the efficacious COVID-19 treatment. World Journal of Engineering. DOI: 10.1108/WJE-01-2021-0062

[22] Srinivasulu A, Barua T. COVID-19 virus prediction using CNN and logistic regression classification strategies. Journal of Data Analysis and Information Processing. 2022;**10**:78-89. DOI: 10.4236/jdaip.2022.101005

[23] Srinivasulu A, Barua T. Early prediction of Covid-19 using modified convolutional neural networks. International Journal of Advanced Computational Engineering and Networking. 2022;**9**(4). DOI: 10.1007/978-981-16-5090-1_6

Chapter 13

Perspective Chapter: Blockchain-Enabled Trusted Longitudinal Personal Health Record

Yibin Dong, Seong K. Mun and Yue Wang

Abstract

In the United States, longitudinal personal health record (LPHR) adoption rate has been low in the past two decades. Patients' privacy and security concern is a major roadblock. Patients like to control the privacy and security of their own LPHR distributed across multiple information systems at various facilities. However, little is known how a scalable and interoperable LPHR can be constructed with patient-controlled security and privacy that both patients and providers trust. As an effort to increase LPHR adoption rate and improve the efficiency and quality of care, we propose a blockchain-enabled trusted LPHR (BET-LPHR) design in which security and privacy are protected while patients have full control of the access permissions. Two limitations associated with the proposed design are discussed. Options and practical resolutions are presented to stimulate future research.

Keywords: longitudinal personal health record, security, privacy, confidentiality, permissioned blockchain

1. Introduction

LPHR is "an electronic, lifelong resource of health information needed by individuals to make health decisions" [1] and to "improve the quality and efficiency of their own health care" [2, 3]. Building electronic health record (EHR) was required by "the Health Insurance Portability and Accountability Act of 1996 (HIPAA)" [4]. The first "HIPAA Privacy Rule was released" [4] to "improve privacy standards and to restrict the disclosure of Protected Health Information (PHI) and personal identifiers to unauthorized individuals" [4]. In 2009, "the Health Information Technology for Economic and Clinical Health Act (the HITECH Act) was enacted" [5] to remediate a loophole in HIPAA Privacy Rule and promote personal health record (PHR). Untethered (cross-organizational) [6] PHR has been a preferred choice of building LPHR [6]. LPHR is attractive to patients because patients can have holistic view of their health information that are scattered in multiple information systems at various facilities. Federal agencies and local governments have been promoting PHR adoption in the past two decades with numerous "laws and regulations, incentives, and penalties" [3, 7, 8]. However, "the LPHR adoption rate has been low" [3, 8] in the United States. A.A. Abd-alrazaq et al. found out that "patients' privacy and security concerns is a major negative factor impacting LPHR adoption" [3, 7]. Patients like to fully control privacy and security of their own LPHR [3, 7]. "However, little is known how to model and construct a scalable and interoperable LPHR with patient-controlled security and privacy that both patients and providers trust" [3]. Solving this problem is "considered important to increase LPHR adoption rate and improve the efficiency as well as the quality of care" [3].

To protect the security and privacy of LPHR, encryption is an intuitive and good choice of solution [9–16]. Encryption can prevent external security attack, however, it cannot defend against insider threat [3, 17]. We argued that insider threat can be remediated via a secure access control model that is implemented correctly at user or session or process level [3, 18, 19]. Combining access control model with encryption is a better resolution. Traditional access control model, in which users are well known, has been used to couple with attribute-based encryption (ABE) as an approach. However, in PHR systems, users can be known or unknown. To overcome this problem, we proposed next generation access control which offers "open access surroundings" where "users can be centrally known or unknown" [3]. We chose the "National Institute of Standards and Technology (NIST)" "Next Generation Access Control (NGAC)" [20], a type of "attribute-based access control (ABAC)" model [3, 21]. Nevertheless, NGAC suffers a race condition in distributed system. This led to our proposal of a "novel Blockchain-enabled Next Generation Access Control (BeNGAC) model" [3] using permissioned blockchain that can ease the race condition. We explained the merits of the new model with additional benefits brought by blockchain technology. We offered the freedom of choice of encryption methods to PHR generators. With BeNGAC as the core of the LPHR access control mechanism, we designed the BET-LPHR that both patients and providers can trust. We discussed two application limitations of the design: a) when the secret private key is lost; b) when the patient cannot directly authorize the access. Possible solutions are offered to solve the limitations. We also compared our approach with prior works.

2. LPHR requirements

The LPHR requirements are summarized in **Table 1**.

3. BeNGAC

Encryption is a good choice of protecting privacy and confidentiality of PHR on premises and in the cloud. Invented by Amit Sahai and Brent Waters [9], ABE and its extensions have been researched extensively and applied to PHR confidentiality and privacy protection [9–16]. However, encryption alone cannot prevent insider threat [3, 17]. Miller & Tucker [17] suggested applying access control to remediate insider threat [3, 17]. Akshay Tembhare et al. combined role-based access control (RBAC) model with ABE to protect PHR in the cloud [25]. However, RBAC is a type of traditional access control method that users are known. PHR users can be centrally known or unknown, in which next generation access control model is a better fit.

Requirements	Interpretation
Security: Availability	LPHR "is accessible and usable on demand by authorized persons" [22]
Security: Integrity	LPHR "is not altered or destroyed in an unauthorized manner" [22]
Security and Privacy: Confidentiality	Disallow unauthorized use and disclose of LPHR
Privacy	Use or disclose of LPHR either requires authorization from patients or is obligated to local, state, or federal laws [23, 24]
Authorization	Patients have full control of authorizing LPHR access to other health care providers.
Tamper resistance	Automatically detect and prevent any unauthorized modification.
Access Auditability	Access to LPHR is auditable.
Scalability	LPHR system is enterprise scalable.
Distributedness	LPHR is distributed in EHR vendors and patient's PHR.
Interoperability	LPHR allows sharing information with other EHRs and stakeholders.
Integration	LPHR allows integrating with smart and wearable personal devices.

Table 1.
LPHR requirements.

This led us to choose the NIST's NGAC as an authorization model. Furthermore, NGAC meets the LPHR distributedness requirement because NGAC provides unified access control policies and resources reinforcing the policies are distributed [3, 21]. Moreover, NGAC is scalable at enterprise level [21] which "fulfills the LPHR scalability requirement" [3].

Nevertheless, in distributed system, NGAC suffers a "race condition" when access control policies are centralized while decisions making processes are localized [26]. To solve this problem, we proposed a decentralized yet distributed access control policy expression unit by using permission blockchain technology Hyperledger Fabric (HF) [3]. We introduced a novel BeNGAC model [3]. In LPHR, patients and providers trust each other, which matches the property of permissioned blockchain. The "race condition" in NGAC is eased by HF "concurrency control" [3, 27] contributed by "HF consensus" [3, 27]. The access control policy information is immutable by inheriting HF's immunizability property. The blockchain transaction audit logs are on chain while the access control policies are stored in private off-chain database [3]. Furthermore, NGAC access control policies compensate HF's weak confidentiality protection. The BeNGAC architecture is sketched in **Figure 1**. "Policy enhancement point (PEP), policy decision point (PDP), event processing point (EPP), and resource access point (RAP)" [3, 26] are distributed and act locally. The policy administration unit (PAU) consists of blockchain-enabled policy administration point (BePAP) and blockchain-enabled policy information point (BePIP) that are decentralized. An application requests to access BET-LPHR. The request is processed by PEP. PEP relays the request to decision maker PDP. PDP queries the policy database BePIP via BePAP. The request is processed and a grant or deny decision is sent to the application via PEP. If the decision is to "allow", the application will send a request to access the BET-LPHR through RAP.

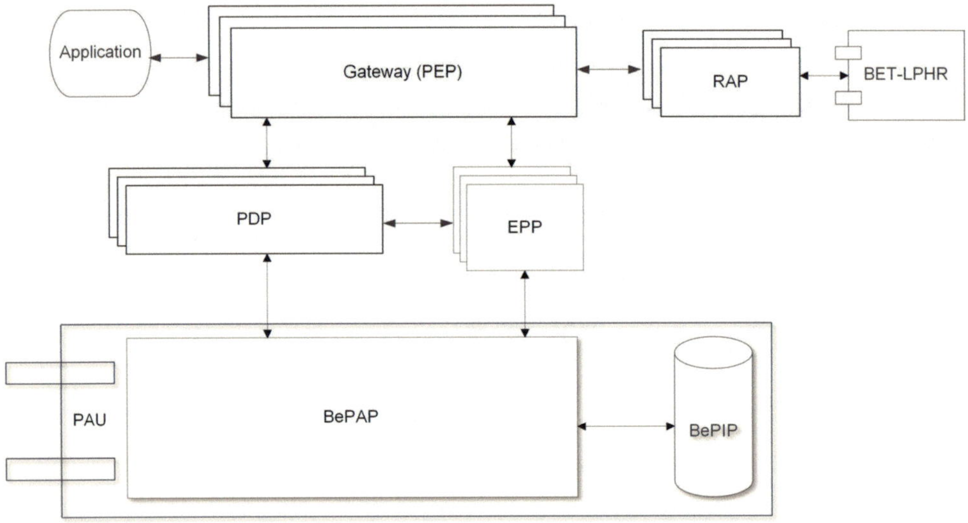

Figure 1.
BeNGAC architecture [3, 26]

4. BET-LPHR

The BET-LPHR consists of two partitions of data: 1) patient self-generated PHR such as data from personal wearable devices; 2) PHR data replicated from the EHRs the patient has visited.

Figure 2 illustrates the BET-LPHR model. VistA_EHR_A and VistA_EHR_B represent the EHR providers that the patient has visited. The patient, VistA_EHR_A, and VistA_EHR_B form a trusted network and are connected by HF BeNGAC policy secure channel. The patient (considered as one organization in this setting) and the two EHR organizations share the same access control policies. "The patient has full

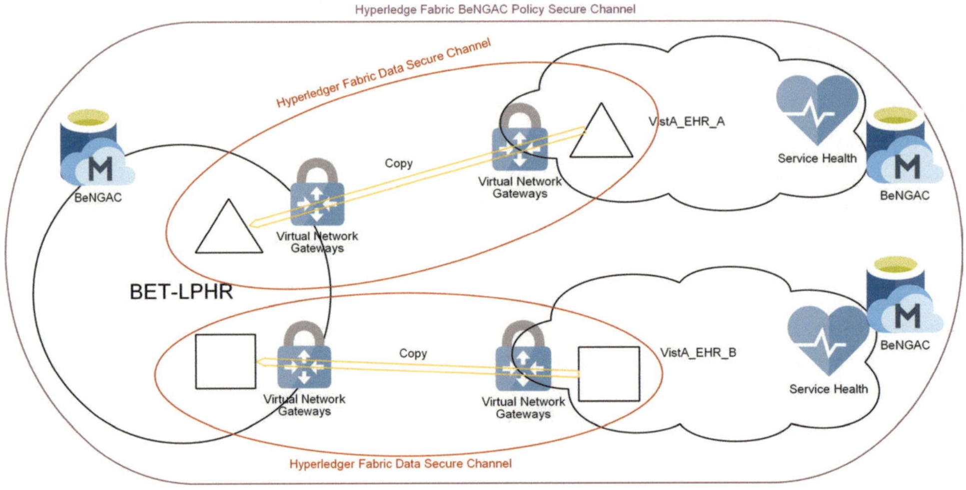

Figure 2.
BET-LPHR model.

control of granular permissions on his or her own LPHR" [3] using the shared access control policies that are realized via a Web-based interface presented to the patients. "The blockchain based peer-to-peer BeNGAC database avoids racing condition during policy reinforcement" [3]. The data sharing operates on a different type of HF communication channel, HF data secure channel. The patient and VistA_EHR_A constitute a trusted HF data secure channel. The patient's PHR data in VistA_EHR_A (triangle shape) is copied (disclosed) to the patient's BET-LPHR. Similarly, BET-LPHR has a PHR copy (square shape) from VistA_EHR_B on a different HF data secure channel. BET-LPHR is distributed to three organizations. Among the trusted HF data sharing channel, BET-LPHR is decentralized (or peer-to-peer).

At a high level, BET-LPHR data flow is summarized in **Figure 3**. The patient and the EHR organizations are communicated via Fast Healthcare Interoperability Resources (FHIR) interface to ensure interoperability. "Digital certificate guarded secure authentication and BeNGAC policies meet the LPHR security and privacy requirements" [3]. Both NGAC and HF are enterprise scalable, so BET-LPHR is enterprise scalable to meet LPHR scalability requirement. The authentication to BET-LPHR relies on asymmetric cryptography private keys. Data in transit is encrypted by the public key and protected by transport layer security (TLS). Data at rest in EHR's silos are protected by the encryption methods the EHR organizations freely choose.

BET-LPHR offers unique event processing capabilities, such as prohibition or obligation, that is inherited from BeNGAC. This fills a gap in traditional access control model when handling insider threat. For example, in RBAC, a doctor is authorized to read a patient's record, but a nurse is not. The RBAC does not prohibit the doctor copy and paste a record to a file that the nurse has permission to read. In BET-LPHR, this is remediated by the prohibition policy as the following [3]:

> "Configuration Rule 1:
> When process p performs (r, o) where $o \rightarrow med\_rec$ do create $u\_deny(p, \{w\}, \neg o)$.
> Configuration Rule 2:
> When process p performs $(copy, o)$ do create $u\_deny(p, \{w\}, \neg o)$"

The doctor authenticates with a user session and the session launched a read (r) process (p). When the process p performs a *read* or *copy* operation on an object which is assigned to the patient's medical record (med_red) attribute, it triggers a prohibition condition to deny writing to an object that is not the object being read.

The LPHR requirements are fulfilled by the BET-LPHR design and are summarized in **Table 2**.

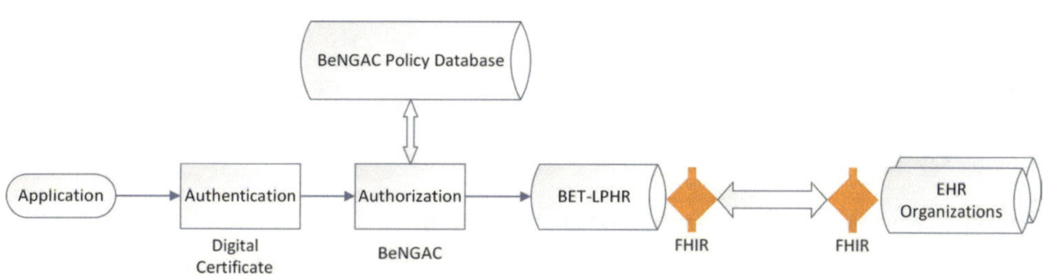

Figure 3.
BET-LPHR architecture.

Requirements	NGAC	HF Blockchain	FHIR
Security	X	X	
Privacy, Confidentiality	X	Some	
Integrity: changes to a LPHR - Tamper Resistance	X	X	
Access Audit		X	
Scalability	X	X	
Distributedness	X	X	
Interoperability and Integration			X

Table 2.
BET-LPHR requirements fulfillment.

5. Limitations and alternatives

There are few limitations of the BET-LPHR design. Some are inherited from EHR and PHR, others are from blockchain technology itself.

5.1 When BET-LPHR owner lost the secret private key

From authentication to authorization, owner's secret private key is essential to unlock the patient's BET-LPHR as a passport to securely administrate BET-LPHR. Being a problem inherited from the blockchain technology itself, losing the secret private key presents a barrier.

We propose a separate blockchain-enabled SecureKey recovery process with a secure login portal using the "BeNGAC and RBAC Separation of Duty (SoD) capability" [3]. Accessing to this portal requires strong multi-factor authentication (MFA) with Fast Identity Online (FIDO) 2.0 biometrics [28]. The patient secret key pair is generated by the key administrator and delivered to the owner in a secure manner. At the same time, a recovery key pair is also generated and sent to the owner. The key administrator and the owner possess the recovery public key, but only the owner has the BET-LPHR patient secret key pair and recovery private key. The keys roles and permissions are described in **Table 3**.

At the BET-LPHR patient key pair generation time, a copy of the secret private key (*SecPrivKey*) is encrypted with the recovery public key (*RecPubKey*). The encrypted secure private key (*SecPrivKey*⊙*RecPubKey*) and recovery private key (*RecPrivKey*) are stored in a *Blockchain-Enabled SecureKey* database with read only permission.

Role	Key administrator	BET-LPHR patient/owner
BET-LPHR Patient Secret Key Pair		Secure Public Key (*SecPubKey*)
		Secure Private Key (*SecPrivKey*)
Recovery Key Pair	Recovery Public Key (*RecPubKey*)	Recovery Public Key (*RecPubKey*)
		Recovery Private Key (*RecPrivKey*)

Table 3.
Roles of key administrator and BET-LPHR patient/owner.

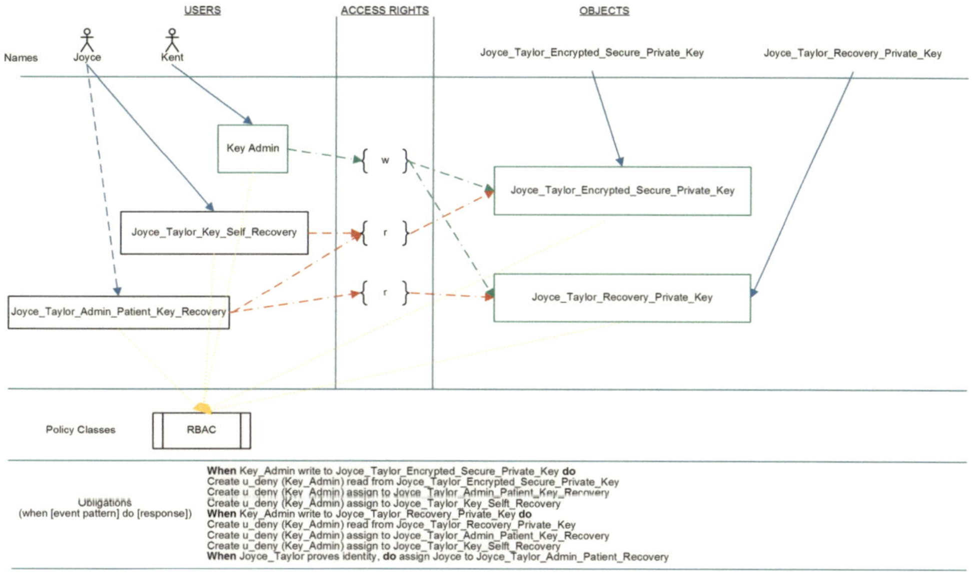

Figure 4.
Secure private key recovery.

When the BET-LPHR owner lost the secure private key (*SecPrivKey*), there are two scenarios for the BET-LPHR owner to recover the key (**Figure 4**):

- **Scenario #1**. If the BET-LPHR owner has the recovery private key (*RecPrivKey*)., the BET-LPHR owner can send a key recovery request through a private key recovery login Web page. The patient will be authenticated with the recovery private key (*RecPrivKey*). Once authenticated, the patient will go through a multi-factor authentication process to answer some secure questions to prove the identity and then retrieve the encrypted secure private key ($SecPrivKey \odot RecPubKey$). The secure private key can be recovered by using ($SecPrivKey \odot RecPubKey \odot RecPrivKey$)

- **Scenario #2**. If the BET-LPHR owner lost both secure private key (*SecPrivKey*) and recovery private key (*RecPrivKey*), the owner can request the Key Administrator to recover the encrypted private key. The owner must go through an identity proof process. Once the owner is identified as the owner of the secure private key (*SecPrivKey*) and recovery private key (*RecPrivKey*), the policy administrator will create a one-time login credential for the owner to login to the private key recovery login Web page, and assign the owner to a user attribute which has read access to the encrypted secure private key ($SecPrivKey \odot RecPubKey$) and the recovery private key (*RecPrivKey*). For example, Joyce Taylor is assigned to Joyce_Taylor_Admin_Patient_Recovery attribute, which has capability of reading the Joyce_Taylor_Encypted_ Secure_Private_Key and Joyce_Taylor_Recovery_ Private_Key. The following BeNGAC obligation rules apply:

  When Key_Admin write to Joyce_Taylor_Encrypted_Secure_Private_Key **do**

  Create u_deny (Key_Admin) read from Joyce_Taylor_Encrypted_Secure_Private_Key

Create u_deny (Key_Admin) assign to Joyce_Taylor_Admin_Patient_Key_Recovery

Create u_deny (Key_Admin) assign to Joyce_Taylor_Key_Selft_Recovery

When Key_Admin write to Joyce_Taylor_Recovery_Private_Key **do**

Create u_deny (Key_Admin) read from Joyce_Taylor_Recovery_Private_Key

Create u_deny (Key_Admin) assign to Joyce_Taylor_Admin_Patient_Key_Recovery

Create u_deny (Key_Admin) assign to Joyce_Taylor_Key_Selft_Recovery

When Joyce_Taylor proves identity, **do** assign Joyce to Joyce_Taylor_Admin_Patient_Recovery

5.2 When the BET-LPHR owner cannot directly authorize the access to the third party

"In general, the BET-LPHR owner can grant permission to a legitimate third party, for instance a specialty doctor he or she will visit during a referral encounter. There are situations such as emergency departments visit, where BET-LPHR access is desired by the ER physicians to make better decision of a care plan by using the patient's BET-LPHR information such as medications taken, allergy conditions, recent doctors' visits, chronic diseases, and recent laboratory test results [29]. However, as a limitation that the patient need to directly grant the permission of BET-LPHR to the doctors in ER, it is not uncommon that the patient is unconscious and cannot authorize the access of his or her BET-LPHR to the doctors in ER facility" [3].

We propose a viable solution using BeNGAC RBAC and Discretionary Access Control (DAC) policies with obligations in the following example (**Figure 5**):

In this scenario, an object attribute Joyce_Taylor_SharedWith_ER_Doctor is created during user profile initialization. A user attribute ER_Doctor is also configured with read only permission on Joyce_Taylor_SharedWith_ER_Doctor. The patient's subset of BET-LPHR that are required during an emergency department visit are assigned to the object attribute Joyce_Taylor_SharedWith_ER_Doctor. The information includes the current medications taken, patient's chronic allergy conditions, doctors' visits in the past 3 months, patient's chronic diseases information, and laboratory test results in the past 90 days. The information is automatically assigned to the Joyce_Taylor_SharedWith_ER_Doctor when a new BET-LPHR record is added either by the encounter during doctor's visit in VistA_EHR_A or when patient records a new self-generated health record to the BET-LPHR. Additionally, a retrospective process can run on demand to assign or unassign patient's BET-LPHR to the Joyce_Taylor_-SharedWith_ER_Doctor.

When the patient is in an emergency department and lost consciousness, the ER physician can send a request to the patient's BET-LPHR administrator along with doctor's identification for a temporary account in patient's BET-LPHR. The doctor (Edward) with this account is assigned to the ER_Doctor role. The account will be disabled after 72 hours of account creation. The following obligation rules are reinforced:

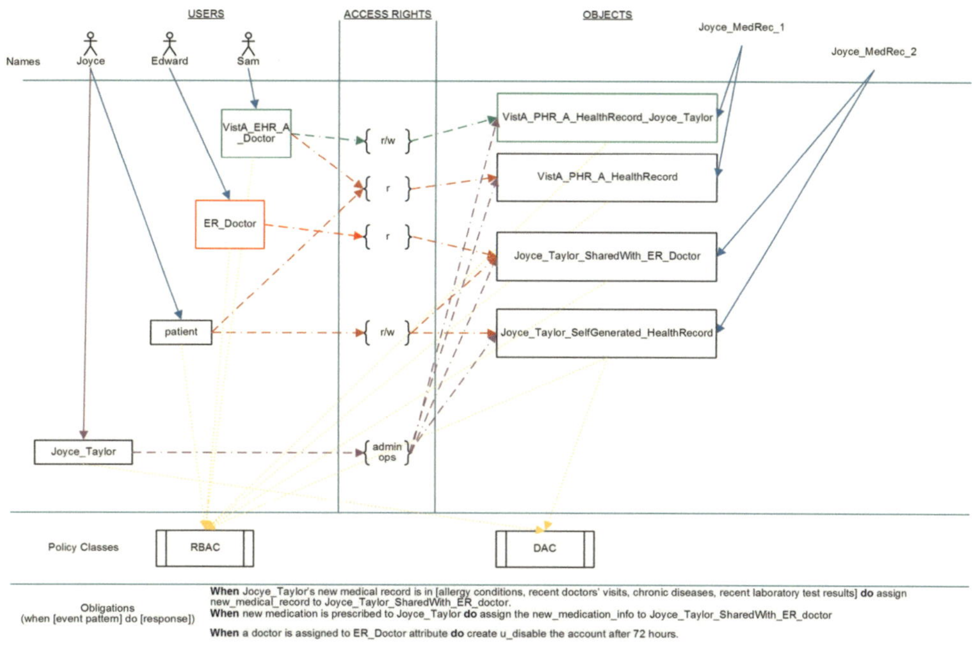

Figure 5.
Indirectly authorize BET-LPHR access to ER physicians.

- **When** Jocye_Taylor's new medical record is in ["allergy conditions, recent doctors' visits, chronic diseases, recent laboratory test results"] [3] **do** assign new_medical_record to Joyce_Taylor_SharedWith_ER_doctor.

- **When** new medication is prescribed to Joyce_Taylor **do** assign the new_medication_info to Joyce_Taylor_SharedWith_ER_doctor

- **When** a doctor is assigned to ER_Doctor attribute **do** create u_disable the account after 72 hours.

5.3 Comparison with prior work

5.3.1 HealthChain

Hylock & Zeng presented HealthChain [30], a proof-of-concept study of patient-centric health record management framework based on blockchain technology. The authors argued patients' health information in current tethered EHRs are inclined to fragmentation due to distributedness of patient records, which leads to poor care coordination. Hylock & Zeng pointed out ONC information blocking discouraged patients' engagement. Therefore, the authors proposed a mixed-block permissioned blockchain solution coupled with FHIR, the interoperability standard. The mixed-block blockchain consists of immutable logs blocks and editable patient blocks, while large size multimedia data are kept off chain at EHR's silos and only reference pointers are stored in patient blocks. HealthChain acts as "an interface between patients and providers or payers" [30]. The authors claimed security was ensured by implementing permissioned blockchain, and only trusted parties including patients could make changes. The privacy or confidentiality was protected by smart contract and 2-party proxy re-encryption decryption.

In HealthChain, the patient data are on the blockchain, which motivated the authors to redesign the data allocation strategy with patient block data consolidation for a certain patient via Chameleon hashing [31]. The blockchain patient block is redactable (or editable). On one hand, Hylock & Zeng regarded consensus and immutability, which are the core properties of blockchain technology, as shortcomings of computing performance and cost barrier when data blocks are being modified [30]. The authors argued modifications to existing patient blocks could avoid costly consensus. On the other hand, the authors stated using blockchain and smart contacts in HealthChain could meet the HIPAA privacy and security rules requirements. Therefore, if core features of blockchain technology, consensus and immutability, were identified as roadblocks to HealthChain sketch, the applicability of blockchain technology to such a design should be reconsidered. Furthermore, keeping patients' data on-chain is not practical at enterprise level. This design could not scale when number of patients and providers are increasing.

Compared with HealthChain, BET-LPHR patient data is off-chain in a "private database while the audit logs are on-chain" [3]. BET-LPHR is enterprise scalable. BET-LPHR provides privacy and confidentiality protection via BeNGAC.

5.3.2 MedRec

In August 2016, Ekblaw, Azaria, Halamka, & Lippman prototyped an Ethereum blockchain based MedRec 1.0 [32, 33] for EHR and medical research data to engage patients as agencies to their own health records. MedRec acts as an interface between providers' EHRs and patients. The patients EHRs data are siloed at providers' data centers, while patients are presented a local cached database to patients' records. Through MedRec patient-provider relationship (PPR) smart contract, a certain degree of fine granular access control to patients' health records is achieved by checking on or off fields of medical records steered by patients via a graphical application portal. The MedRec Summary Smart Contract (SSC) weaves PPR smart contracts together to form a holistic view of a patient's medical records from all providers by integrating the reference points to PPRs. The SSC is persistent on the blockchain, which offers flexibility to patients or providers to re-join the network and recover from a system disaster. The MedRec Registrar Smart Contract (RSC) links a patient's existing EHR participant ID to an Ethereum cryptographic public key identifier. The identification registration process is controlled by limited authorization institutions. In MedRec, any changes to a patient's records on a provider's EHR requires an acknowledgment of acceptance or rejection from patient's client. In MedRec, the authors argued the authentication, confidentiality, and data sharing accountability are managed by blockchain smart contracts.

There are few drawbacks in MedRec design. Firstly, when creating a new medical record in provider's EHR, MedRec requires the provider to compose a query string that retrieves that part of data and associate a hash of the query output to guarantee data integrity. However, before a patient accepts this new change, this new record is not in patient's holistic view nor treated as patient's genuine data by the patient. At this point, a hacker (either internal or external) can disclose this new record to a third party without notifying the patient, which violates privacy and confidentiality of the patient record. Secondly, since any change to a patient's records on provider's side requires patient's communication to accept or deny the change, if for some reason, the patient cannot respond to the communication, the authors did not explain the results of those affected records. Thirdly, Proof-of-Work (PoW) was implemented as a mining approach, which consumes excessive computing energy.

In 2019, Nchinda, Cameron, Retzepi, & Lippman introduced a new architectural design of MedRec 2.0 [34]. The MedRec 2.0 replaced PoW with computing cost saving Proof-of-Authority (PoA) based on the trusted participants of EHR data providers on the blockchain network. The MedRec 2.0 is an open-source solution, claimed by authors to be a robust approach with small system resource consumption overhead to the existing EHRs. However, the scalability needs to be tested when more health care community users adopt the solution.

In contrast to MedRec, BET-LPHR does not require the patient to confirm denying or accepting changes when adding a health record. Furthermore, BET-LPHR uses HF consensus so it eliminated the consensus overhead of PoW or PoA.

6. Conclusions

In this chapter, we presented a scalable and interoperable BET-LPHR solution to solve a longstanding PHR problem. Patients get benefits of controlling the security and privacy of their own LPHR when sharing the information with trusted health care providers. The permission control autonomy is achieved via BeNGAC. The BET-LPHR is built on top of BeNGAC network with a FHIR interface so it is interoperable with other EHRs. Both BeNGAC and BET-LPHR are enterprise scalable. The BET-LPHR is distributed yet decentralized and tamper resistant with auditable changes. We discussed two limitations of the solution when owner lost the private key or cannot directly authorize the access to BET-LPHR. Also, the current HF version supports up to 100 organizations [27] on the same policy or data secure channel, which can present a limit when a patient wants to include more than 99 EHR organizations to BET-LPHR on the same secure policy channel. We leave this for future research.

Acknowledgements

Authors are grateful for technical assistance provided by Mr. Peter Li while he was with Open Source Electronic Health Record Alliance, Arlington, VA.

Author details

Yibin Dong, Seong K. Mun and Yue Wang*
Virginia Polytechnic Institute and State University, Arlington VA, USA

*Address all correspondence to: yuewang@vt.edu

References

[1] AHIMA. Defining the personal health record. Journal of AHIMA. 2005;**76**(6): 24-25

[2] Personal health records and the HIPAA privacy rule [Internet]. Available from: https://www.hhs.gov/sites/default/files/ocr/privacy/hipaa/understanding/special/healthit/phrs.pdf. [Accessed: April 23, 2022]

[3] Dong Y, Mun SK, Wang Y. Blockchain-enabled next generation access control. In: PA PJ, Leitão P, Pinto A, editors. BLOCKCHAIN 2021. Cham: Springer; 2021. Lecture Notes in Networks and Systems. p. 2022

[4] HIPAA history [Internet]. Available from: https://www.hipaajournal.com/hipaa-history/. [Accessed: May 23, 2022]

[5] H.R.1. American Recovery and Reinvestment Act of 2009 [Internet] [cited 2022 Apr 23]. 2009. Available from: https://www.congress.gov/bill/111th-congress/house-bill/1/text

[6] Key considerations, venesco and personal health records community of practice. Venesco LLC, (ONC) OotNCfHI; 2015 Report No.: Contract # 14-233-SOL-00533

[7] Abd-Alrazaq AA, Bewick BM, Farragher T, Gardner P. Factors that affect the use of electronic personal health records among patients: A systematic review. International Journal of Medical Informatics. 2019;**126**: 164-175

[8] ONC. Office-based physician electronic health record adoption [Internet]. 2017. Available from: https://dashboard.healthit.gov/quickstats/pages/physician-ehr-adoption-trends.php. [Accessed: May 23, 2022]

[9] Sahai A, Waters B. Fuzzy identity-based encryption. In: Advances in Cryptology – EUROCRYPT 2005. Berlin, Heidelberg: Springer Berlin Heidelberg; 2005. pp. 457-473

[10] Goyal V, Pandey O, Sahai A, Waters B. Attribute-based encryption for fine-grained access control of encrypted data. In: Proceedings of the 13th ACM conference on Computer and communications security. 2006. pp. 89-98

[11] Li M, Yu S, Ren K, Lou W. Securing personal health records in cloud computing: patient-centric and fine-grained data access control in multi-owner settings. Berlin, Heidelberg: Springer Berlin, Heidelberg; 2010

[12] Zheng Y. Privacy-preserving personal health record system using attribute-based encryption. Worcester Polytechnic Institute; 2011

[13] Li M, Yu S, Zheng Y, Ren K, Lou W. Scalable and secure sharing of personal health records in cloud computing using attribute-based encryption. IEEE Transactions on Parallel and Distributed Systems. 2013;**24**(1):131-143

[14] Debnath MK, Samet S, Vidyasankar K. A secure revocable personal health record system with policy-based fine-grained access control. Thirteenth Annual Conference on Privacy, Security and Trust; 2015

[15] Au MH, Yuen TH, Liu JK, Susilo W, Huang X, Xiang Y, et al. A general framework for secure sharing of personal health records in cloud system. Journal of Computer and System Sciences. 2017;**90**:46-62

[16] Sookhak M, Yu FR, Khan MK, Xiang Y, Buyya R. Attribute-based data access control in mobile cloud computing: Taxonomy and open issues. Future Generation Computer Systems. 2017;**72**:273-287

[17] Miller AR, Tucker CE. Encryption and the loss of patient data. Journal of Policy Analysis and Management. 2011; **30**(3):534-556

[18] Samarati P, de Vimercati SC, editors. Access control: Policies, models, and mechanisms. Foundations of Security Analysis and Design. Berlin, Heidelberg: Springer; 2001

[19] Landwehr CE. Formal models for computer security. ACM Computing Surveys (CSUR). 1981;**13**(3):247-278

[20] INCITS. Information technology - Next Generation Access Control - Functional Architecture (NGAC-FA). ANSI/INCITS 499-2018. American National Standards Institute; 2018

[21] Hu VC, Ferraiolo DF, Kuhn DR. Assessment of access control systems. Gaithersburg, MD 20899-8930. NIST; 2006

[22] HIPAA security rule. U.S. Department of Health & Human Services [Internet]. 2013. Available from: https://www.hhs.gov/hipaa/for-professionals/security/laws-regulations/index.html. [Accessed: May 23, 2022]

[23] CMS interoperability and patient access final rule, 42 CFR Parts 406, 407, 422, 423, 431, 438, 457, 482, and 485 (March 9, 2020)

[24] ONC Cures Act Final Rule, 45 CFR Parts 170 and 171 RIN 0955-AA01 (March 9, 2020)

[25] Tembhare A, Sibi Chakkaravarthy S, Sangeetha D, Vaidehi V, Venkata RM. Role-based policy to maintain privacy of patient health records in cloud. The Journal of Supercomputing: An International Journal of High-Performance Computer Design, Analysis, and Use. 2019;**75**(9):5866-5881

[26] Ferraiolo DF, Gavrila SI, Jansen W, Stutzman PE. Policy machine: features, architecture, and specification. NIST; 2015

[27] IBM. Hyperledger fabric a blockchain platform for the enterprise [Internet]. Available from: https://hyperledger-fabric.readthedocs.io/en/latest/. [Accessed: May 23, 2022]

[28] FIDO alliance [Internet]. Available from: https://fidoalliance.org/. [Accessed: May 21, 2022]

[29] Wilcox AB, Shen S, Dorr DA, Hripcsak G, Heermann L, Narus SP. Improving access to longitudinal patient health information within an emergency department. Applied Clinical Informatics. 2012;**3**(3):290-300

[30] Hylock RH, Zeng X. A blockchain framework for patient-centered health records and exchange (healthchain): evaluation and proof-of-concept study. Journal of Medical Internet Research. 2019;**21**(8):e13592

[31] Ashritha K. MS, KVL, th International Conference on Advanced C, Communication Systems Coimbatore IMM. In: Redactable blockchain using enhanced chameleon hash function. 2019 5th International Conference on Advanced Computing & Communication Systems (ICACCS). IEEE; 2019. pp. 323-328

[32] Ekblaw A, Azaria A, Halamka JD, Lippman A. In: Lab MM, BIDM C,

editors. A case study for blockchain in healthcare: "MedRec" prototype for electronic health records and medical research data. 2016

[33] Azaria A, Ekblaw A, Vieira T, Lippman A. MedRec: using blockchain for medical data access and permission management. In: 2016 2nd International Conference on Open and Big Data (OBD). 2016. pp. 25-30

[34] Nchinda N, Cameron A, Retzepi K, Lippman A. MedRec a network for personal information distribution. In: 2019 International Conference on Computing, Networking and Communications (ICNC). Honolulu, HI, USA; 2019. pp. 637-641

Section 6

Blockchain Monitoring and Security

Chapter 14

Methodology of the Blockchain Monitoring Framework

Dominique Bernard Kanga, Mohamed Azouazi, Mohammed Yassine El Ghoumrari and Abderrahmane Daif

Abstract

A blockchain is a technology that allows the storage and transmission of information without a control body. Technically, it is a distributed database in which the information sent by users is verified and grouped into blocks, thus forming a chain. Thanks to the secure encryption of the data and the fact that new transactions are linked to the previous ones, it is almost impossible to modify the old records without modifying the following ones. On the other hand, the control of the blockchain by more than half of the nodes in the network (by consensus) makes it impossible to falsify the data in the blockchain. However, this public/private, anonymous, and unforgeable ledger that is the blockchain contains a set of information (metrics, logs, etc.) that can provide clues for an efficient monitoring and allow the reinforcement of the security of the blockchain that could be discussed in the future with the advent of quantum machines.

Keywords: blockchain, monitoring, framework, blockchain security, smart contract, big data, metrics

1. Introduction

Blockchain is now one of the most important technologies to have emerged in recent years. Many experts believe that this technology has the potential to change the world over the next two decades. Although it is still in its infancy, corporate giants are interested in its applications in several areas. So far, venture capitalists have invested billions of dollars in this field, with several applications [1]

Indeed, the applications of blockchain seem close to infinite [2]. While one immediately thinks of its financial applications—international payments, money transfers, complex financial products—blockchain can also solve problems and create new opportunities in healthcare, defense, management, supply chains, luxury, and other industries. At more advanced stages, blockchain could give rise to what Gartner calls the "programmable economy" [3], powered by entirely new business models that eliminate all kinds of middlemen. Given the importance of blockchain in the technological evolution of society, including across industries, especially the financial sector, researchers have undertaken considerable work to further strengthen the security level of this technology.

Indeed, work on the methodologies for encrypting the data preceding and following each block has made the data of the blockchain virtually unbreakable. In addition, the security of the blockchain is also due to the fact that several computers called nodes to store the blockchain. In addition to that, to modify the ledger, one would have to take control of at least 50% of the nodes in the network and their computing power in order to modify the data in the blockchain [4]. This is a difficult feat to accomplish, especially for a public blockchain such as the one behind bitcoin. With the advent of the quantum machine, this unparalleled security within the blockchain may be challenged in the future. Therefore, blockchain monitoring [1] could add an important layer of security to the blockchain. However, our work will consist in discussing and studying this topic while proposing, at the end of this study, an efficient and exploitable blockchain monitoring methodology [5] in order to allow a good understanding of the topic.

1.1 Blockchain technology

A blockchain is a technology that allows information to be stored and transmitted without a control body [6]. Technically, it is a distributed database in which the information sent by users is verified and grouped at regular intervals into blocks, thus forming a chain. The whole is secured by cryptography. By extension, a blockchain is a distributed database that manages a list of records that are protected against alteration or modification by storage nodes [7]. Not all blockchains work in the same way. For example, they may differ in their consensus mechanisms, whose rules prevail depending on the technology that updates the ledger [1]. But fundamentally, a blockchain is a distributed and secure record of all transactions made since the beginning of the distributed system [8]. By extension, a blockchain constitutes a database that contains the history of all exchanges made between its users since its creation as shown in **Figure 1** [9].

However, there are public blockchains, open to all, and private blockchains, whose access and use are limited to a certain number of actors. A public blockchain can therefore be likened to a public, anonymous, and unforgeable accounting ledger. As the mathematician Jean-Paul Delahaye writes, one must imagine "a very large notebook, which everyone can read freely and for free, on which everyone can write, but which is impossible to erase and indestructible" [10], which is well illustrated in the figure above. Today, blockchains cover several aspects of computer security because of the numerous researches made around this technology.

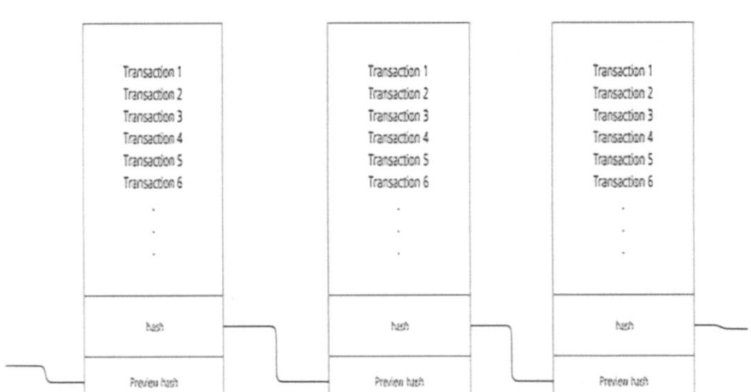

Figure 1.
Block in blockchain technology.

1.2 Security aspects of blockchain information

The basic security properties of the blockchain stem from both advances in cryptography and the design and implementation of bitcoin. Theoretically, the first secure blockchain was formulated using cryptography in 1991.

A proposal to improve the efficiency of the cryptographic blockchain was published in 1993, incorporating Merkle trees and placing multiple documents in a block. The blockchain is designed to ensure a number of inherent security attributes, such as consistency, proof of forgery, resistance to a distributed denial of service (DDoS) attack, pseudonymity resistance to a double attack, and OWASP resistance. However, to use the blockchain for secure distributed storage, additional security and privacy properties are required. In this section, we describe the fundamental security and privacy properties of blockchains before addressing the topic of monitoring blockchain systems with the aim of proposing a methodology for exploring, analyzing, and visualizing the behaviors of blockchain actors. In this chapter, we will only deal with the monitoring of applications.

1.2.1 Controlling data consistency

The criteria that are required for a functional blockchain are the following:

- a replicated register that only allows the irreversible addition of data;

- a data protection.

In a traditional database, it is possible to guarantee these properties by controlling access to the registry, which implies having confidence in the entity that maintains it.

The blockchain solution is to decentralize and replicate the maintenance of the registry between several locations. Thus, the participating entities do not need to trust each other, and it works as long as enough entities are actually trustworthy and do not form coalitions (of more than 51%). This honesty is motivated by a reward for producing blocks that are cryptographically protected. All of these blocks are replicated in a P2P network (with no central node), avoiding a single point of failure [11].

Satoshi Nakamoto's initial blockchain was permissionless, meaning that anyone could participate in maintaining the registry, without the need to register first. This meant that it would work efficiently regardless of the number of participating entities.

Later, a variation more suited to certain applications emerged: consortium blockchains, where participating entities are pre-registered. The registry can be faster and more reliable, while still being controlled by the majority of participants.

1.2.2 Defending against DDoS ("denial service attack")

A denial of service attack is called a DoS attack on a host. It is a type of cyberattack that disrupts hosted Internet services, making the host machine or network resource on the host unavailable to intended users. DoS attacks attempt to overload the host system or network resource on the host by flooding it with unnecessary requests, thereby blocking the execution of legitimate services. The DDoS attack refers to a "distributed" DoS attack, meaning that the flooding attack of incoming traffic to a victim comes from many disparate sources spread across the Internet [11]. A DDoS attacker can compromise and use one person's computer to attack another computer by taking advantage of security vulnerabilities or weaknesses. By taking advantage of a set of compromised computers in this way, a DDoS attacker can send huge amounts of data to a hosting website or send spam to specific email addresses. Therefore, it is very difficult to prevent the attack by simply jamming the individual sources one by one. The arm wrestling depends on the rate of repair of these compromised nodes versus the success rate of compromising computer nodes in the network. The major concern in a DDoS attack is the availability of the blockchain and is related to the question of whether a DDoS attacker can make the blockchain unavailable by taking down part or all networks. The answer to this question is no, thanks to the fully decentralized construction and maintenance of the blockchain, particularly the bitcoin system that has a large network (interconnected node), as well as the consensus protocol for generating new blocks and adding them to the blockchain, which ensures that the processing of blockchain transactions can continue even if several blockchain nodes are offline. For a cyberattacker to successfully take the blockchain offline, he or she must gather sufficient computing resources to compromise a very large portion of the blockchain nodes on the entire blockchain network. The larger the blockchain network, the harder it is to pull off such a large DDoS attack [11]. This is the case with the bitcoin blockchain network, which continues to grow and now has 14719 nodes worldwide retrieved on Sat Dec 18 16:24:33 2021 +01 in bitnodes.io and as shown in **Figure 2**.

1.2.3 Resistance to double spending attacks

The double-spending attack in the context of the bitcoin blockchain refers to a specific problem unique to digital currency transactions. It should be noted that the double-spending attack can be considered a general security problem due to the fact that digital information can be replicated relatively easily. Specifically, in digital token exchange transactions, such as electronic money, there is a risk that the holder will duplicate the digital token and send multiple identical tokens to multiple recipients. If inconsistency can be incurred due to duplicate digital token transactions (e.g., spending the same bitcoin token twice), then the problem of double spending becomes a serious security threat. To prevent duplication, bitcoin evaluates and verifies the authenticity of each transaction using the transaction logs on its blockchain with a consensus protocol. By ensuring that all transactions are included in the blockchain, in which the consensus protocol allows everyone to publicly verify the transactions in a block before committing the block to the global blockchain, this ensures that the sender of each transaction only spends the bitcoins he or she legitimately owns.

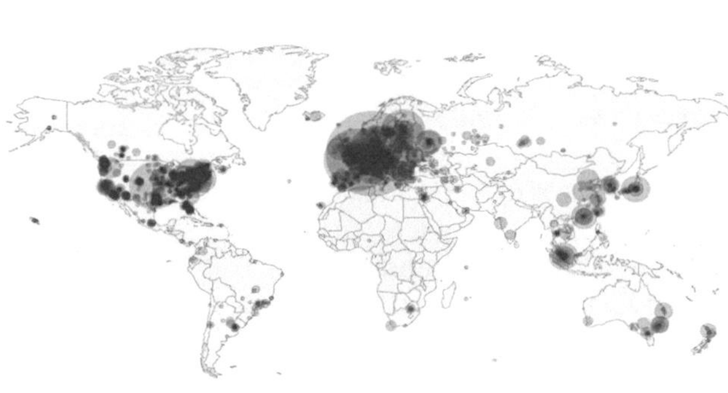

Figure 2.
Live Map shows concentration of reachable bitcoin nodes found in countries around the world.

In addition, each transaction is signed by its sender using a secure digital signature algorithm. This ensures that if someone forges the transaction, the verifier can easily detect it. The combination of transactions signed using digital signatures and public verification of transactions using majority consensus ensures that the blockchain can withstand the double-spending attack [11].

1.2.4 Defying the attacks of the majority consensus (51%)

A 51% attack is an attack that targets the so-called Proof of Work (PoW) or Proof of Stake (PoS) blockchains.

This attack refers to the risks of cheating in the majority consensus protocol. One of these risks is often referred to as the 51% attack, particularly in the context of double spending.

An example of a 51% attack can occur if one cooperative becomes too large relative to the others, which can allow for a 51% attack, when they agree to carry out a conspiracy, such as in vote counting or illegally transferring cryptocurrencies to one or more target wallets, reversing authentic transactions as if they never happened, etc.

Today, measures have been taken on large-scale blockchains like bitcoin to resist this type of attack, but it is still exploitable.

1.2.5 Resistance to OWASP

The Open Web Application Security Project (OWASP) is an online community working on web application security [12]. While OWASP's top ten vulnerabilities list is designed to describe vulnerabilities faced by web application developers, nine of OWASP's ten vulnerabilities also apply to blockchain systems. Even though the blockchain ecosystem has been designed to solve most of the security issues faced in web application and information system design due to the use of advanced cryptographic mechanisms and 51% consensus in blockchain, it is worth noting that the avenues for monitoring remain unexplored, and it is worth considering the possibilities of investigating the implementation of the blockchain monitoring mechanism (**Table 1**).

Top 10 OWASP	Applicable / resistance
injection	Yes/Yes
Broken Authentication	Yes/Yes
Sensitive Data Exposure	Yes/Yes
XML External Entities (XXE)	No/Yes
Broken Access Control	Yes/Yes
Security Misconfiguration	Yes/Yes
Cross-Site Scripting (XSS)	Yes/Yes
Insecure Deserialization	Yes/Yes
Using Components with Known Vulnerabilities	Yes/Yes
Insufficient Logging & Monitoring	Yes / Not explore

Table 1.
OWASP Top 10 application security risks—2017 and blockchain resistance [12].

1.3 A blockchain system to monitor

Figure 3 summarizes the blockchain layout that we see the need to include monitoring. A typical blockchain network consists of a set of interconnected nodes that act in pairs. These nodes are typically hosted in a cloud or on-premises infrastructure, where the blockchain execution engine is configured natively on a virtual machine (VM) or using containerization technologies such as Docker or a physical machine. Transactions submitted to the blockchain network are broadcast to all pairs and newly created blocks are propagated, so that all pairs have an up-to-date copy of the shared ledger. To get a snapshot of the block, in terms of transaction events and associated metadata, all you need to do is monitor one of the pairs. And this is typically done using blockchain explorer, which listens for events and provides some visualization of the number of transactions received, queued, processed, and finally consolidated into a new block. However, this level of monitoring does not provide any clues about the resource utilization of that node, the health of other nodes, or the latency experienced within the blockchain network.

Another key element that must be monitored to achieve end-to-end visibility of a blockchain-based solution is the off-chain components that include the application layer (decentralized application). The DAPP layer [1] includes a user interface, storage, and SDK (Software Development Kit) API (Application Program Interface) components, through which interaction with a blockchain node is possible.

1.4 Blockchain framework monitoring

Effective monitoring and management of a blockchain system require a framework that can integrate data, process generated events, and provide adequate visualization of blockchain-related metrics. This framework [13] must be flexible and support deployment configurations of off-blockchain applications and blockchain nodes individually. As shown in **Figure 4**, the diagram describes a proposed blockchain monitoring framework, which includes the following:

- A monitoring agent [14, 15], which is deployed on each blockchain node (blockchain network agent) and associated applications (agent provider), can read logs generated as part of the transaction process and relay data about processor, memory, and device usage. I / O.

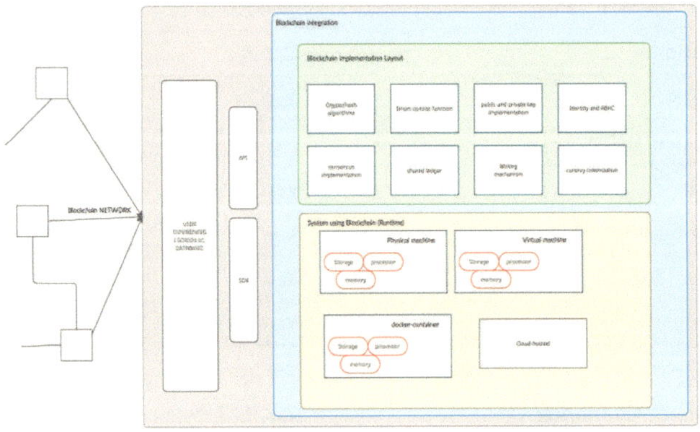

Figure 3.
Blockchain system to monitor.

DOI: http://dx.doi.org/10.5772/intechopen.109550

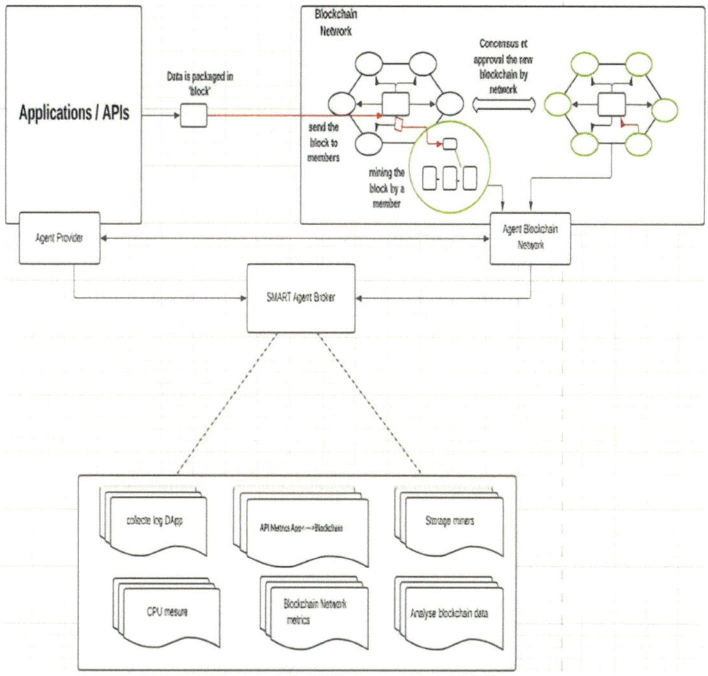

Figure 4.
Blockchain system monitoring framework.

- A log collection engine that continuously manages log information and sends it for further processing.
- Collects metrics that will allow us to sort and filter information in the blockchain network (agent blockchain network) and the DAPP (agent Provider) such as:
    - The time and date the transaction was initiated.
    - The time it took for the request to succeed or fail.
    - The size of the request or response.
    - The endpoint for which the transaction was sent (distributed network).
    - The entries of the endpoint.
    - The execution details of the environment.
    - Status of the request, whether it failed or not.
    - The network code of the request. This will be one of the standard HTTP/HTTP status codes.
    - The origin of the request.

- Monitor the creation of smart contracts.
- The agent-broker [16] collects from the two smart agents a large amount of log data to organize and index it into corresponding documents, which are shared and stored for analysis [1].
- A visualization platform connected to the broker-agent consumes the data collected by the nodes and provides statistics on the efficiency of the blockchain nodes and the network overview [17].
- It allows parties to conduct analytical research and generate reports.

Based on the proposed monitoring framework, there are important indicators that can be extracted and help strengthen the security of the blockchain while ensuring the protection of funds in the case of cryptocurrencies and personal data:

- Analyze how the blockchain's transaction processing and consensus mechanism uses the resources of the underlying infrastructure [1].
- Provide visibility into an end-to-end business transaction presented is initiated by a dApp user and captured in the blockchain.
- Allow miner pools to integrate/remove specific machines (CPU, graphics card, etc.) based on machine performance needs.
- Visualize the 51% of attack attempts that can occur when a group of miners attempt to perform a conspiracy.
- Combine and correlate block and transaction events from each node and determine the performance and throughput of the blockchain network.
- Configure a noninvasive monitoring solution that can be dynamically enabled for each embedded pair and also supports a common network provider model.

2. Conclusions

Today, billions of dollars have been invested in cryptocurrencies whose core technology is blockchain, while solutions and techniques to effectively monitor existing blockchain networks are not well thought out and are almost nonexistent. The main reason is that few commercial use cases have not yet translated into blockchain production systems or that most smart contracts or cryptocurrencies are of the "fire-forge" type (i.e., designating a crypto or smart contract whose postlaunch monitoring no longer requires the intervention of the platform operator). Furthermore, the decentralized nature of blockchain raises the following question: is monitoring of the entire blockchain network really necessary [18, 19]? What are the indicators that could be obtained if it were possible to propose an effective tool or methodology for monitoring the blockchain? Could these indicators be used for big data analysis [20]? Therefore, we proposed a framework for monitoring a blockchain system in a general way, based on several existing applications and system monitoring solutions applicable to the blockchain ecosystem. In addition, the implementation of a blockchain monitoring system could detect anomalies or fraud throughout the system and, for example, reject transactions even before the blockchain records are updated.

The next step in our work would be to follow an approach that would allow us to design a model on a private blockchain to see the possibilities of exploitation, and to list all the information, logs, and statistics that can be used in a larger (public) blockchain.

Author details

Dominique Bernard Kanga*, Mohamed Azouazi, Mohammed Yassine El Ghoumrari and Abderrahmane Daif
Faculty of Sciences Ben M'sik Casablanca University Hassan II, Morocco

*Address all correspondence to: kangadominiquebernard@gmail.com

References

[1] Kanga DB, Azzouazi M, el Ghoumrari MY, Daif A. Management and monitoring of blockchain systems. Procedia Computer Science. 2020;**177**:605-612. DOI: 10.1016/j.procs.2020.10.086

[2] Reijers W, Coeckelbergh M. The blockchain as a narrative technology: Investigating the social ontology and normative configurations of cryptocurrencies. Philosophy & Technology. 2018;**31**(1):103-130. DOI: 10.1007/s13347-016-0239-x

[3] Herraiz-Faixó F, Arroyo-Cañada FJ, López-Jurado MP, Lauroba-Pérez AM. Digital and programmable economy applications: A smart cities congestion case by fuzzy sets. Journal of Intelligent and Fuzzy Systems. 2020;**38**(5):5391-5404. DOI: 10.3233/JIFS-179632

[4] Li Y, Ouyang K, Li N, Rahmani R, Yang H, Pei Y. A blockchain-assisted intelligent transportation system promoting data services with privacy protection. Sensors. 2020;**20**(9):2483. DOI: 10.3390/s20092483

[5] Li X, Jiang P, Luo X, Chen T, Wen Q. A survey on the security of blockchain systems. Future Generation Computer Systems. 2017. DOI: 10.1016/j.future.2017.08.020

[6] Camilleri AF, Grech A, Inamorato dos Santos A, European Commission. Blockchain in Education. Joint Research Centre; 2022. Available from: https://www.pedocs.de/volltexte/2018/15013/pdf/Grech_Camilleri_2017_Blockchain_in_Education.pdf

[7] Labbi M, Kannouf N, Chahid Y, Benabdellah M, Azizi A. Blockchain-Based PKI for Content-Centric Networking. 2019. pp. 656-667. DOI: 10.1007/978-3-030-11196-0_54

[8] Umeh J. Blockchain double bubble or double trouble? ITNOW. 2016;**58**(1):58-61. DOI: 10.1093/itnow/bww026

[9] Saleh I. Internet of Things (IoT): Concepts, issues, challenges and perspectives. In: Challenges of the Internet of Things. John Wiley & Sons; 2018. pp. 1-26. DOI: 10.1002/9781119549765.ch1

[10] Desplebin O, Lux G, Petit N. L'évolution de la comptabilité, du contrôle, de l'audit et de leurs métiers au prisme de la Blockchain : une réflexion prospective. Management & Avenir. 2018;**103**(5):137. DOI: 10.3917/mav.103.0137

[11] Zhang R, Xue R, Liu L. Security and privacy on blockchain. ACM Computing Surveys. 2019;**52**(3):52. DOI: 10.1145/3316481

[12] Poston H. Mapping the OWasp top ten to blockchain. Procedia Computer Science. 2020;**177**:613-617. DOI: 10.1016/j.procs.2020.10.087

[13] Monitoring a Blockchain Network. 2022. Retrieved September 1, 2020, from: https://cloud.ibm.com/docs/blockchain?topic=blockchain-monitor-blockchain-network

[14] Hernantes J, Gallardo G, Serrano N. IT infrastructure-monitoring tools. IEEE Software. 2015;**32**(4):88-93. DOI: 10.1109/MS.2015.96

[15] Ward R, Ward R. Cognitive conflict without explicit conflict monitoring in a dynamical agent. Neural Networks. 2006;**19**(9):1430-1436. DOI: 10.1016/j.neunet.2006.08.003

[16] Faci N, Bernard C, Lyon U, Meneguzzi F, Modgil S, Oren N, Miles S, Luck M. A framework for monitoring agent-based normative systems. Cloud Computing View Project RiskTrack-Tracking Tool Based on Social Media for Risk Assessment on Radicalisation View Project: A Framework for Monitoring Agent-Based Normative Systems. 2009. DOI: 10.1145/1558013.1558034

[17] Khan KM, Arshad J, Iqbal W, Abdullah S, Zaib H. Blockchain-enabled real-time SLA monitoring for cloud-hosted services. Cluster Computing. 2021a. DOI: 10.1007/s10586 021 03116 y

[18] Avatangelou E, Dommarco RF, Klein M, Müller S, Nielsen CF, Soriano MPS, Schmidt A, Tazari MR, Wichert R. Conjoint PERSONA – SOPRANO Workshop. 2008. pp. 448-464. DOI: 10.1007/978-3-540-85379-4_51

[19] Rathee G, Balasaraswathi M, Chandran KP, Gupta SD, Boopathi CS. A secure IoT sensors communication in industry 4.0 using blockchain technology. Journal of Ambient Intelligence and Humanized Computing. 2021;**12**(1):533-545. DOI: 10.1007/s12652-020-02017-8

[20] Chang C-L, McAleer M, Wong W-K. Big data, computational science, economics, finance, marketing, management, and psychology: Connections. SSRN Electronic Journal. 2018. DOI: 10.2139/ssrn.3117386

Section 7

Blockchain in the Tourism Industry

Chapter 15

Perspective Chapter: Prospects of Using Blockchain Technology in the Tourism Industry

Leyla Gamidullaeva, Ivan Karelin and Svetlana Zinchenko

Abstract

This article is devoted to an actual research problem in conditions of increased uncertainty and the need for a theoretical rethinking of the tourism phenomenon in the society, when geopolitical upheavals and a decrease in international security are taking place, as well as the pandemic COVID-19, which currently poses new challenges for the tourism industry. Today, it is more relevant and appropriate than ever to think about them from the perspective of blockchain technology, using a fundamental approach to the digitalization of decentralized management of the life cycle of an internal regional tourism product. It is required to develop an integrated scientific and methodological approach to modeling and designing cyber-physical systems for monitoring and managing tourism products, objects, and processes based on blockchain technologies in order to maximize the contribution of the tourism industry to the socioeconomic development of regions.

Keywords: digitalization, life cycle management, internal tourism product, economic efficiency, management optimization, tourism

1. Introduction

Modern conditions for the development of socioeconomic systems, characterized by high uncertainty of geopolitical, macroeconomic, and epidemiological processes, lead to the need to form new organizational and economic mechanisms that ensure innovative economic growth of the Russian economy. In the current conditions, a serious reboot and a comprehensive transformation of all sectors of the economy are required. Great challenges and threats pose new challenges to the tourism industry as the most vulnerable to external shocks and threats. At the same time, the importance of the tourism industry in terms of developing the country's interior and mobilizing the existing natural, cultural, human capital, etc., is difficult to overestimate.

Today, influenced by digitalization, the global world tourism industry is undergoing very significant changes. The digitalization of the activities of enterprises and industries is happening at a tremendous speed. This dynamic is the result of many factors, such as increasing the speed and breadth of internet coverage, optimizing business processes through integrated automation, and many others. The global COVID-19 pandemic has become a significant driver of the digitalization process, creating an objective need for online services and services.

The problem is that in the face of growing uncertainty that threatens the economic security of the country, and the rapid digital transformation of the economy, the need for rational use and optimization of the management of the tourism industry is becoming more urgent. It is required to develop an integrated scientific and methodological approach to modeling and designing systems for monitoring and managing tourism products, facilities, and processes in order to maximize the contribution of the tourism industry to the socioeconomic development of regions.

The need to develop a tourist destination is primarily related to such problems as security [1], the presence of intermediaries, high demand for a national product, etc. The development of information technology has made it possible to solve some of the identified problems using distributed registry technologies.

Distributed ledger technologies are at the initial stage of development, but their influence is observed in many areas: science, manufacturing, economics, etc. They set new development trends, as they offer data anonymity, the absence of a third party in commodity-money relations [2], and low commission for economic transactions [3].

Since 2020, distributed registry technologies have been gaining ground in the travel and tourism industry [4]. The main direction was the development of smart platforms for the implementation of new systems and methods of conducting tourism activities [5].

The significance of this system in the field of tourism was confirmed in the study of a trial product. To demonstrate the usability and effectiveness of the developed approach, [6] present a hotel booking case, the analysis of which shows the significance of the proposed system. A similar result was achieved by other researchers [7].

It is important that the development also concerns such a fairly new phenomenon as medical tourism, which is travel abroad to receive medical services. Medical tourists as well as healthcare providers can benefit from technology that makes it easier to find a healthcare provider, make fast and secure payments, and keep data secure and private [3].

The further development of the tourism industry is closely related to the development of smart platforms, which are similar in technical characteristics to technological platforms. The term "technology platform" was introduced in 2004 by the European Commission to identify prioritized scientific and technical areas for the development of the European Union to ensure Europe's technological independence [8].

According to the authors [8], the tourism technological platforms "must provide the solution to the following main tasks:

- the influx of private investment in the tourism sector;
- improving the technology level of the reproduction process of both local tourism products and regional tourism products in general;
- expansion of high-tech exports, including tourism technology exports (tourism products are referred to as so-called hidden exports, when the currency or financial resources directly enter the region from other countries and regions together with tourists);
- providing the conditions for business growth, the formation of new high-tech companies in the region's economy (a smart resort city, digital technologies);
- increasing the efficiency of the use of tourism and other resources, as well as preventing their deterioration and even more so exhaustion, since otherwise the sustainable reproduction of regional tourism products will be undermined;

- solving significant social problems (preserving and building up the human potential of a region, industry, corporation, country, and civilization; improving the environmental situation; ensuring safety)."

Tyan et al. [5] highlighted the following positive aspects of developing a smart platform for tourism:

- enhancing tourism experience;
- rewarding sustainable behavior;
- ensuring the benefits for local community;
- reducing privacy concerns.

As noted in the study [9], the most important advantage of such a system is the creation of new forms of communication between the supplier and the consumer. In decentralized systems, there is no need for outside support to complete a transaction. The transaction occurs due to a consensus mechanism that protects the network and provides a mechanism for the interaction of its participants. The development of this technology will especially affect the SME (Small and Medium Enterprises) sector [10], where there is the greatest pressure from large companies [7].

Developers of smart platforms for various industries, including the tourism industry, are faced with the task of choosing the optimal technology that provides a mechanism for interaction between participants, maintaining confidentiality, eliminating intermediaries, etc. This study also aims to substantiate the need for a tourism smart platform based on distributed ledger technology as a tool and mechanism for optimizing tourism industry development *via* digital transformation under the influence of current challenges.

2. Tourism and sustainable development

It is extremely important for tourism to develop in the context of the sustainable development goals proclaimed by the UN in 2015 (Agenda 2030). This poses new challenges for the industry, the solution of which is possible with the use of the latest information technologies. The growth of sustainable tourism implies a long-term development, in which a balance is achieved between economic, social, environmental, and cultural goals, and the interests of all stakeholders are taken into account. In the period of transition to a digital and smart economy, the introduction of tourist technological platforms is a necessary condition for increasing the competitiveness of regional tourism products in Russia. Tourism within the framework of sustainable development acquires environmentally friendly tourism products, and reduces the financial costs of users. The constant and long-term development of tourism information platforms is expected through technology transfer and capacity building, both public and private [8]. The need to develop this area is emphasized by the study by [11], which notes that sustainable development in the field of tourism is relevant not only for the consumer but also for the service provider, the state, etc. The level of use of advanced manufacturing technologies; and the number of organizations performing research and development. The demand for sustainable tourism is constantly growing, and studies are being conducted on the relationship between sustainable and urban tourism [12], where the authors focus on improving the technological environment in this area, introducing solutions to optimize existing ones, as well as creating new, relevant solutions in this area. Studies conducted in the field of tourism show a high correlation between tourism development and STP, this research has also confirmed the influence of absorption, adaptation, and innovation capabilities in the sustainable development of the tourism sector [13].

With the development of IT technologies and the globalization of our world, the attention of companies is increasingly focused on the "green image" (charity, support for those in need, social development, and culturalization). Producers tend to the location of the client, to reapply for the services provided. Research shows that there is a positive relationship between sustainability practices and return visit intent, and between a green image, both directly and indirectly through trust [14]. This shows the relevance of the development of the system, as well as the development of new solutions for the development of the tourism region.

3. Features and limitations for the introduction of distributed ledger technologies in the tourism industry

Tourism business models include accommodation services, food & beverage services, travel transportation services, transportation supporting services, rental services, recreation & sports services, and souvenir retailers, as well as online & offline travel agencies and reservation services [15]. Support for such a number of solutions in the system determines a number of limitations. To obtain an acceptable result when introducing TRR into the tourism environment, it is necessary to solve three main blocks of problems [16]:

1. Legal - There are no norms, rules, or laws that regulate the activities of network participants. Norms on the form, conditions, and procedure for concluding smart contracts are not provided. There are no legal norms to protect consumer rights, there is no mechanism for bringing intruders to justice, and the issue of taxation when making a transaction has not been resolved.

2. Technical - In case of loss of personal data, there is no way to return the profile. Failures, errors, malfunctions, and attacks on the network are possible. There may be problems associated with the processing or transmission of data, which is unacceptable in financial systems, an insufficient level of scientific and technical progress, and the absence of cyber-physical devices that allow recording and collecting the necessary data [17]. The confidentiality of information, at the moment, can be violated by any physical purchase, for example, renting a car or buying a plane ticket, which is impossible without presenting personal data, which will link information on the network with a specific individual [18].

3. Economic - Innovative transition is associated with large financial investments. It is necessary to constantly maintain a stable state of the system, as well as organize sufficient computing power to protect against intruders. It is necessary to train and train specialists to work with TRR, develop the system, modernize technology, and develop new solutions to maintain relevance [16], and it is also necessary to overcome the low level of consumer awareness [5].

The development of the digital tourism area is achieved through the creation of a favorable business environment, which is achieved by combining the efforts of the state, science, and business [8]. Overcoming these barriers will open the C2C market within the system, new forms of investment will appear, and there will also be disintermediation in the tourism sector [18].

4. Comparative analysis of distributed ledger technologies

On a schematic representation of the blockchain systems (**Figure 1**) and the Hashgraph (**Figure 2**), circles indicate blocks of transactions, lines indicate their relationship, and dashed lines indicate unapproved transactions. When depicting the TraceChain technology (**Figure 3**), circles represent network nodes, lines show their interconnection.

4.1 Blockchain

In 2008, the first technology was introduced, based on the concept of a decentralized network called blockchain. If in a centralized network, the security of a transaction is assumed by the central authority, then in this concept, security is determined by the interaction of its participants. The classic representative of the system is Bitcoin, based on this technology. Bitcoin is a chain of immutable blocks (**Figure 1**), the integrity of which is confirmed by the Proof-of-Work consensus protocol. To add a new block to the network, it is necessary to perform some "work," during which the block information is confirmed [19]. This protocol provides sufficient network security, but the structure in the form of series-connected blocks and the use of an expensive protocol for finding consensus do not allow unlimited scaling. In our

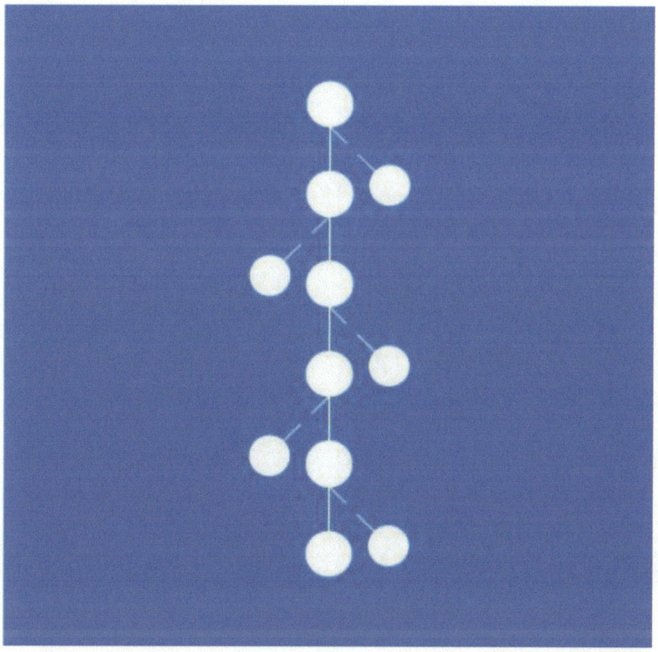

Figure 1.
Visualization of blockchain technology.

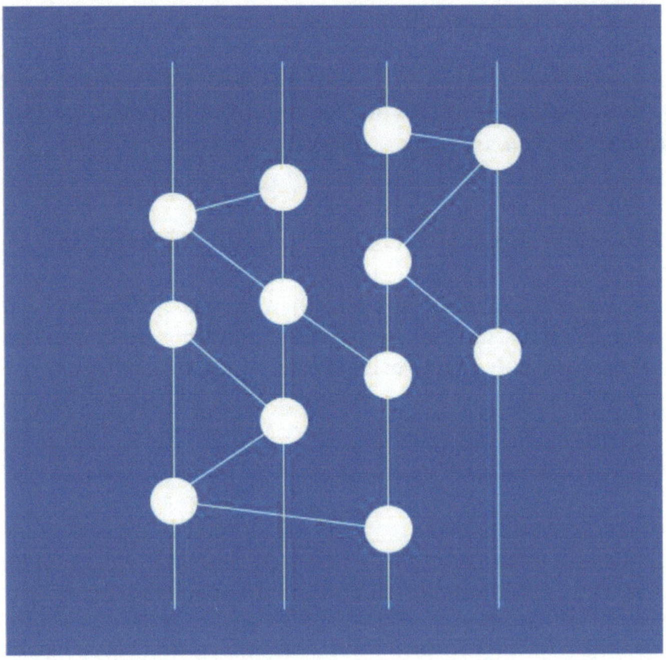

Figure 2.
Visualization of Hashgraph technology.

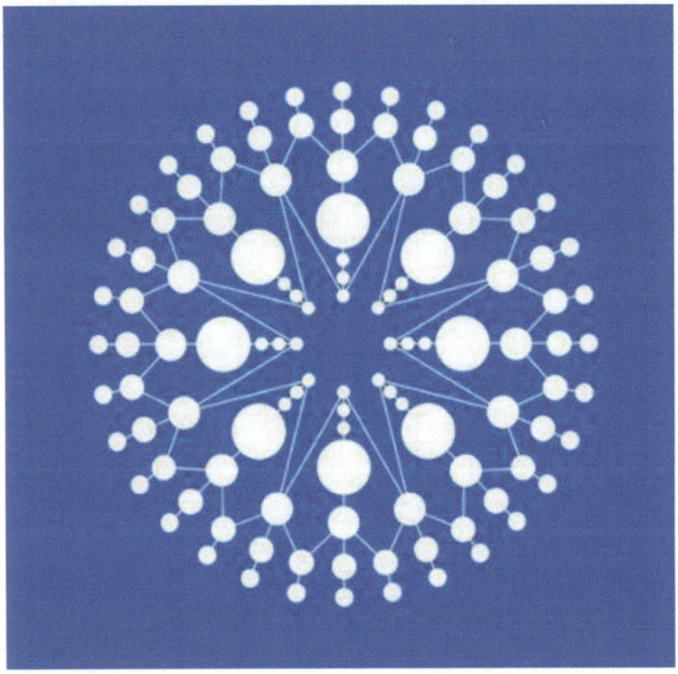

Figure 3.
Visualization of TraceChain technology.

opinion, at the moment, blockchain technology has almost exhausted its potential and is in a state of alternative to the gold exchange system.

During further development, Blockchain technology was able to increase its throughput, for example, Ethereum, but this problem still remains relevant.

4.2 Hashgraph

In 2017, the Hashgraph technology was released, the rights to the algorithms of which belong to Swirlds (HEDERA) [20]. The technology was supported by large companies that formed a platform called "Hedera Hashgraph Platform." The technology is currently patented, and the only authorized ledger is Hedera Hashgraph (**Figure 2**).

The Hashgraph technology does not use the blockchain in its structure; directed acyclic graph (DAG) was taken as the basis. The structure of the technology is the ledger in which no transaction is discarded. The interaction of network nodes is a constantly expanding tree [21] (**Figure 2**). Hashgraph technology uses other ways to achieve consensus between network participants when interacting, one of which is called "gossip about gossip." When sending information, a network participant randomly selects another participant to whom it transfers all the information it has accumulated in the form of a hash (Already encrypted data is transmitted, which contributes to confidentiality). This algorithm is repeated several times, resulting in a tree that has the general direction, which named Hashgraph [19]. The rate of dissemination increases exponentially until a consensus is reached.

The network structure and consensus technology allow Hashgraph to be fair, fast, compliant, efficient, inexpensive, timestamped, and resistant.

This project is supported by large companies such as LG, Google, IBM, and Ubisoft.

The HEDERA platform supports the following areas: enterprise, gaming, healthcare, and case studies.

4.3 TraceChain

In 2017, TraceChain technology was developed. The technology is based on AI, which determines the structure of the network, and distributes nodes to ensure maximum speed [19]. Checking the correctness of transactions is extremely fast, as it is performed immediately on all devices at the same time. The outer radius of the network keeps connections with many clients, the middle one—the load from slow intercontinental connections, and the central parts of the network are responsible for the complete synchronization of lost transactions. This forms a new type of consensus called multi-PoS (**Figure 3**). The average transaction processing speed is up to 3 seconds [21].

This technology at this moment belongs to MetaHash AG (METAHASH) [22], which develops and promotes the product on the technology market. The transaction approval rate of more than 50,000 transactions per second is achieved by the optimal location of the nodes in the network. Like competitors, this technology has vulnerabilities, the main of which is the cybersecurity problem, which is being optimized by the developer company.

Distributed ledger technologies may become new growth points in the tourism industry (**Table 1**).

Criteria	Technologies		
	Blockchain	**Hashgraph**	**TraceChain**
Transactions per second	Up to 30	10,000	50,000
Consensus mechanism	Proof-of-Work	Gossip about gossip	multi-PoS
Transaction confirmation	15 seconds	3–5 seconds	<3 seconds
Structure	Chain	Tree	Rings
Safety	High	High	Being tested

Table 1.
Comparative analysis of the existing technologies of distributed ledger.

The introduction of distributed ledger technologies in the tourism sector opens up new opportunities, which determines the importance of technology in this area [23]. Decentralization, fee reduction, smart platform applications, elimination of intermediaries, and security will allow tourism to move to the next level. The choice of a system for the transition depends on the starting conditions (financial resources, the required speed of transactions, etc.). For example, blockchain is reliable and low in cost, while limited in the speed of transactions, Hashgraph is expensive, while providing a high transaction speed, TraceChain is budgetary, and provides a high transaction speed, while the security of this option, at this moment, is inferior to competitors. Studies conducted by other research and development organizations [18, 24] have the same results.

5. Conclusion

The formation and development of smart platforms can take place both in an evolutionary way in the conditions of market relations, and by a decision from above, that is, at the initiative of state bodies. At the same time, a smart platform, as a rule, unites all participants in business processes within a single sector of the economy (manufacturing, trading, service enterprises, their customers, public authorities, and other economic entities), creating and structuring information flows between all stakeholders in a digital format and market participants within the industry. The digital industry platform in order to improve management efficiency allows you to form a holistic objective picture of the state of the industry.

In technological terms, an industry smart platform based on blockchain is an information system for accumulating, exchanging, and managing data in a structured form, as well as for creating an ecosystem of services with information systems of smart platform participants connected to it. Thus, with the help of a smart platform, horizontal integration of information systems of market participants in a specific sector of the economy is ensured.

Obviously, for interested participants, the formation of a smart platform for the tourism industry has many positive consequences, as it opens up access to a wide range of opportunities and allows the implementation of a sustainable model for the development of the industry [25]. Therefore, the essence of the formation of a business model should be to organize work with all interested parties (external partners, contractors, government, and control bodies) without creating a rigid vertical hierarchical management system. On the other hand, the quality and efficiency of the control function are improving to assess the latent factors of growth or decline in the development of both individual business entities and the tourism industry as a whole.

A promising direction for further research is the development of universal mechanisms for modeling and synthesizing the architecture of intelligent cyber-physical systems of a new generation based on the blockchain, which is a digital ecosystem. The latter is designed to provide decision support processes based on the monitoring of tourism products and processes at distributed cyber-physical objects of the regional tourism industry.

The solution to this problem requires the development of a new scientific and methodological approach to modeling and designing cyber-physical systems for monitoring and managing distributed objects and processes in order to optimize management at all stages of the life cycle of a regional tourism product. This will eliminate the gaps between the stages of the life cycle (design, planning, promotion and commercial implementation, implementation on the ground, and optimization), ensure the control of the life cycles of all tourism products and regulate the processes of changing the shape and duration of the life cycle of each individual product at all stages in order to rationally use of tourist and recreational resources of the region. The results of the study create a methodological and practical basis for the creation of new artificial intelligence technologies, which corresponds to the priority direction of the Strategy for Scientific and Technological Development of the Russian Federation until 2035 “Transition to advanced digital, intelligent manufacturing technologies, robotic systems, new materials and design methods, creation of processing systems big data, machine learning and artificial intelligence. In future research, we are planned to continue our research work on the sustainable development of tourism, as well as methods of implementing distributed ledger technologies in the tourism areas.

Acknowledgements

The research was funded by the Russian Science Foundation (RSF) and Penza Region, grant number 22-28-20524, https://rscf.ru/en/project/22-28-20524/.

Author details

Leyla Gamidullaeva[1*], Ivan Karelin[2] and Svetlana Zinchenko[1]

1 Department of Economics and Management, Penza State University, Russia

2 K.G. Razumovsky Moscow State University of Technologies and Management (FCU), Russia

*Address all correspondence to: gamidullaeva@gmail.com

References

[1] Tarlow P. Tourism Security. 1st ed. Butterworth-Heinemann; 2014. DOI: 10.1016/C2012-0-06812-3

[2] Rashideh W. Blockchain technology framework: Current and future perspectives for the tourism industry. W. Rashideh. Tourism Management. 2020;**80**:104125. DOI: 10.1016/j.tourman.2020.104125

[3] Tyan I, Guevara-Plaza A, Yagüe MI. The benefits of blockchain technology for medical tourism. Sustainability. 2021;**13**(22):12448. DOI: 10.3390/su132212448

[4] Thees H. The application of blockchain in tourism: Use cases in the tourism value system. European Journal of Tourism Research. 2020;**26**:1-21. DOI: 10.54055/ejtr.v26i.1933

[5] Tyan I, Yagüe MI, Guevara-Plaza A. Blockchain technology for smart tourism destinations. Sustainability. 2020;**12**(22):9715. DOI: 10.3390/su12229715

[6] Zhang L, Hang L, Jin W, Kim D. Interoperable multi-Blockchain platform based on integrated REST APIs for reliable tourism management. Electronics. 2021;**10**(23):2990. DOI: 10.3390/electronics10232990

[7] Erceg A, Damoska Sekuloska J, Kelić I. Blockchain in the tourism industry—A review of the situation in Croatia and Macedonia. Informatics. 2020;7(1):5. DOI: 10.3390/informatics7010005

[8] Sharafutdinov V, Onishchenko E, Nakonechnyi A. Tourism technology platforms as a tool for supporting competitiveness of regional tourism products. Regional Research of Russia. 2020;**10**:48-55. DOI: 10.1134/S2079970520010104

[9] Raluca-Florentina T. The utility of Blockchain Technology in the Electronic Commerce of tourism services: An exploratory study on Romanian consumers. Sustainability. 2022;**14**(2):943. DOI: 10.3390/su14020943

[10] Nuryyev G, Wang Y-P, Achyldurdyyeva J, Jaw B-S, Yeh Y-S, Lin H-T, et al. Blockchain technology adoption behavior and sustainability of the business in tourism and hospitality SMEs: An empirical study. Sustainability. 2020;**12**(3):1256. DOI: 10.3390/su12031256

[11] Giorgi E, Cattaneo T, Ni M, Enríquez AR. Sustainability and effectiveness of Chinese outline for National Tourism and leisure. Sustainability. 2020;**12**(3):1161. DOI: 10.3390/su12031161

[12] Grah B, Dimovski V, Peterlin J. Managing sustainable urban tourism development: The case of Ljubljana. Sustainability. 2020;**12**(3):792. DOI: 10.3390/su12030792

[13] Rodríguez AJG, Barón NJ, Martínez JMG. Validity of dynamic capabilities in the operation based on new sustainability narratives on nature tourism SMEs and clusters. Sustainability. 2020;**12**(3):1004. DOI: 10.3390/su12031004

[14] Mercadé Melé P, Molina Gómez J, Sousa MJ. Influence of sustainability practices and green image on the Re-visit intention of small and medium-size towns. Sustainability. 2020;**12**(3):930. DOI: 10.3390/su12030930

[15] Joo J, Park J, Han Y. Applications of Blockchain and Smart Contract for Sustainable Tourism Ecosystems. 2021. DOI: 10.1007/978-981-15-5258-8_71

[16] Nikitina AA, Tishchenko SV. Blockchain technologies are an innovative breakthrough in tourism. Problems of Economics and Legal Practice. 2018;**2**:218-220. DOI: 10.33693/2541-8025

[17] Mkrttchian V, Vertakova Y, Symonyan A. Data integrity Management for Laboratory of the control of lifecycle of domestic Russian tour products. In: Data Integrity and Quality. London, UK: IntechOpen; 2021. DOI: 10.5772/intechopen.96071

[18] Treiblmaier H. Blockchain and tourism. 2020. DOI: 10.1007/978-3-030-05324-6_28-1

[19] Karelin IV, Akimova IV, Grosheva ES, Artyukhin VV. Comparison of distributed ledger technologies. Modern Science-Intensive Technologies. 2021;**12**(2):226-230. DOI: 10.17513/snt.38979

[20] HEDERA [Internet]. Available from: https://hedera.com [Accessed: March 12, 2022]

[21] Pomazkova EE. Comparative analysis of blockchain and alternative distributed ledger technologies. International Journal of the Humanities and Natural Sciences. 2019;**4**(2):43-50. DOI: 10.24411/2500-1000-2019-10749

[22] METAHASH [Internet]. Available from: https://metahash.org [Accessed: March 12, 2022]

[23] Vasyuta EA, Ovakimyan MA, Podolskaya TV. Analysis of the prospects for the development of blockchain tourism: a study of existing practices and an assessment of existing problems. Innovation and Investment. 2020;**7**:259-261. DOI: 10.24411/2307-180X-2020-00025

[24] Chowdhury M et al. A comparative analysis of distributed ledger technology platforms. IEEE Access. 2019;**7**:167930-167943. DOI: 10.1109/ACCESS.2019.2953729

[25] Cooper S. Corporate Social Performance: A Stakeholder Approach. 1st ed. Routledge; 2004. DOI: 10.4324/9781315259239

Section 8

Blockchain Algorithms and Data Analysis

Chapter 16

Perspective Chapter: Matching-Based Clustering Algorithm for Categorical Data

Ruben Gevorgyan and Yenok Hakobyan

Abstract

Blockchain technology allows confidential data to remain strictly confidential and, at the same time, can be used for machine learning with external researchers. Blockchain enables valuable datasets to be reliably processed and speeds up the process of developing valid data mining applications. Blockchain can make it much easier to share datasets, machine learning models, decentralized intelligence, and trustworthy decision-making, which is very important in anomaly detection and fraud detection. This chapter presents a new framework for partitioning categorical data, which does not use the distance measure as a key concept. The matching-based clustering algorithm is designed based on the similarity matrix and a framework for updating the latter using the feature importance criteria. The experimental results show this algorithm can serve as an alternative to existing ones and can be an efficient knowledge discovery tool, especially in anomaly detection using blockchain technologies. While the algorithms for continuous data are relatively well studied in the literature, there are still challenges to address in case of categorical data. Based on the similarity matrix and a novel method for updating it using the feature importance, a matching-based clustering algorithm is designed.

Keywords: categorical data, clustering, similarity matrix, feature importance, anomaly detection

1. Introduction

In the academic literature, one can find many publications in which data mining methods have been applied to solve problems in relation to blockchain systems. In their detailed review article, Liu et al. [1] divide these tasks into the following three categories: cryptocurrency price prediction, blockchain address deanonymization, and anomaly detection. This chapter presents a new clustering method for anomaly detection.

Blockchain platforms are often subject to a variety of malicious attacks [2]. Such actions can potentially be detected by analyzing patterns in transactions. Since the number of anomalous transactions is small, this problem was solved using either empirically derived rules or cluster analysis [3, 4]. As noted by Liu et al. [1], more

IntechOpen

research is required in this area, since most of the models obtained were characterized by a low proportion of identified anomalies.

Clustering is one of the "super problems" in data mining. Generally speaking, clustering is partitioning data points into intuitively similar groups [5]. This definition is simple and does not consider the challenges that occur while applying cluster analysis to real-world datasets. Nevertheless, this type of analysis is common in different areas such as text mining, marketing research, customer behavior analysis, financial market exploration, and so on. Nowadays various clustering algorithms are introduced in the literature, each of them with its advantages and disadvantages. Moreover, as the data come in different forms such as text, numeric, categorical, image, and so on, they perform differently in different scenarios. In other words, the performance of a particular clustering algorithm depends on the structure of the data under consideration.

Cluster analysis of numeric data is relatively well studied in the literature. Various approaches are implemented such as representative base, hierarchical, density base, graph base, probabilistic, and so on [6]. Recently, increasing attention has been paid to clustering none numeric types of data. An important topic is the clustering of categorical data. The problem is that the algorithms for categorical data clustering are mainly modifications of the ones introduced for numeric data. For instance, the most common algorithm is K-modes [7] which is a prototype of the K-means [8] algorithm. However, several researchers have developed algorithms specifically for categorical data, but there is still much room for new approaches.

The main problem with partitioning categorical data is that the standard operations used in clustering algorithms are not applicable. For instance, the definition of distance between two objects with categorical attributes is not as straightforward as with numeric attributes. The main problem is that categorical data takes only discrete values, which do not have any order, unlike continuous data. Thus, the definition of the distance in case of categorical data is ambiguous. Therefore, researchers have developed and used similarity measures [9–11] or have applied different types of transformation [12]. Another problem is the assessment of cluster representatives because many mathematical operations are not applicable to categorical data. For instance, it is impossible to assess the mean of the categorical feature. Taking into account the limitation of existing algorithms one may consider developing an algorithm, which is not using predefined distance/similarity measures as a key concept and is not based on representatives for assigning data points to clusters.

This idea motivated us to develop the matching-based clustering algorithm. In this paper, we are not interested in improving the similarity measure or modifying existing algorithms. The key concept of the algorithm introduced is that two objects with categorical features are similar only if all the features match. Thus, the algorithm is based on the similarity matrix. Besides, we employ a feature importance framework to choose which features to drop on each iteration until all the objects are clustered. The tests on the soybean disease dataset show that the algorithm is highly accurate and possesses much better results.

The rest of the paper is organized as follows. We briefly review the common categorical data clustering algorithms in Section 2. In Section 3, we discuss the categorical data and its limitations. In Section 4 we introduce the general framework of the matching-based clustering algorithm. Section 5 presents the experimental results on the soybean disease dataset. Finally, we summarize our work and describe our future plans in Section 5.

2. Categorical data clustering literature review

Researchers have proposed various methods and algorithms for clustering categorical data. The most common approach is the transformation of the data into binary dataset and then implementation of the standard algorithms with some modification if required. Nevertheless, scholars have developed a wide variety of algorithms for clustering categorical data in recent years. These algorithms can be grouped into five main classes: model-based, partition-based, density base, hierarchical, and projection-base [12]. The main difference between these algorithms is how the similarity or distance is defined for the data points, and according to what criteria the clusters are formed.

Model-based clustering is based on the notion that data come from a mixture model. The most common models used are statistical distributions. Based on the user-specified parameters the prior models are assessed. Then the algorithm aims at recovering the latent model by changing it on each iteration. The main disadvantage of this type of clustering is that it requires user-specified parameters. Hence, if the assumptions are false the results will be inaccurate. At the same time, models may oversimplify the actual structure of the data. Another disadvantage of model-based clustering is that it can be slow on large datasets. Some model-based clustering algorithms are AutoCLass [13], SVM clustering [14], BILCOM Empirical Bayesian [15], etc.

Partition-based clustering algorithms are the most common ones. The main advantage of them is the fast processing time on large datasets. The main concept is defining representatives of each cluster, allocating objects to the cluster, redefining representatives, and reassigning objects based on the dissimilarity measurements. This is repeated until the algorithm converges. The main drawback of this type of algorithm is that they require the number of clusters to be predefined by the user. Another disadvantage is that several algorithms of this type produce locally optimal solutions and are dependent on the structure of the dataset. Several partition-based algorithms are K-modes, Fuzzy K-modes [16], Squeezer [17], COOLCAT [18], etc.

Density-based algorithms define clusters as subspaces where the objects are dense and are separated by subspaces of low density [19]. The implementation of density-based algorithms for categorical data is challenging as the attributes values are unordered. Even though they can be fast in clustering, they sometimes may fail to cluster data with varying densities [20].

Hierarchical algorithms represent the data as a tree of nodes, where each node is a possible grouping of data. There are two possible ways of clustering categorical data using hierarchical algorithms: in an agglomerative (bottom-up) and divisive (top-down) fashion. However, the latter is less common. The main concept of the agglomerative algorithm is using a similarity measure to gradually allocate the objects to the nodes of the tree. The main disadvantage of hierarchical clustering is its slow speed. Another problem is that the clusters may merge thus these algorithms might lead to information distortion. Several categorical data hierarchical clustering algorithms are ROCK [21], LIMBO [22], COBWEB [23], etc.

Projected clustering algorithms are based on the fact that in high-dimensional datasets clusters are formed based on specific attribute subsets. In other words, each cluster is a subspace of high-dimensional datasets defined by a subset of attributes only relevant to that cluster. The main issue with projected clustering algorithms is that it requires user-specified parameters. If the defined parameters are inaccurate, the clustering will be poor. Projected cluster algorithms include CACTUS [24], CLICKS [25], STIRR [26], CLOPE [27], HIERDENC [28], MULIC [29], etc.

More detailed presentations and comparisons of the existing algorithm can be found in [30–32]. Summarizing the existing algorithms, we can conclude that most of them find some tradeoff between accuracy and speed. However, considering the growing interest in analyzing categorical data in social, behavioral, and biomedical science we are more interested in highly accurate algorithms. Furthermore, as one can see the majority of the algorithms uses some distance/similarity metrics and defines representatives of clusters as subroutine of the algorithms. At the same time, they also require user-specified parameters. These factors can be seen as limitations in case of clustering categorical data. Therefore, we purpose another approach to partitioning the categorical data, which tries to avoid these features. Therefore, in the next section, we discuss the main characteristics of categorical data.

3. Categorical data overview

Data comes in various forms such as numeric, categorical, mixture, spatial, and so on. The analysis of each type of data possesses unique challenges. The categorical data is not an exception. This type of data is widely used in political, social, and biomedical science. For instance, the measures of attitudes and opinions can be assessed with categorical data. The performance of medical treatments can also be categorical. Even though the mentioned areas of science have the largest influence on the development of the methods for categorical data, this type of data commonly occurs in other areas of science such as marketing, behavior science, education, psychology, public health, engineering, and so on.

Generally speaking, categorical data is the data where objects are defined by categorical features. Categorical features can have two types of scales: nominal and ordinal. In the case of nominal scale, they have unordered categories. In contrast, ordinal scale possesses ordered categories, but the interval between categories is unknown. In this paper, we focus only on categorical features with nominal scales.

For sake of notation, consider a multidimensional dataset **D** containing **n** objects, such that each object is described by **m** categorical features each with **k** = 1, 2, 3, ... unique categories. Thus, the dataset **D** can be viewed as a matrix below:

$$D_{n,m} = \begin{pmatrix} a_{1,1} & a_{1,2} & \cdots & a_{1,m} \\ a_{2,1} & a_{2,2} & \cdots & a_{2,m} \\ \vdots & \vdots & \ddots & \vdots \\ a_{n,1} & a_{n,2} & \cdots & a_{n,m} \end{pmatrix} \tag{1}$$

where each object is described by a $\boldsymbol{a}$ set of categories $\boldsymbol{O_i} = [\boldsymbol{a_{i,1}}, \boldsymbol{a_{i,2}}, \boldsymbol{a_{i,3}} \cdots \boldsymbol{a_{i,m}}]$. As the categorical attributes have discrete values with no order values, the application of distance measures such as ***lpnorm*** will produce inaccurate results. However, the most common approach to overcome this limitation is the implementation of data transformation techniques. For instance, one can use binarization to transform the data into binary data and then apply the distance measure. On the other hand, the traditional way of comparing two objects with categorical features is to simply check if the categories coincide. If the categories of all the features under consideration match, the objects can be viewed as similar. This does not mean they are the same, because they can be distinguished by other features. Thus, researchers have proposed various similarity measures instead of requiring all the features to match. The most popular approach is the overlap. According to it, the similarity between two objects $\mathbf{O_x} = [\mathbf{a_{x,1}}, \mathbf{a_{x,2}}, \mathbf{a_{x,3}} \cdots \mathbf{a_{x,m}}]$ and $\mathbf{O_y} = [\mathbf{a_{y,1}}, \mathbf{a_{y,2}}, \mathbf{a_{y,3}} \cdots \mathbf{a_{y,m}}]$ is assessed by:

$$Ov(O_x, O_y) = \frac{1}{m}\sum_{i=1}^{m} \gamma_i$$
$$\text{where } \gamma_i = \begin{cases} 1 & \text{if } a_{x,i} = a_{y,i} \\ 0 & \text{otherwise} \end{cases}, \tag{2}$$

It can take values from [0, 1]. The closer value gets to one, the higher the similarity between the objects.

While implementing overlap, one can notice that the probability of finding another object with the same categories rapidly decreases as the number of features and the number of unique categories of each feature increases. To illustrate this, one can calculate the probability of finding another object with the same categories as object $\mathbf{O_x}$ using the formula below:

$$P = \prod_{i=1}^{m} \frac{f(a_{x,i}) - 1}{k_m(n-1)} \tag{3}$$

where $\boldsymbol{f}(\mathbf{a_{x,i}})$ is the frequency of category $\mathbf{a_{x,i}}$ in the dataset.

If we consider the number of objects constant, the probability of finding another similar object depends on the number of features and the number of unique categories of each. It can be seen from the formula above that as the number of attributes or the number of categories increases the probability of finding another object rapidly decreases. The problem is that the overlap measure gives equal weights to the features and does not take into account the importance of each feature in partitioning the data. However, the researchers have proposed more efficient ways of assessing similarity, which takes into account the frequency of each category in the dataset. There are various types of similarity measures that are based on this concept. For instance, Goodall [33], Lin [34], and so on:

$$\text{Goodall}(O_x, O_y) = \frac{1}{m}\sum\nolimits_{i=1}^{m} S_i$$
$$\text{where } \ S_i = \begin{cases} 1 - \dfrac{f(a_{x,i})(f(a_{x,i}) - 1)}{n(n-1)} & \text{if } a_{x,i} = a_{y,i} \\ 0 & \text{otherwise} \end{cases}, \tag{4}$$

$$\text{Lin}(O_x, O_y) = \frac{1}{m}\sum\nolimits_{i=1}^{m} S_i$$
$$(\text{where}) \ \ S_i = \begin{cases} 2\log f(a_{x,i}) & \text{if } a_{x,i} = a_{y,i} \\ 2\left(\log\left(\dfrac{f(a_{x,i})}{n} + \dfrac{f(a_{y,i})}{n}\right)\right) & \text{otherwise} \end{cases}, \tag{5}$$

Nevertheless, there are still cases when the use of similarity measures can be misleading. For instance, consider the dataset below with four objects and two categorical attributes with $[\mathbf{c_1}, \mathbf{c_2}]$, $[\mathbf{b_1}, \mathbf{b_2}]$ categories, respectively:

$$D_{4,2} = \begin{pmatrix} c_1 & b_1 \\ c_2 & b_2 \\ c_1 & b_2 \\ c_2 & b_1 \end{pmatrix} \tag{6}$$

According to the measures presented above, the similarity between each unique pair of these objects will be:

Object	Overlap	Lin	Goodall
(O_1, O_2)	0.00	0.00	0.00
(O_1, O_3)	0.50	0.69	0.33
(O_1, O_4)	0.50	0.69	0.33
(O_2, O_3)	0.50	0.69	0.33
(O_2, O_4)	0.50	0.69	0.33
(O_3, O_4)	0.00	0.00	0.00

As one can see, these measures can be misleading. For instance, one can group $\mathbf{O_3}$, $\mathbf{O_4}$ to either $\mathbf{O_1}$ or $\mathbf{O_2}$ as the similarity measures are the same. Therefore, similarity measures are powerful tools, but they should be used with caution. In this regard, one may consider using a quantitative measure to compare the features and choose relatively important ones. Then the objects will be similar if the categories of the selected features match. This is the main motivation for our approach.

Therefore, we employ several feature importance measures. We define the partial grouping power of a feature in dataset **D** as the number of unique matching pairs on the feature divided by the total number of unique matching pairs in the dataset. This is based on the notion that if the feature has a relatively higher number of matching pairs than others, it is more likely to group objects. The **PGPI l** can be assessed by:

$$PGPI_l = \frac{\sum_{s=1}^{k_l} \frac{f(c_s)(f(c_s)-1)}{2}}{\sum_{i=1}^{m} \sum_{j=1}^{k_m} \frac{f(c_j)(f(c_j)-1)}{2}} \tag{7}$$

where $\boldsymbol{c_s}$ is the unique category of the feature, and $\boldsymbol{f(c_s)}$ is the frequency of the category in the dataset. This measure can take values from $[0, 1]$. The closer the value to one the higher the importance of the feature in aggregating the objects.

We also define a measure for the partitioning power of a feature. We define the partial partitioning power of a feature in dataset **D** as the number of unique mismatching pairs on the feature divided by the total number of unique mismatching pairs in the dataset. The **PPPIl** can be assessed by:

$$PPPI_l = \frac{\sum_{s=1}^{k_l} \frac{n(n-1)}{2} - \frac{f(c_s)(f(c_s)-1)}{2}}{\sum_{i=1}^{m} \sum_{j=1}^{k_m} \frac{n(n-1)}{2} - \frac{f(c_j)(f(c_j)-1)}{2}} \tag{8}$$

This measure can take values from $[0, 1]$. The closer the value to one the higher the importance of the feature in partitioning the objects. Both methods can be used in any analysis. However, as the objects can be relatively grouped or separated depending on the features under consideration, one of the measure may perform better.

We also present another measure for assessing the feature importance. This one is based on the similarity matrix. The similarity matrix is defined as the matrix below:

$$SM_{n,n} = \begin{pmatrix} m_{1,1} & m_{1,2} & \cdots & m_{1,m} \\ m_{2,1} & m_{2,2} & \cdots & m_{2,m} \\ \vdots & \vdots & \ddots & \vdots \\ m_{n,1} & m_{n,2} & \cdots & m_{n,m} \end{pmatrix} \tag{9}$$

where $m_{i,j}$ is a similarity measure between object **i** and **j** such as Overlap, Lin, and Goodall. Throughout this paper, we will use the count of matches between two objects as a similarity measure:

$$m_{i,j} = \sum_{i=1}^{m} \gamma_i, \quad \text{where } \gamma_i = \begin{cases} 1 & \text{if } a_{x,i} = a_{y,i} \\ 0 & \text{otherwise} \end{cases} \tag{10}$$

This measure is also known as the hamming distance. The similarity matrix is symmetrical, thus only the upper triangular matrix is used in the calculations. Furthermore, the diagonal will also be ignored. For another measure of the feature importance, based on the similarity matrix we define the general influence matrix as:

$$IM_{n,n} = \begin{pmatrix} I_{1,1} & I_{1,2} & \cdots & I_{1,m} \\ I_{2,1} & I_{2,2} & \cdots & I_{2,m} \\ \vdots & \vdots & \ddots & \vdots \\ I_{n,1} & I_{n,2} & \cdots & I_{n,m} \end{pmatrix}, \quad \text{where } I_{i,j} = \begin{cases} 1 & \text{if } m_{i,j} > \alpha \\ 0 & \text{otherwise} \end{cases} \tag{11}$$

where α is a threshold, which is bounded by the values similarity measure can take. In this chapter, we set α to **0**. After the construction, the features or the subset of features under consideration are dropped, and the influence matrix is updated. The matrix after the drop is defined as the partial influence matrix of corresponding feature or subset of features **l**. In this case, the partial grouping power of a feature or subset of features is assessed by dividing the count of the ones in the partial influence matrix by the count of ones in general influence matrix:

$$PGPI2_j = \frac{\eta_{PIM_l}}{\eta_{GIM}} \tag{12}$$

where η_{PIM_l} is the count of ones in the $\boldsymbol{PIM_l}$ and η_{GIM} is the count of ones in the $\boldsymbol{GIM}$. One can notice that these measures of feature importance depend only on the number of unique matches in the dataset, and the number of categories of each feature does not influence them. In the next section, we present the matching-based clustering algorithm, which combines the importance measures of the features and the similarity matrix to partition categorical data into homogeneous groups.

4. Matching-based clustering algorithm

Similar to any clustering algorithm the main objective of matching-based clustering is partitioning the data into relatively similar groups. The algorithm is defined for categorical data only. However, one can modify it to work with other types also, but it is out of the scope of this paper. The main idea is, while there are still objects without clusters, the algorithm will choose features to drop based on their importance. Then it will update the similarity matrix and try to cluster the objects based on the new SM. It uses the similarity matrix where the similarity measure between two objects is defined by formula (10). We also use either PGPI or PPPI measure to choose the features to drop on each iteration. For the sake of notations, we define $\boldsymbol{\theta p}$ as the count of the remaining features on iteration **p**. The initial value of $\boldsymbol{\theta_0} = \mathbf{m}$. We consider two objects to belong to the same cluster if $\boldsymbol{m_{i,j}} = \boldsymbol{\Theta p}$. In other words, they are grouped if their categories coincide for all the remaining features on iteration **p**.

The algorithm consists of the following steps:

1. Construct the similarity matrix $SM_{n,n}$.

2. Calculate the PGPI or PPPI of each feature.

3. Allocate the objects to clusters based on the similarity matrix. In other words, group two objects (and j), if $m_{i,j} = \theta_p$. If one of them is already allocated to a cluster, assign the second one to the same cluster.

4. Check if there are still objects not assigned to any cluster, if yes continue to next step, otherwise terminate.

5. Remove the features with the lowest *PGPI* or the highest *PPPI*. If there are more than one feature, one may consider either dropping all of them or use the PGPI2 to choose which one to drop.

6. Update the similarity matrix. Furthermore, to avoid the merging of existing clusters, additionally update the SM using: ∀ existing cluster i and j, if $m_{i,j} = \theta_p$, then the values, equal to θ_p, of rows and columns i and j are set to zero.

7. Return to step 3.

The algorithm stops if all the objects are clustered or the importance of remaining features is the same. To illustrate how the algorithm works, we will apply it to the dataset below:

Objects	A	B	C	D	E
O_1	a_2	b_1	c_2	d_3	e_2
O_2	a_2	b_1	c_2	d_3	e_2
O_3	a_2	b_1	c_2	d_3	e_1
O_4	a_2	b_1	c_2	d_3	e_4
O_5	a_1	b_2	c_4	d_2	e_3
O_6	a_1	b_2	c_3	d_4	e_4
O_7	a_1	b_2	c_4	d_2	e_2
O_8	a_1	b_2	c_3	d_4	e_1
O_9	a_1	b_2	c_1	d_1	e_3
O_{10}	a_1	b_2	c_4	d_2	e_2

In this dataset 10 objects are defined by 4 categorical features A, B, C, D, and E with $[a_1, a_2]$, $[b_1, b_2]$, $[c_1, c_2, c_3, c_4]$, $[d_1, d_2, d_3, d_4]$, and $[e_1, e_2, e_3, e_4]$ unique categories, respectively. Thus, we initialize the algorithm by constructing the similarity matrix:

$$S_{10,10} = \begin{pmatrix} - & 5 & 4 & 4 & 0 & 0 & 1 & 0 & 0 & 1 \\ - & - & 4 & 4 & 0 & 0 & 1 & 0 & 0 & 1 \\ - & - & 4 & 0 & 0 & 0 & 1 & 0 & & 0 \\ - & - & - & - & 0 & 1 & 0 & 0 & 0 & 0 \\ - & - & - & - & 2 & 4 & 2 & 3 & & 4 \\ - & - & - & - & - & 2 & 4 & 2 & & 2 \\ - & - & - & - & - & - & 2 & 2 & & 5 \\ - & - & - & - & - & - & - & 2 & & 2 \\ - & - & - & - & - & - & - & & & - \\ - & - & - & - & - & - & & & & - \\ - & - & - & & & & & & & - \end{pmatrix} \tag{13}$$

Then, the importance of each feature is assessed. In this example, we will use the PGPI measure. For instance, the PGPI_A will be:

$$PGPI_A = \frac{21}{21 + 21 + 10 + 10 + 9} = 0.296 \tag{14}$$

respectively $PGPI_B = 0.296, PGPI_C = 0.14, PGPI_D = 0.14$, and $PGPI_E = 0.127$. Then as the $\theta_0 = m = 5$, all the objects i, j with $m_{i,j} = 5$ are grouped. As we can see we have two clusters $[O_1, O_2]$ and $[O_7, O_{10}]$, respectively. As we still have some objects left without cluster allocation, we continue to next step. In particular, as the feature E has the lowest $PGPI$, we drop it and the similarity matrix is updated. Also to avoid the merging of already existing clusters, we additionally update SM according to step 6. As similarity between clusters $[O_1, O_2]$ and $[O_7, O_{10}]$ is not equal to $\theta_1 = 4$, we will not make additional changes, and the new data view and corresponding similarity matrix will be:

Objects	A	B	C	D
(O_1, O_2)	a_1	b_0	c_1	d_2
O_3	a_1	b_0	c_1	d_2
O_4	a_1	b_0	c_1	d_2
O_5	a_0	b_1	c_3	d_1
O_6	a_0	b_1	c_2	d_3
(O_7, O_{10})	a_0	b_1	c_3	d_1
O_8	a_0	b_1	c_2	d_3
O_9	a_0	b_1	c_0	d_0

(15)

$$S_{8,8} = \begin{pmatrix} -4 & 4 & 0 & 0 & 0 & 0 & & 0 \\ - & - & 4 & 0 & 0 & 0 & 0 & 0 \\ - & - & 0 & 0 & 0 & 0 & & 0 \\ - & - & - & 2 & 4 & 2 & & 2 \\ - & - & - & - & 2 & 4 & & 2 \\ - & - & - & - & - & 2 & & 2 \\ - & - & - & - & - & - & & 2 \\ - & - & - & - & - & - & & -. \end{pmatrix} \tag{16}$$

As $\theta_1 = 4$, we will have $[O_1, O_2, O_3, O_4]$, $[O_5, O_7, O_{10}]$, and $[O_6, O_8]$ clusters. However, we still have one more object to assign to a cluster, thus we drop C and D. We update the similarity matrix:

Objects	A	B
(O_1, O_2, O_3, O_4)	a_1	b_0
(O_5, O_7, O_{10})	a_0	b_1
(O_6, O_8)		$a_0 b_1$
O_9		$a_0 b_1$

$$S_{4,4} = \begin{pmatrix} - & 0 & 0 & 0 \\ - & - & 2 & 2 \\ - & - & - & 2 \\ - & - & - & - \end{pmatrix}$$

But we also check the statement in step 6 between any pair of the existing clusters. As $\theta_2 = 2$, the statement is true in the case of $[O_5, O_7, O_{10}]$ and $[O_6, O_8]$. Thus, the values corresponding rows and columns, which are equal $\theta_2 = 2$, are set to zero. The purpose of this modification is that as we are dropping features with the low grouping power, the cluster is more likely to merge. Therefore, we may lose important local partitioning of data points. Thus the finally updated similarity matrix will be:

Objects	A	B	C	D	E	Cluster
O_1	a_1	b_0	c_1	d_2	e_2	1
O_2	a_1	b_0	c_1	d_2	e_2	1
O_3	a_1	b_0	c_1	d_2	e_1	1
O_4	a_1	b_0	c_1	d_2	e_4	1
O_5	a_0	b_1	c_3	d_1	e_3	2
O_6	a_0	b_1	c_2	d_3	e_4	3
O_7	a_0	b_1	c_3	d_1	e_2	2
O_8	a_0	b_1	c_2	d_3	e_1	3
O_9	a_0	b_1	c_0	d_0	e_3	4
O_{10}	a_0	b_1	c_3	d_1	e_2	2

Table 1.
Final form of the clustering.

$$\begin{pmatrix} - & 0 & 0 & 0 \\ - & - & 0 & 0 \\ - & - & - & 0 \\ - & - & - & - \end{pmatrix} \tag{17}$$

However, the second iteration does not group the O_9. At the same time as the importance of the remaining features A and B is the same the algorithm terminates and the object O_9 forms the fourth cluster. Thus, the final form of the clustering is presented in **Table 1**.

The algorithm has some unique characteristics worth mentioning. First, to achieve better performance one can notice that all the changes required in each step should be done only on the similarity matrix and there is no need to update the dataset. Second, there is no need for user-defined parameters. However, one may consider one, for instance, the required number of clusters to be created. Third, even though we introduced step 6 to avoid the merging of clusters to achieve higher accuracy, one can avoid this step. In this case, the algorithm will create a tree were each leaf is a possible cluster, like in the hierarchical cluster. Furthermore, based on the user-defined parameter the algorithm can create the required number of clusters if required. For instance, in case of our example above the dendrogram will be (**Figure 1**).

Forth, as the algorithm is based on either feature grouping or partitioning power, this information can be used to understand the data better. For instance, this algorithm can serve as a subroutine for selecting features for other clustering algorithms. The main disadvantage is that may create too many clusters.

5. Experimental results and discussion

To test how the matching-based algorithm performs on real-world dataset, we have employed it to the soybean disease dataset [23]. It is one of the standard test data

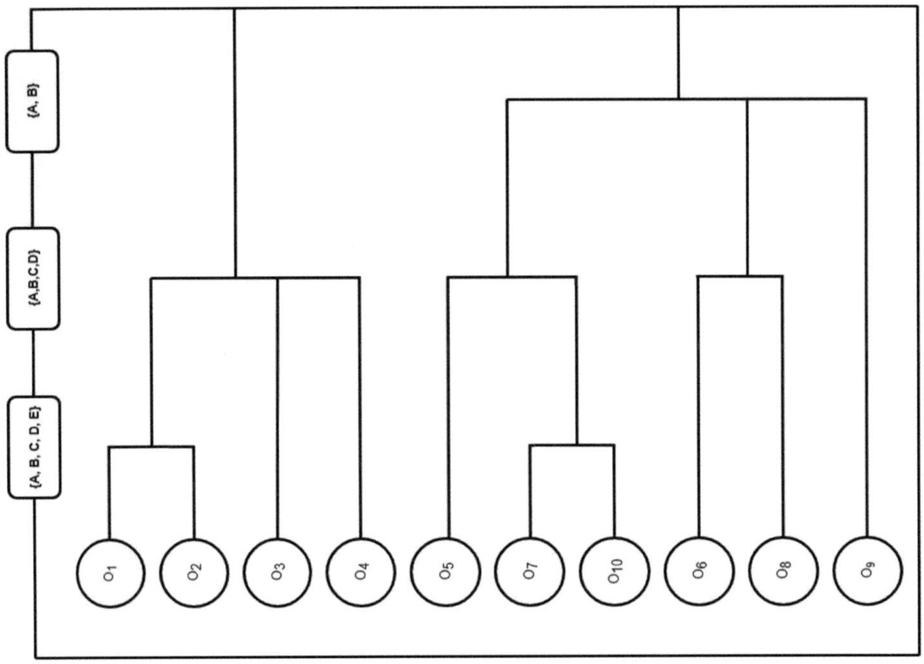

Figure 1.
Dendrogram, based on the user-defined number of clusters.

sets used in the machine learning community. It has often been used to test conceptual clustering algorithms. We chose this data set to test our algorithm because of its publicity and because all its attributes can be treated as categorical without categorization. The soybean data set has 47 observations, each being described by 35 attributes. Each observation is identified by one of the four diseases – Diaporthe Stem Canker, Charcoal Rot, Rhizoctonia Root Rot, and Phytophthora Rot. Which are used as indicators of the efficiency of the algorithm.

After applying the MBC algorithm to the soybean disease dataset, we got 18 different clusters.

DT	1	2	3	4	5	6	7	8	9	10	11	12	13	14	15	16	17	18
*D*1	−3	2	–	–	–	–	–	−2	2	–	–	–	–	–	–			1
*D*2	–	–	−3	–	−2	2	3	–	–	–	–	–	–	–	–			–
*D*3	–	–	–	−2	–	–	–	–	–	–	5	2	–	–	–	–		1
*D*4	5	–	–	–	−3	–	–	–	–	–	–	2	3	2	2			–

However, as we can see from the table above all the clusters except for one entirely belong to one of the groups mentioned above. In other words, we have only one possible misclassification. However, as already mentioned one may require specific number of clusters. In this case, one can use the dendrogram (**Figure 2**).

Furthermore, we can compare the performance of the algorithm with K-modes [35]. In that paper, the algorithm was also applied to the soybean disease dataset. The author emphasized the fact that K-modes depend on the data order and the user should also give the number of clusters. In case of MBC, we do not have these

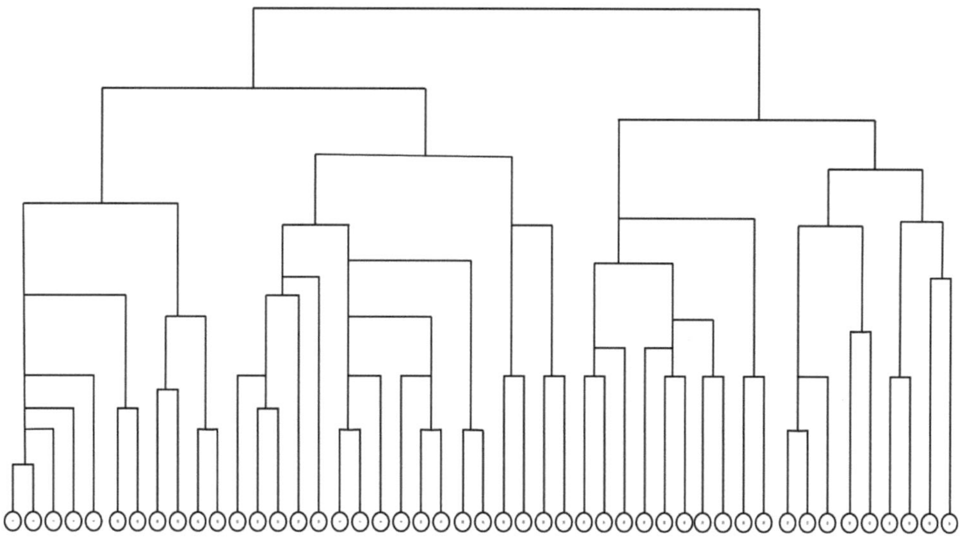

Figure 2.
Dendogram for the soybean disease dataset after applying the MBC algorithm.

limitations. Thus the application of the MBC algorithm may result in more accurate outcome. However, we still should consider how to lower the number of clusters in case of it.

6. Conclusion

The vital issue in clustering categorical data is the notion of distance/similarity between the observations. The best practice approaches are limited to numeric values. Hence, specific models are being developed for categorical data. The algorithm introduced in this paper can serve as an alternative to existing ones. It is based on the main characteristics of categorical data. It presents a new framework for clustering categorical data, which is not based either on distance or on similarity measure. The main concept of the algorithm is the assessment of the similarity matrix, updating latter based on the important criteria of each feature or subset of feature and grouping only if the categories of objects entirely match. These allow to cluster of categorical data without conversion. Another advantage is the description of the features, as the algorithm allows to identify the feature which causes the partitioning of the data. These can be very important in interpreting clustering results. The main advantage of the algorithm is high accuracy and few initial parameters.

Our future work plan is to develop and implement a modification of the algorithm to cluster mixture data. Furthermore, overcome its limitation and adopt it to clustering big datasets. Such an algorithm is required in a number of data mining applications, such as partitioning very large sets of objects into a number of smaller and more manageable subsets that can be more easily modeled and analyzed.

Mathematics subject classification (2010):

62H30, 62H17, 62H20

Author details

Ruben Gevorgyan* and Yenok Hakobyan
Faculty of Economics and Management, Yerevan State University, Yerevan, Armenia

*Address all correspondence to: rubengevorgyan@ysu.am

References

[1] Liu XF et al. Knowledge discovery in cryptocurrency transactions: A survey. IEEE Access. 2021;**9**:37229-37254

[2] Rahouti M, Xiong K, Ghani N. Bitcoin concepts, threats, and machine-learning security solutions. IEEE Access. 2018;**6**: 67189-67205

[3] Camino RD, State R, Montero L, Valtchev P. Finding suspicious activities in financial transactions and distributed ledgers. In: 2017 IEEE International Conference on Data Mining Workshops (ICDMW). New Orleans: IEEE; 2017. pp. 787-796

[4] Pham T, Lee S. Anomaly detection in the bitcoin system-a network perspective. 2016. arXiv preprint arXiv: 1611.03942

[5] Jain A, Murty M, Flynn P. Data clustering: A review. ACM Computing Surveys (CSUR). 1999;**31**(3):264-323

[6] Aggarwal CC. Data Mining: The Textbook, 154–259. Switzerland: Springer International Publishing; 2015

[7] Chaturvedi A, Green P, Carroll JD. K-modes clustering. Journal of Classification. 2001;**18**(1):35-55

[8] MacQueen JB. Some methods for classification and analysis of multivariate observations. In: Proceedings of the 5th Berkeley Symposium on Mathematical Statistics and Probability. Berkley and Los Angeles: University of California Press; 1967. pp. 281-297

[9] Boriah S, Chandola V, Kumar V. Similarity measures for categorical data: A comparative evaluation. In: SIAM Conference on Data Mining. Atlanta: SIAM; 2008

[10] Gouda K, Zaki MJ. Genmax: An efficient algorithm for mining maximal frequent itemsets. Data Mining and Knowledge Discovery. 2005;**11**(3): 223-242

[11] Burnaby T. On a method for character weighting a similarity coefficient employing the concept of information. Mathematical Geology. 1970;**2**(1):25-38

[12] Hubert MR, Segaert P, Pieter. Multivariate and functional classification using depth and distance. Adv. Data Anal. Classif. 2017; **11**:445-466

[13] Stutz J, Cheeseman P. Bayesian classification (AutoClass): Theory and results. In: Advances in Knowledge Discovery and Data Mining. Menlo Park, CA: AAAI Press; 1995. pp. 153-180

[14] Winters-Hilt S, Merat S. SVM clustering. BMC Bioinformatics. 2007; **8**(7):S18

[15] Andreopoulos B, An A, Wang X. Bi-level clustering of mixed categorical and numerical biomedical data. International Journal of Data Mining and Bioinformatics (IJDMB). 2006;**1**(1): 19-56

[16] Huang Z. Clustering large data sets with mixed numeric and categorical values. In: Knowledge Discovery and Data Mining. Techniques and Applications. Singapore: World Scientific; 1997

[17] He Z, Xu X, Deng S, et al. Squeezer: An efficient algorithm for clustering categorical data. Journal of Computer Science and Technology. 2002;**17**(5): 611-624

[18] Barbara D, Li Y, Couto J. COOLCAT: An entropy-based algorithm for categorical clustering. In: Proceedings of CIKM 2002. Vol. 2002. McLean, VA, USA: ACM Press; pp. 582-589

[19] Gionis A, Hinneburg A, Papadimitriou S, Tsaparas P. Dimension induced clustering. In: Proceedings of KDD'05. Chicago: Association for Computing Machinery; 2005. pp. 51-60

[20] Grambeier J, Rudolph A. Techniques of cluster algorithms in data mining. Data Mining and Knowledge Discovery. 2002;**6**:303-360

[21] Guha S, Rastogi R, Shim K. ROCK: A Robust clustering algorithm for categorical attributes. Information Systems. 2000;**25**(5):345-366

[22] Andritsos P, Tsaparas P, Miller R, et al. LIMBO: Scalable clustering of categorical data. In: Proceedings of the 9th International Conference on Extending Database Technology EDBT'04. Heraklion, Greece: Springer; 14–18 March 2004. pp. 123-146

[23] Fisher D. Knowledge acquisition via incremental conceptual clustering. Machine Learning. 1987;**2**: 139-172

[24] Ganti V, Gehrke J, Ramakrishnan R. CACTUS-clustering categorical data using summaries. In: Proceedings of KDD'99. San Diego, CA, USA: SIGMOD; 1999. pp. 73-83

[25] Zaki MJ, Peters M. CLICKS: Mining subspace clusters in categorical data via K-partite maximal cliques. In: Proceedings of the 21st International Conference on Data Engineering (ICDE). Tokyo, Japan: IEEE Computer Society; 2005. pp. 355-356

[26] Gibson D, Kleiberg J, Raghavan P. Clustering categorical data: An approach based on dynamical systems. In: Proceedings of 24 the International Conference on Very Large Databases (VLDB'98). New York City, USA: IEEE Computer Society; 24–27 August 1998. pp. 311-323

[27] Yang Y, Guan S, You J. CLOPE: A fast and effective clustering algorithm for transactional data. In: Proceedings of KDD 2002. Edmonton, Alberta, Canada: Association for Computing Machinery; 2002. pp. 682-687

[28] Andreopoulos B, An A, Wang X. Hierarchical density-based clustering of categorical data and a simplification. In: In: Proceedings of the 11th Pacic-Asia Conference on Knowledge Discovery and Data Mining (PAKDD 2007), Springer LNCS 4426/2007. Nanjing, China: Springer; 22–25 May 2007. pp. 11-22

[29] Andreopoulos B. Clustering algorithms for categorical data. [PhD thesis], Dept of Computer Science Engineering, York University, Toronto, Canada. 2006

[30] Anderberg MR. Cluster Analysis for Applications. New York: Academic Press; 1973

[31] Jain AK, Dubes RC. Algorithms for Clustering Data. New Jersey: Prentice-Hall, Inc.; 1998

[32] Kaufman L, Rousseeuw PJ. Finding Groups in Data: An Introduction to Cluster Analysis. Brussels: Wiley; 2009

[33] Goodall DW. A new similarity index based on probability. Biometrics. 1966; **22**(4):882-907

[34] Lin D. An information-theoretic definition of similarity. In: Proc ICML

1998 15th International Conference on Machine Learning. San Francisco, CA, USA: Morgan Kaufmann Publishers Inc; 1998. pp. 296-304

[35] Zhexue H. A fast clustering algorithm to cluster very large categorical data sets in data mining. In: Research Issues on Data Mining and Knowledge Discovery. Arizona, USA: SIGMOD; 1997. pp. 1-8

Section 9

Blockchain in the Food Supply Chain

Chapter 17

Perspective Chapter: Blockchain Adoption in Food Supply Chain

Jerome Verny and Wei Guan

Abstract

Modern food supply chain involves numerous stakeholders that are geographically dispersed. This scattered and complex structure impedes the free flow of information throughout the supply chain. Low transparency lays the groundwork for food fraud and delays the implementation of necessary countermeasures when a food contamination incident occurs. Moreover, customers nowadays are increasing demanding in terms of product provenance and sustainability. Under this circumstance, it is urgent to identify effective solutions that can mitigate the concerns of food safety, quality, and fraud. Blockchain has been identified as a promising technology for enabling end-to-end supply chain traceability. This study investigates the main challenges of agri-food supply chain and how blockchain attributes can address these challenges. We also propose an integrative framework of key factors that influence blockchain adoption in the food industry.

Keywords: blockchain, adoption, agri-food supply chain, critical synthesis, critical success factors

1. Introduction

Global agri-food supply chain involves a myriad of functionally and geographically diverse stakeholders. This fragmented structure limits the free flow of information among supply chain participants. Due to the low information transparency, modern food supply chain often confronts challenges in production, processing, storage, distribution, and raises concerns about food safety and fraud. As a matter of fact, nearly 10% of the world population (600 million people) have suffered foodborne illness every year, and 110 billion USD is lost each year in productivity and medical expenses resulting from unsafe food in low- and middle-income countries [1]. According to European Commission on food fraud [2], olive oil, milk, honey, saffron, orange juice, apple juice, grape wine, vanilla extract, and fish are on the list of the most common sources of food fraud. Scandals such as the milk adulterated with melamine in China, horse meat in beef products and sold in Europe, fipronil in eggs, and the slaughter of sick cows for meat in Poland have drawn attention worldwide.

Under these circumstances, consumers around the globe are demanding detailed information in terms of product provenance and what parties are involved in each stage of the food supply chain. However, the lack of supply chain transparency severely inhibits the capacity of food supply chain stakeholders to provide such information. Moreover, when food contamination incidents occur, low transparency also delays the effective implementation of countermeasures. In this context, academics and practitioners have begun to investigate the potential of technological innovations (e.g. artificial intelligence, big data analytics, and internet of things) to improve food supply chain transparency. In this paper, we examine how blockchain technology can be used to address food supply chain challenges and the factors that influence its effective implementation in food supply chain. Our analysis is based on a critical synthesis of a wide range of sources in the recent years from major businesses who are leading the digitalization efforts in agri-food industry, world leading supply chain consulting firms, global international organizations, trade magazines, and research articles. The contributions of this paper are twofold. First, it provides a holistic overview of the merits of blockchain adoption in the food supply chain. Second, we propose an integrative framework of factors that affect blockchain adoption in this specific context.

2. Agri-food supply chain challenges

Traditional agri-food supply chains encompass diverse actors from the raw material suppliers, processors, wholesalers, and retailers to end customers. Over time, modern agri-food supply chains have transitioned from autonomous, independent, and local actors to globally interconnected systems of multiple actors that affect the way food is produced, sourced, processed, transported, and delivered to the final consumer. Complexities emerge due to the need of real-time information sharing, mutual scheduling, product quality guarantee, and timely fulfillment of delivery promises. Current supply chain transactions are based on complex, paper-based settlement process. Not only do these transactions lack transparency and efficiency, but they are also vulnerable to fraud.

Most of the food supply chain participants still adopt the "one-up-one-down approach" that is laid out by CAC/GL 60-2006: participants can identify at any given stage of the food chain (from production to distribution) where the food comes from (one step back) and where it goes (one step further). By doing so, the visibility of focal firms in the movement of agricultural products is limited to the level of their direct suppliers and customers. This tracking method is largely inadequate, especially for multi-ingredient foods that include elements from different sources in different countries. In case of suspected contamination, the entire shipment will be discarded as a precautionary measure in accordance with the one-up-one-down approach, resulting in heavy economic losses. In addition, the currently deployed food traceability systems are neither integrated with each other nor linked among all participants in the supply chain. This disconnection generates information asymmetry between supply chain parties, resulting in poor supply chain visibility.

Nowadays, consumers are increasingly demanding in terms of product provenance and sustainability. They need to know where and how their food is produced and delivered. The current traceability and provenance systems for food supply chain can no longer fulfill such demands. In addition to the pressure on the demand side, regulations such as the food safety modernization act and general food law regulation profoundly impact the global food supply chain by mandating hazard analysis and end-to-end traceability. Unpredictable incidences of food safety or health hazards can significantly reduce a company's brand value, erode consumer trust, and lead to lawsuits and product recalls.

There are some technological innovations that are applied for agri-product traceability purposes, such as radio-frequency identification tags, electronic data interchange, and internet of things. However, current food traceability systems are built on top of centralized infrastructures, which leave room for unresolved issues, including data integrity, tampering, and single points of failure [3]. To address these issues and enable end-to-end traceability, supply chain professionals envision the use of blockchain technology.

3. Blockchain technology

Blockchain is a distributed database that allows the storage and transmission of information in a transparent and secure manner. It operates without a centralized control body because it is managed by a network of computers/users, also called nodes, on a peer-to-peer basis. This database is constituted by a growing list of digital records of validated transactions, known as blocks, which are chained to each other in chronological order through hashing function [4]. The validation of a new block of transactions involves all the relevant nodes. They execute algorithms to evaluate and verify the authenticity and accuracy of the transactions. If the majority of nodes agree that these transactions are valid, then the new block of transactions is accepted into the database. Each block is attributed with a unique hash number – a digital fingerprint of data, and it also carries the hash of the previous block. Once these blocks are chained to each other, they become immutable. **Figure 1** illustrates the main properties and operations mode of a blockchain.

Blockchain contains the following key features: immutability, automation, and security [6, 7]. First, transaction information stored on the blockchain can be seen by all participants and cannot be altered by any single node, as each node possesses a complete record of all the information within the blockchain. Data immutability ensures data accuracy, increases trust, and reduces fraud. This feature enables the tracking of the provenance of assets, which means that for any asset it is possible to tell where it is, where it has been, and what has happened throughout its lifetime [8].

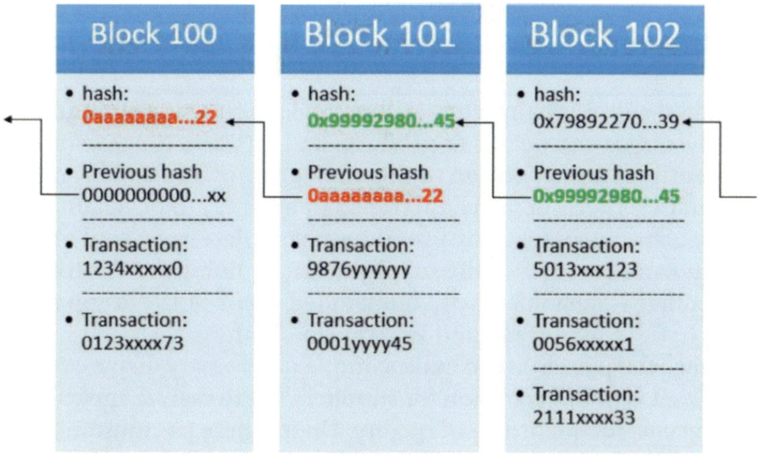

Figure 1.
The key properties of blockchain. Source: Queiroz & Wamba [5].

Second, blockchain technology enables smart contracts, which is a consensus agreement based on a specific computer protocol [9]. The smart contract is executed automatically when conditions are proven to have been met [10]. In this way, digital assets could be operated without the need for third-party interventions, but by a program executing automatically certain rules defined by users [11, 12]. The trust issue in the network is resolved as everyone is playing by the rules and operation could be done more efficiently. Third, the distributed and encrypted nature of blockchain technology makes it resilient to different kinds of cyberattacks [13].

Blockchain has broadly been used according to one of the two models: public and private. Based on peer-to-peer network and consensus mechanisms, a public blockchain is a decentralized or distributed network that uses individual node to record transactions and distribute the data directly to each connected node. All the data and clusters of transactions are organized in a group, or "block." By adding a set of new blocks, the chain is hence formed. Every block is generated by a specific consensus algorithm to assure that all the participants involving in the chain agree upon a specific state of the system as its true state [14–16]. A public blockchain network is totally open to every creator of a block, which suggests that anyone can access to the blockchain and allow them to read its contents. The network typically has an encouraging mechanism to encourage more participants to join the network. Bitcoin is a typical public blockchain networks today [17, 18].

A private blockchain is similar as a public blockchain. They are both decentralized network connected by different blocks generated by a specific consensus mechanism. The only distinction between public and private blockchain is related to who is allowed to participate within the network, execute the consensus protocol, and maintain the shared ledger [18]. A private blockchain calls for invitation proof that should be validated by either the network starter or by a group of rules. Private blockchain is usually operated by one organization, which defines limited visibility rights to chosen participants within a permissioned network [13]. In another word, participants must obtain a permission to join the blockchain. The access control mechanism could vary according to organization's choice: existing participants could decide future entrants; a regulatory agency could issue licenses for participation or a consortium could make the choices instead [19].

4. Why blockchains are important for agri-food supply chain

Currently, food supply chain actors still struggle to find a reliable and effective way to verify the origin and details of products and services. More than 70,000 consumers recently signed a petition urging major companies and brands, including Walmart, to enhance their supply chain transparency [20]. Blockchain has the potential to improve such transparency by providing a complete audit trail of transaction data collected at various stages of the supply chain. By doing so, this technology offers customers and other stakeholders with undisputed proof of the origin and authenticity of products to fight food fraud and improve food safety.

Blockchain has the possibility to make complex and costly dispute settlement a thing of the past. It is not uncommon for suppliers fail to deliver agricultural products on time or in the correct quantity and quality. Under these circumstances, relevant parties need to identify the source of the problem and resolve the dispute, usually through fines or compensation. However, supply chain disputes are often tedious and costly to handle and manage. Blockchain's ability to record ownership transfers and legal and security requirements in real time can help to reduce the likelihood of disputes. Smart contracts enabled by blockchain technology can automatically trigger compensation or fines at low cost if predetermined terms are violated.

There are many requirements to monitor and comply with in the agri-food industry, including product safety and integrity, ethical sourcing, technical regulations, and social and environmental responsibility of suppliers [21]. Failing to comply with these requirements can lead to potential regulatory scrutiny and have a negative impact on firm's reputation. Blockchain can address agri-food supply chain compliance issues. By providing real-time visibility and data auditability into the supply chain, blockchain ensures that all contractual terms are met and compels supply chain participants to work with each other within regulatory requirements. In case of the violation of the required environmental conditions detected by sensors in a container, smart contracts can help avoid agricultural product degradation by sending real-time warnings for inspection. Data immutability enabled by blockchain provides a reliable means for supply chain actors to protect the interests of final customers.

There are many interesting use cases of blockchain in the following area: agri food distribution, food origin and sourcing, and food safety and quality [22]. For instance, Walmart, the world's largest retailer, uses IBM blockchain service to quickly pinpoint the culprit in future food safety scares. In 2016, it partnered with BLU-82 to form IBM Food Trust. The main objective is to increase the supply chain transparency in responding to the increasing customer demand in terms of food provenance and safety. Walmart has cooperated with IBM to develop a blockchain-based food traceability system and completed two tests: trace pork sold in China and mangoes in the Americas [18]. For pork sold in its Chinese stores, this food traceability system allowed uploading certificates of authenticity to the blockchain, bringing more trust to a system where that used to be a serious issue. For mangoes in the US, the time needed to trace their provenance went from 7 days to 2.2 seconds [23]. In September 2018, Walmart required all its lettuce and spinach suppliers to log their shipments on the blockchain. Today, Walmart can now track the origins of 25 products (e.g. strawberries, yogurt, and chicken) using the aforementioned system.

Another example of blockchain-enabled transparency in food origin and sourcing is the export of beef from Australia to China. China's growing demand for beef and the difficulty of meeting domestic demand have led to the import of beef mainly from Australia. BeefLedger, an Australian blockchain company, has developed a token-driven platform for food sourcing and monitoring, where beef supply chain members take part in the network by purchasing beef tokens. Importers, wholesalers, and retailers can use these tokens as payment for beef shipments. BeefLedger stores all information associated with cattle (cow feeding and health history), meat processing, transportation, and storage conditions. Chinese consumers can thus easily access to the information about the origins of the beef, how was it sourced, and how authentic the claims on the label are.

Blockchain is also used in combination with other digital technologies, such as internet of things, to address fraudulent practices that lead to food safety and quality issues. Downstream Beer is a pioneer in the beer industry to jointly use blockchain technology and internet of things to provide full transparency of beer ingredients and brewing techniques. By using sensors, every aspect of the beer-making process (e.g. location, temperature, and humidity monitoring) is recorded safely on the blockchain. Consumers can use their smartphones to scan the QR codes marked on the bottles where they can access information related to the ingredients, processing methods, bottling process, storage, temperature conditions, etc.

Blockchain along with big data analytics, automation, internet of things, and artificial intelligence are rapidly becoming pillars of supply chain digitalization. The adoption of blockchain helps to foster a paperless agri-food supply chain [24]. It will help break down bureaucracy, radically reduce transaction times and administrative costs, and speed up the flow of goods by eliminating tedious paperwork and using automated data storage process.

5. Factors influencing blockchain adoption in agri-food supply chain

The application of blockchain in agri-food supply chain is still in its infancy. While some use cases can be found, the scope of most blockchain pilots in agri-food industry is still quite limited. Therefore, it is urgent to carefully examine the factors that affect the implementation of blockchain in this specific industry. In this paper, we adopt Technology-Organization-Environment (TOE) framework [25] to comprehensively analyze blockchain adoption from the technological, organizational, and environmental perspective. TOE framework is widely used by previous studies to investigate the adoption of technological innovations, such as social commerce [26], customer relationship management systems [27], and software as a service [28]. We believe that this integrative framework is also suitable to understand blockchain adoption in agri-food industry. **Figure 2** shows the integrative adoption model proposed in this paper.

5.1 Technological considerations

5.1.1 Scalability

Each node within the blockchain network has a complete copy of all the information stored on the blockchain. When a new block is added, the system needs to update the copy at each node so that a single version of truth is ensured among all participants. As the network expands with more members and data, this update process slows down accordingly, and latency becomes an important issue. Agri-food supply

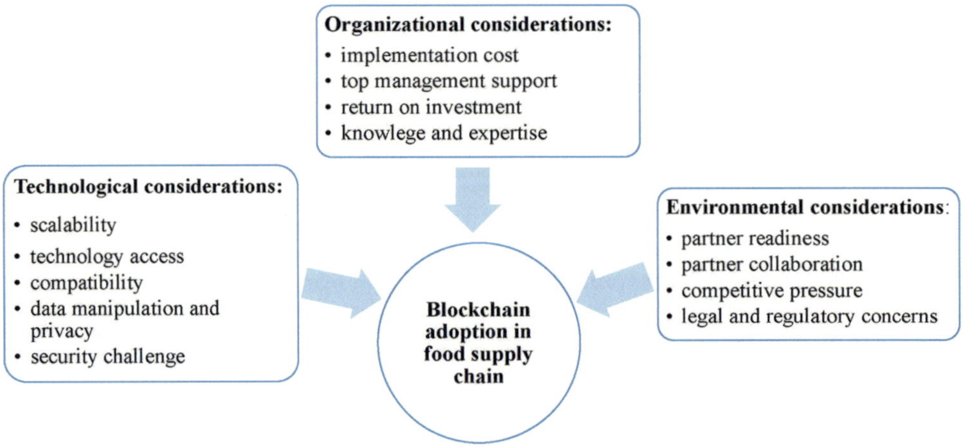

Figure 2.
An integrative framework for blockchain adoption in food supply chain.

chain involves many actors and generates a very large amount of information every day. In the pilot stage, the latency issue described before may not be pronounced, but this issue will be exacerbated when a blockchain initiative scales up beyond the pilot stage. This is the reason why many supply chain blockchain initiatives remain proof of concept for years [24].

5.1.2 Technology access

Internet access and IT infrastructure are prerequisites for the implementation of blockchain in agri-food supply chain. However, most of the food chain actors are farmers and small- and medium-sized enterprises, and they are not likely to have the adequate IT infrastructure to adopt blockchain. Thus, incapacity to access supply chain information in a timely manner due to technology access limitations could be a barrier for blockchain adoption.

5.1.3 Compatibility

The traceability systems developed by different blockchain pilots in the agri-food industry are not based on the same standard. They are designed and operated with different governance rules and consensus models. The question of how one blockchain will be used alongside other blockchains or other available systems remains unanswered [13]. Lack of interoperability can make large-scale blockchain adoption a nearly impossible mission. Recently, there are increasing number of interoperability projects that aim to connect various private blockchain with common rules and understanding. This is the only way to ensure smooth data transfer between different blockchain systems.

5.1.4 Data manipulation and privacy

Data integrity issue is one of the major technological hurdles for blockchain adoption. Data may be manipulated or compromised before it is recorded into blockchain. To address this issue, authenticated data entry system needs to be developed to automatically convert physical events to digital inputs for the blockchain. By doing so, the information inconsistency between physical reality and data stored on blockchain can be eliminated. The transparency enabled by blockchain technology also raises concerns in terms of data privacy, especially some participants may be competitors. Therefore, participants may be reluctant to use blockchain to avoid the leakage of their proprietary information.

5.1.5 Security challenge

Blockchain is usually viewed as a reliable technology with characteristics, such as encryption, data immutability, and decentralization. However, this does not mean that blockchain is not vulnerable to cyberattacks and security fraud. People with malicious intent can exploit blockchain security vulnerabilities for their own benefit. There are already many cybersecurity incidents related to blockchain usage. For example, a cyberattack on Bithumb, a major cryptocurrency exchanges for bitcoin, and Ethereum caused an economic loss of 870,000 USD and the exposure of more than 30,000 users' information. The potential security threats could discourage firms willingness to use blockchain.

5.2 Organizational considerations

5.2.1 Top management support

Top management support and commitment play a pivotal role in the successful implementation of any technological innovations. The higher the support and commitment, the more likely the firm will develop new organizational policies to clarify the usage of blockchain and reduce the resistance from employees to embrace the new system. Lack of technology awareness and involvement from upper management would impede the allocation of adequate human, financial, and technological resources for technology adoption projects.

5.2.2 Cost

The adoption of blockchain incurs diverse costs, such as recruiting blockchain expert, investment in additional equipment (e.g. sensors and radio-frequency identification tags). In addition, adopters need to spend a lot of time and resources to master the complexity of blockchain. The cost may be justifiable for the leading firms that initiate blockchain implementation projects, because they are usually large firms with bountiful resources. It is financially viable for them to invest in costly project that promises long-term returns. However, this is not the case for small- and medium-sized food supply chain actors with low margins. Therefore, high implementation cost may hinder the adoption intention of these actors.

5.2.3 Return on investment (ROI)

According to [21] global blockchain survey, more than 33% of the respondents claims that their current ROI in blockchain technology remains uncertain. The lack of clear ROI is recognized as the top barrier for blockchain adoption [29]. With the increasingly usage of blockchain in agri-food industry, it is likely that the financial benefits of blockchain implementation will be clarified in the foreseeable future.

5.2.4 Lack of knowledge and expertise

Blockchain is still largely an emerging technology, and the skills required to develop and use it are in short supply. The willingness to invest in blockchain requires a certain degree of technological awareness and knowledge. The technical complexity of blockchain makes it a challenge for individual users to understand, accept, and confidently participate in it. Lack of in-house technological capabilities is a serious barrier for the wide adoption of blockchain. One of the solutions to mitigate the low in-house technological capabilities is to purchase blockchain as a service, which allows firms to access the benefits of blockchain usage without having to make a substantial investment in the technical talent behind it.

5.3 Environmental considerations

5.3.1 Trading partner readiness

If only a few nodes in the food supply chain are ready to join the network, the ability of blockchain to enable end-to-end traceability and improve transparency will be largely compromised. As a matter of fact, most of the food supply chain actors are financially constrained small- and medium-sized companies with limited technical expertise. They are unlikely to have sufficient capability to adopt blockchain alone. While large companies are able to initiate blockchain projects, the participation of smaller companies is needed to achieve fruitful results. Therefore, if the other trading partners in the food supply chain are not ready to adopt blockchain, the focal firm may also hesitate to invest in this technology.

5.3.2 Collaboration among trading partners

Blockchain implementation in food supply chain requires all relevant parties to have the similar level of technological awareness and maturity. The large company that initiates the blockchain project can provide necessary help to the other firms with lower digital readiness. But they first need to have the same conception about blockchain technology, and they must achieve a consensus on the benefits of blockchain-based food supply chain transformation. Mutual commitment and shared vision are the foundations of any supply chain collaboration. If the other firms share the same vision with the focal firm and they are willing to engage in collaboration to facilitate the co-adoption of blockchain, then this paves the way for the focal firm to adopt blockchain.

5.3.3 Competitive pressure

Modern food supply chain actors can leverage technological innovations to gain a competitive advantage over competitors. By adopting innovative technologies, firms may change the rules of competition and exploit new ways to outperform competitors, thus changing the competitive structure of the industry. Blockchain technology has promising applications in the field of agri-food distribution, food origin and sourcing, and food safety and quality. Many firms are looking to reap such benefits in an increasingly competitive marketplace. Therefore, it can be assumed that companies will adopt blockchain if their competitors are also considering it.

5.3.4 Legal and regulatory concerns

The distributed nature of blockchains gives rise to some unique legal concerns. Since different nodes of the distributed ledger may be located in different regions of the world, it is a complex and even conflicting task to decide which laws should be complied with and which courts have the authority to decide what matters in blockchain-related issues [29]. The blockchain pilot initiators should collaborate with legal professionals to define rules and a detailed set of contingencies to anticipate potential legal issues [13].

6. Conclusion

In this paper, we attempt to provide answers to the following questions: how blockchain technology can be used to address food supply chain challenges and what are the factors that influence its effective implementation in food supply chain? Based on a critical synthesis of the state of the art, we first identify the main challenges of food supply chain, such as food fraud, food safety, customer, and regulatory pressure to ensure transparency and responsiveness in case of product contamination. We then demonstrate how the key attributes of blockchain technology, such as immutability, traceability, transparency, automation, and security, can address the identified challenges. Finally, we propose an integrative framework to help scholars and practitioners better understand the critical success factors of blockchain adoption in the agri-food industry.

Author details

Jerome Verny[1*] and Wei Guan[2]

1 NEOMA Business School, France

2 HIGHFI, France

*Address all correspondence to: jerome.verny@neoma-bs.fr

References

[1] World Health Organization. Food Safety. 2020. Available from: https://www.who.int/news-room/fact-sheets/detail/food-safety

[2] European commission. Food fraud. Available from: https://knowledge4policy.ec.europa.eu/food-fraud-quality/topic/food-fraud_en

[3] Longo F, Nicoletti L, Padovano A. Estimating the impact of blockchain adoption in the food processing industry and supply chain. International Journal of Food Engineering. 2020;**16**(5-6):1-18

[4] Nakamoto S.. Bitcoin: A Peer-to-Peer Electronic Cash System. 2008

[5] Queiroz MM, Wamba SF. Blockchain adoption challenges in supply chain: An empirical investigation of the main drivers in India and the USA. International Journal of Information Management. 2019;**46**:70-82

[6] Ganne E. Can Blockchain Revolutionize International Trade? Geneva: World Trade Organization; 2018

[7] Iansiti M, Lakhani KR. The truth about blockchain. Harvard Business Review. 2017:118-127

[8] Cole R, Stevenson M, Aitken J. Blockchain technology: implications for operations and supply chain management. Supply Chain Management: An International Journal. 2019;**24**(4):1-34

[9] Zheng Z, Xie S, Dai HN, Chen X, Wang H. Blockchain challenges and opportunities: A survey. International Journal of Web and Grid Services. 2018;**14**(4):352-375

[10] Pilkington M. Blockchain technology: Principles and applications. In: Research Handbook on Digital Transformations. Edward Elgar Publishing; 2016

[11] Li Z, Barenji AV, Huang GQ. Toward a blockchain cloud manufacturing system as a peer to peer distributed network platform. Robotics and Computer-Integrated Manufacturing. 2018;**54**(January):133-144

[12] Zhao JL, Fan S, Yan J. Overview of business innovations and research opportunities in blockchain and introduction to the special issue. Financial Innovation. 2016;**2**(1):1-7

[13] Wang Y, Hugh Han J, Beynon-Davies P. Understanding blockchain technology for future supply chains: A systematic literature review and research agenda. Supply Chain Management: An International Journal. 2019;**24**(1):62-84

[14] Christidis K, Devetsikiotis M. Blockchains and smart contracts for the internet of things. IEEE Access. 2016;**4**:2292-2303

[15] Eyal I, Gencer AE, Sirer EG, Van Renesse R. Bitcoin-ng: A scalable blockchain protocol. In: Paper Presented at the NSDI, Santa Clara, CA. 2016

[16] IBM. Making Blockchain Real for Business. New York, NY: IBM; 2016

[17] Guo Y, Liang C. Blockchain application and outlook in the banking industry. Financial Innovation. 2016;**2**:24

[18] IBM. IBM Announces Major Blockchain Collaboration with Dole, Driscoll's. Unilever and Walmart to Address Food Safety Worldwide: Golden State Foods, Kroger, McCormick and Company, McLane Company, Nestlé, Tyson Foods; 2017. Available from:

https://www-03.ibm.com/press/us/en/pressrelease/53013.wss

[19] Yu T, Lin Z, Tang Q. Blockchain: The introduction and its application in financial accounting. The Journal of Corporate Accounting & Finance. 2018;**29**(4):37-47

[20] Scarano G. Walmart, primark urged by 70k consumers to boost supply chain transparency. Sourcing Journal. 2018. Available from: https://sourcingjournal.com/topics/compliance/walmart-primark-urged-by-70k-consumers-to-boost-supply-chain-transparency-77062/

[21] Deloitte. New Tech on the Block: Planning for blockchain in the retail and consumer packaged goods industries. 2018. Available from: https://www2.deloitte.com/content/dam/Deloitte/uk/Documents/ConsumerIndustrialProducts/deloitte-uk-blockchain-in-retail-and-cpg.pdf

[22] Menon S, Jain K. Blockchain technology for transparency in agri-food supply chain: Use cases, limitations, and future directions. IEEE Transactions on Engineering Management. 2021:1-15

[23] Hyperledger, Case study: How walmart brought unprecedented transparency to the food supply chain with Hyperledger Fabric. 2018. Available from: https://www.hyperledger.org/learn/publications/walmart-case-study

[24] Gartner. Gartner Top Strategic Predictions for 2019 and Beyond. 2018. Availale from: https://www.gartner.com/smarterwithgartner/gartner-top-strategic-predictions-for-2019-and-beyond

[25] Tornatzky LG, Fleischer M, Chakrabarti AK. The Process of Technological Innovation. Lexington, MA: Lexington Books; 1990

[26] Abed SS. Social commerce adoption using TOE framework: An empirical investigation of Saudi Arabian SMEs. International Journal of Information Management. 2020;**53**:102118

[27] Cruz-Jesus F, Pinheiro A, Oliveira T. Understanding CRM adoption stages: Empirical analysis building on the TOE framework. Computers in Industry. 2019;**109**:1-13

[28] Oliveira T, Martins R, Sarker S, Thomas M, Popovič A. Understanding SaaS adoption: The moderating impact of the environment context. International Journal of Information Management. 2019;**49**:1-12

[29] Chang Y, Iakovou E, Shi W. Blockchain in global supply chains and cross border trade: A critical synthesis of the state-of-the-art, challenges and opportunities. International Journal of Production Research. 2020;**58**(7):2082-2099